Heme Oxygenase (HO)-1 as an Immunoregulator in Health and Disease

Heme Oxygenase (HO)-1 as an Immunoregulator in Health and Disease

Editors

Elias A. Lianos
Maria G. Detsika

MDPI • Basel • Beijing • Wuhan • Barcelona • Belgrade • Manchester • Tokyo • Cluj • Tianjin

Editors
Elias A. Lianos
Carilion School of Medicine
Salem, VA, USA

Maria G. Detsika
National and Kapodistrian
University of Athens
Athens, Greece

Editorial Office
MDPI
St. Alban-Anlage 66
4052 Basel, Switzerland

This is a reprint of articles from the Special Issue published online in the open access journal *Antioxidants* (ISSN 2076-3921) (available at: https://www.mdpi.com/journal/antioxidants/special_issues/Heme_Oxygenase_Immunoregulator).

For citation purposes, cite each article independently as indicated on the article page online and as indicated below:

LastName, A.A.; LastName, B.B.; LastName, C.C. Article Title. *Journal Name* **Year**, *Volume Number*, Page Range.

ISBN 978-3-0365-8136-1 (Hbk)
ISBN 978-3-0365-8137-8 (PDF)

Contents

antioxidants

MDPI

Editorial

Immune-Related Functions of Heme Oxygenase-1

Elias A. Lianos [1] and Maria G. Detsika [2,*]

1 Veterans Affairs Medical Center and Virginia Tech, Carilion School of Medicine, Salem, VA 24153, USA; elias.lianos@va.gov
2 1st Department of Critical Care Medicine & Pulmonary Services, GP Livanos and M Simou Laboratories, Evangelismos Hospital, National and Kapodistrian University of Athens, 10675 Athens, Greece
* Correspondence: mdetsika@med.uoa.gr

Citation: Lianos, E.A.; Detsika, M.G. Immune-Related Functions of Heme Oxygenase-1. *Antioxidants* **2023**, *12*, 1322. https://doi.org/10.3390/antiox12071322

Received: 13 June 2023
Accepted: 19 June 2023
Published: 22 June 2023

Heme oxygenase (HO)-1 is a well-known cytoprotective enzyme due to its enzymatic action, which involves the catalysis of heme into anti-apoptotic and antioxidant molecules such as bilirubin, biliverdin and CO. Apart from heme, its natural inducer and substrate, HO-1 responds to numerous stimuli and is, therefore, implicated in multiple cellular, molecular and immunological pathways. Its role as a potential immunoregulator has been studied both via its by-products and via the direct involvement of HO-1 in immune-related cellular functions and responses. The *Antioxidants* journal Special Issue titled "Heme Oxygenase (HO)-1 as an Immunoregulator in Health and Disease" aims to highlight its role in immune-related pathways. This Special Issue includes twelve articles, eight of which are research articles reporting recent advances in research on the role of HO-1 in various immune-related mechanisms, and four of which are review articles that provide detailed, up-to-date information and evidence on its role as an immunoregulator in different diseases or health systems.

Cannedo-Marroquin et al. described the increase in the lung HO-1 expression of mice primarily infected with human respiratory syncytial virus (hRSV) and subsequently with a *Mycobacterium bovis* (*M. bovis*) strain referred to as *Bacillus Calmette-Guerin* (BCG) [1]. In a series of elegantly performed experiments, the study showed that a pre-infection with hRSV promoted lung pathology such as increased infiltration of innate immune cells, including interstitial and alveolar macrophages, during a subsequent mycobacterial challenge, thus impairing pulmonary immune responses and promoting secondary mycobacterial colonization, which was also associated with increased lung tissue HO-1 expression.

The involvement of HO-1 in viral infection was also explored in the study by Detsika M. G. and colleagues, who investigated changes in the expression of HO-1 in COVID-19 [2]. The study assessed HO-1 mRNA levels in COVID-19 patients with severe and critical illness and found that HO-1 levels increased in critically ill patients. This increase was accompanied by an increase in HO-1 expression at the tissue level in critically ill COVID-19 patients and was associated with poor prognosis.

The role of HO-1 in COVID-19 infection was further analyzed by Toro et al. [3]. Their review article consists of a thorough description of the anti-inflammatory and antiviral properties of HO-1 and its consequent potential for combating clinical manifestations of the disease.

Fitzgerald H.K. et al. explored the potential of *Trypanosoma brucei* (*T. brucei*)-derived ketoacids indole pyruvate (IP) and hydroxyphenylpyruvate (HPP) to induce HO-1 and modulate $CD4^+$ T cell responses [4]. The study reported upregulation of HO-1 by IP and HPP through Nrf2 in human dendritic cells (DC) associated with decreased DC maturation and pro-inflammatory cytokine production, thus affecting $CD4^+$ T cell differentiation.

The role of HO-1 as an immunomodulator was also assessed in a model of parasitic disease [5]. Costa et al. showed that mice infected with *Fasciola hepatica* (*F. hepatica*), a fluke that infects livestock and humans, causing fasciolosis, exhibited increased HO-1 expression in peritoneal antigen-presenting cells (APCs). The later produced decreased levels of reactive oxygen and nitrogen species, and their presence was associated with increased levels of regulatory T cells in an IL-10 activity-dependent manner.

The effect of the HO-1 pathway on inflammation was also assessed by Hirvonen et al., whose work involved healthy volunteers and revealed that the administration of D-glyceric acid activated mitochondrial metabolism and reduced inflammation [6]. Measurement of the HO-1 reaction by-products showed increased blood bilirubin levels, which were associated with ameliorated inflammation and lower levels of blood triglycerides.

Another study utilized CO-releasing molecules (CORMs) on primary tenocytes in order to investigate their potential to reduce inflammation [7]. Appetecchia et al. treated primary tenocytes with various concentrations of specific CORMs previously tested for their efficacy to produce efficient amounts of CO, an important HO-1 reaction by-product. The results of the study showed an improvement in tendon homeostasis and a reduction in PGE2 secretion.

The impact of HO-1 and its reaction by-products, especially bilirubin, is described in detail in the review article by Thomas D. T. and colleagues [8]. The article depicts the function of bilirubin as an antioxidant and metabolic hormone and how the HO-1–BVRA–bilirubin–PPAR axis influences inflammation and metabolic function and interacts with exercise to improve the outcomes of weight management.

An additional role of HO-1 as a potential modulator of the expression of proteins regulating the activation of the complement cascade was investigated by Detsika et al. [9]. Using tissue samples obtained from transgenic rats with complete HO-1 deficiency, it was shown that la ack of HO-1 reduced decay accelerating factor (DAF) expression in kidney, liver and lung tissue and CR1-related gene/protein Y (Crry) expression in kidney and liver tissue. The reduction in kidney DAF and Crry expression resulted in an increase in C3b deposition in kidney tissue, thus indicating an additional immunomodulatory role of HO-1 as a complement cascade regulator.

The important relationship of HO-1 and kidney disease via its potential as an immunomodulator is described in detail in the review article by Athanassiadou and colleagues [10]. The article serves as an up-to-date overview on the involvement of HO-1 in kidney disease through its effect on immune pathways and responses.

Similarly, in the review article by Xia and Zhong, the immunomodulatory role of HO-1 in allergic airway inflammation is described in detail [11]. Specifically, the anti-inflammatory effect of HO-1 in different stages of airway inflammation via the regulation of various immune cells, including dendritic cells, mast cells, basophils, T cells and macrophages, is discussed.

Finally, the anti-tumoral role of HO-1 via different mechanisms, including immune mediated pathways, was demonstrated in the study by Lage-Vickers et al. [12]. The study employed a proteomics approach and analyzed prostate cancer proteomic data in order to assess the significance of HO-1 interactors.

In conclusion, this Special Issue emphasizes the potential of HO-1 to modulate various immune cellular and physiological pathways, thus strengthening the evidence for its promising role as a putative target leading to innovative avenues of therapeutic strategies.

Funding: This work was supported by a Veteran Affairs Merit Award 5I01BX004333-04, titled 'Novel Complement-targeted treatment strategies', awarded to Elias A. Lianos.

Conflicts of Interest: The authors declare no conflict of interest.

References

1. Canedo-Marroquin, G.; Soto, J.A.; Andrade, C.A.; Bueno, S.M.; Kalergis, A.M. Increased Heme Oxygenase 1 Expression upon a Primary Exposure to the Respiratory Syncytial Virus and a Secondary Mycobacterium bovis Infection. *Antioxidants* **2022**, *11*, 1453. [CrossRef] [PubMed]
2. Detsika, M.G.; Nikitopoulou, I.; Veroutis, D.; Vassiliou, A.G.; Jahaj, E.; Tsipilis, S.; Athanassiou, N.; Gakiopoulou, H.; Gorgoulis, V.G.; Dimopoulou, I.; et al. Increase of HO-1 Expression in Critically Ill COVID-19 Patients Is Associated with Poor Prognosis and Outcome. *Antioxidants* **2022**, *11*, 1300. [CrossRef] [PubMed]

3. Toro, A.; Ruiz, M.S.; Lage-Vickers, S.; Sanchis, P.; Sabater, A.; Pascual, G.; Seniuk, R.; Cascardo, F.; Ledesma-Bazan, S.; Vilicich, F.; et al. A Journey into the Clinical Relevance of Heme Oxygenase 1 for Human Inflammatory Disease and Viral Clearance: Why Does It Matter on the COVID-19 Scene? *Antioxidants* **2022**, *11*, 276. [CrossRef] [PubMed]
4. Fitzgerald, H.K.; O'Rourke, S.A.; Desmond, E.; Neto, N.G.B.; Monaghan, M.G.; Tosetto, M.; Doherty, J.; Ryan, E.J.; Doherty, G.A.; Nolan, D.P.; et al. The Trypanosoma brucei-Derived Ketoacids, Indole Pyruvate and Hydroxyphenylpyruvate, Induce HO-1 Expression and Suppress Inflammatory Responses in Human Dendritic Cells. *Antioxidants* **2022**, *11*, 164. [CrossRef] [PubMed]
5. Costa, M.; da Costa, V.; Frigerio, S.; Festari, M.F.; Landeira, M.; Rodriguez-Zraquia, S.A.; Lores, P.; Carasi, P.; Freire, T. Heme-Oxygenase-1 Attenuates Oxidative Functions of Antigen Presenting Cells and Promotes Regulatory T Cell Differentiation during Fasciola hepatica Infection. *Antioxidants* **2021**, *10*, 1938. [CrossRef] [PubMed]
6. Hirvonen, O.P.; Lehti, M.; Kyrolainen, H.; Kainulainen, H. Heme Oxygenase-1 and Blood Bilirubin Are Gradually Activated by Oral D-Glyceric Acid. *Antioxidants* **2022**, *11*, 2319. [CrossRef] [PubMed]
7. Appetecchia, F.; Consalvi, S.; Berrino, E.; Gallorini, M.; Granese, A.; Campestre, C.; Carradori, S.; Biava, M.; Poce, G. A Novel Class of Dual-Acting DCH-CORMs Counteracts Oxidative Stress-Induced Inflammation in Human Primary Tenocytes. *Antioxidants* **2021**, *10*, 1828. [CrossRef] [PubMed]
8. Thomas, D.T.; DelCimmuto, N.R.; Flack, K.D.; Stec, D.E.; Hinds, T.D., Jr. Reactive Oxygen Species (ROS) and Antioxidants as Immunomodulators in Exercise: Implications for Heme Oxygenase and Bilirubin. *Antioxidants* **2022**, *11*, 179. [CrossRef] [PubMed]
9. Detsika, M.G.; Theochari, E.; Palamaris, K.; Gakiopoulou, H.; Lianos, E.A. Effect of Heme Oxygenase-1 Depletion on Complement Regulatory Proteins Expression in the Rat. *Antioxidants* **2022**, *12*, 61. [CrossRef] [PubMed]
10. Athanassiadou, V.; Plavoukou, S.; Grapsa, E.; Detsika, M.G. The Role of Heme Oxygenase-1 as an Immunomodulator in Kidney Disease. *Antioxidants* **2022**, *11*, 2454. [CrossRef] [PubMed]
11. Xia, Z.; Zhong, W. Immune Regulation of Heme Oxygenase-1 in Allergic Airway Inflammation. *Antioxidants* **2022**, *11*, 465. [CrossRef] [PubMed]
12. Lage-Vickers, S.; Sanchis, P.; Bizzotto, J.; Toro, A.; Sabater, A.; Lavignolle, R.; Anselmino, N.; Labanca, E.; Paez, A.; Navone, N.; et al. Exploiting Interdata Relationships in Prostate Cancer Proteomes: Clinical Significance of HO-1 Interactors. *Antioxidants* **2022**, *11*, 290. [CrossRef] [PubMed]

antioxidants

MDPI

Article

Increased Heme Oxygenase 1 Expression upon a Primary Exposure to the Respiratory Syncytial Virus and a Secondary *Mycobacterium bovis* Infection

Gisela Canedo-Marroquín [1,2,†], Jorge A. Soto [1,3,†], Catalina A. Andrade [1], Susan M. Bueno [1] and Alexis M. Kalergis [1,4,*]

1 Millennium Institute of Immunology and Immunotherapy, Departamento de Genética Molecular y Microbiología, Facultad de Ciencias Biológicas, Pontificia Universidad Católica de Chile, Santiago 8331150, Chile; gecanedo@uc.cl (G.C.-M.); jasoto6@uc.cl (J.A.S.); cnandrade@uc.cl (C.A.A.); sbueno@bio.puc.cl (S.M.B.)
2 Centre for Biomedical Research and Innovation, Faculty of Dentistry, Universidad de los Andes, Santiago 7620157, Chile
3 Millennium Institute on Immunology and Immunotherapy, Departamento de Ciencias Biológicas, Facultad de Ciencias de la Vida, Universidad Andrés Bello, Santiago 8370146, Chile
4 Departamento de Endocrinología, Facultad de Medicina, Pontificia Universidad Católica de Chile, Santiago 8330023, Chile
* Correspondence: akalergis@bio.puc.cl
† These authors contributed equally to this work.

Abstract: The human respiratory syncytial virus (hRSV) is the leading cause of severe lower respiratory tract infections in infants. Because recurrent epidemics based on reinfection occur in children and adults, hRSV has gained interest as a potential primary pathogen favoring secondary opportunistic infections. Several infection models have shown different mechanisms by which hRSV promotes immunopathology to prevent the development of adaptive protective immunity. However, little is known about the long-lasting effects of viral infection on pulmonary immune surveillance mechanisms. As a first approach, here we evaluated whether a primary infection by hRSV, once resolved, dampens the host immune response to a secondary infection with an attenuated strain of *Mycobacterium bovis* (*M. Bovis*) strain referred as to Bacillus Calmette-Guerin (BCG). We analyzed leukocyte dynamics and immunomodulatory molecules in the lungs after eleven- and twenty-one-days post-infection with *Mycobacterium*, using previous hRSV infected mice, by flow cytometry and the expression of critical genes involved in the immune response by real-time quantitative reverse transcription polymerase chain reaction (RT-qPCR). Among the latter, we analyzed the expression of Heme Oxygenase (HO)-1 in an immunization scheme in mice. Our data suggest that a pre-infection with hRSV has a conditioning effect promoting lung pathology during a subsequent mycobacterial challenge, characterized by increased infiltration of innate immune cells, including interstitial and alveolar macrophages. Our data also suggest that hRSV impairs pulmonary immune responses, promoting secondary mycobacterial colonization and lung survival, which could be associated with an increase in the expression of HO-1. Additionally, BCG is a commonly used vaccine that can be used as a platform for the generation of new recombinant vaccines, such as a recombinant BCG strain expressing the nucleoprotein of hRSV (rBCG-N-hRSV). Therefore, we evaluated if the immunization with rBCG-N-hRSV could modulate the expression of HO-1. We found a differential expression pattern for HO-1, where a higher induction of HO-1 was detected on epithelial cells compared to dendritic cells during late infection times. This is the first study to demonstrate that infection with hRSV produces damage in the lung epithelium, promoting subsequent mycobacterial colonization, characterized by an increase in the neutrophils and alveolar macrophages recruitment. Moreover, we determined that immunization with rBCG-N-hRSV modulates differentially the expression of HO-1 on immune and epithelial cells, which could be involved in the repair of pulmonary tissue.

Keywords: RSV; mycobacteria; HO-1; tuberculosis; susceptibility

Citation: Canedo-Marroquín, G.; Soto, J.A.; Andrade, C.A.; Bueno, S.M.; Kalergis, A.M. Increased Heme Oxygenase 1 Expression upon a Primary Exposure to the Respiratory Syncytial Virus and a Secondary *Mycobacterium bovis* Infection. *Antioxidants* **2022**, *11*, 1453. https://doi.org/10.3390/antiox11081453

Academic Editors: Elias Lianos and Maria G. Detsika

Received: 2 June 2022
Accepted: 23 July 2022
Published: 26 July 2022

1. Introduction

Human respiratory syncytial virus (hRSV) is the leading cause of acute lower respiratory tract infection (ALRTI) in children under five-year-old, displaying high rates of hospitalization and over 90,000 death-related cases [1,2]. Children under two years of age, people with comorbidities, such as cardiac and pulmonary affections, and the elderly, are the most susceptible to developing hRSV bronchiolitis and pneumonia [3]. HRSV infections cause airway inflammatory hyperresponsiveness characterized by the secretion of pro-inflammatory cytokines, including interleukin (IL)-6, IL-4, IL-13, and tumor necrosis factor (TNF)-α, leading to the recruitment of innate leukocytes, such as neutrophils, as well as adaptive immune cells including T and B cells [4].

The consequences associated with hRSV infection during the early stages of infection include poor adaptive immunity that allows reinfections [5,6] and increased susceptibility to the development of allergies [7], post-bronchiolitis wheeze (PBW), and asthma [8]. Another important consequence of hRSV infections is the frequent establishment of secondary bacterial infections, such as *Streptococcus pneumoniae* (*S. pneumoniae*), which can lead to more severe clinical symptoms of pulmonary pathology, and are considered an essential factor that contributes to mortality rate [9–12]. Different studies have suggested that primary respiratory viral infections can induce host susceptibility to secondary infections, either acute or concomitant bacterial infection (*S. pneumoniae*) or established bacterial infection after viral clearance, such as *Mycobacterium tuberculosis* (*M. tuberculosis*) [13]. However, the susceptibility that an initial infection with hRSV promotes for secondary infections with *M. tuberculosis* has not been evaluated. Here we sought to evaluate whether hRSV infection shapes the lung immune response and promotes pulmonary inflammatory hyperresponsiveness in a chronic mycobacterial infection model.

The *Mycobacterium bovis* (*M. Bovis*) strain, referred to as Bacillus Calmette-Guerin (BCG), is an attenuated version of a known species belonging to the *M. tuberculosis* complex [14]. Although infections with BCG do not recapitulate all pathological features observed with *M. tuberculosis*, BCG is a valuable model for studying anti-mycobacterial immune responses mimicking important aspects of bacilli-host interaction [15,16]. Moreover, BCG provides a functional model for chronic infections as viable bacilli can persist for up to 10 months in the lungs [17].

Heme Oxygenase (HO)-1 is an enzyme that catalyzes a reaction in which the heme group turns into carbon monoxide (CO), biliverdin, and free iron [18]. HO-1 promotes heme group homeostasis and has a cryoprotective role against tissue damage. Additionally, it has been reported that the activity of HO-1 has an antiviral effect on pathogens such as hRSV, among others [19–21]. In the case of mycobacteria, the function of HO-1 in the host is induced by the dormancy mechanism activated by the mycobacteria, promoting an anti-inflammatory response that avoids the immune system and ensures their survival [22,23]. The expression of HO-1 is controlled by a transcription factor named nuclear factor erythroid 2-related factor (Nrf2), which can be activated as a consequence of stimuli such as hypoxia [24]. The OX-2 glycoprotein membrane (CD200) is an important molecule expressed on the surface of alveolar epithelial cell type II, which can bind to the receptor CD200R on the surface of alveolar macrophages [25]. The union of CD200 with its receptor on the alveolar macrophages leads to their inhibition, downregulating the inflammatory response in the pulmonary epithelium [25]. By doing this, CD200 can regulate the airway immunological response against infections.

As mentioned above, hRSV-infection can cause airway inflammation and hyperresponsiveness [4]. However, the link between hyperresponsiveness of the airways and long-term pulmonary sequels regarding *Mycobacterium* infections is unclear. Therefore, we evaluated whether the lungs of hRSV-infected mice are immunologically susceptible to secondary bacterial colonization along with pulmonary pathology. With this aim, we performed long-term bacterial colonization using BCG as the *Mycobacterium* infectious model after a primary infection with hRSV in C57BL/6 mice [26,27]. Additionally, HO-1 induction during hRSV-infection produces an anti-inflammatory environment and reduction of viral loads [20]. Since the HO-1 effect in BCG immunization has not been evaluated, we decided to test whether the effect previously described was increased using a BCG-based vaccine prototype against hRSV. This study aimed to determine the contribution of HO-1 during a mycobacterial infection that followed an infection by hRSV and after an immunization scheme followed by an hRSV challenge.

2. Materials and Methods

2.1. Ethics Statements

All experimental protocols followed guidelines from the Sanitary Code of Terrestrial Animals of the World Organization for Animal Health (OIE, 24. Edition, 2015) and were reviewed and approved by the Scientific Ethical Committee for Animal and Environment Care of the Pontificia Universidad Católica de Chile (Protocol number 160915010 and 160405005). All mouse experiments were conducted in agreement with international ethical standards and according to the local animal protection law number 20,800.

2.2. Viral Propagation and Titration

Human Epidermoid carcinoma strain 2 (Hep-2) cell line (American Type Culture Collection, CCL-23^{TM}) (American Type Culture Collection, CCL-7^{TM}) was used to propagate hRSV serogroup A2, strain 13018–8, a clinical isolate from Instituto de Salud Pública de Chile [28]. Hep-2 monolayers were grown in T75 flasks with Dulbecco's Modified Eagle Medium (DMEM) (Life Technologies Invitrogen, Carlsbad, CA, USA) supplemented with 10% fetal bovine serum (FBS) (Gibco Invitrogen Corp, Carlsbad, CA, USA) until 80–90% confluency. Flasks containing 5 mL of DMEM 1% FBS for infection with hRSV, the viral inoculum with 2×10^5 plaque formation units (PFU), were incubated at 37 °C. After 2 h of virus adsorption, supernatants were replaced with fresh DMEM 1% FBS medium and incubated for 48 h (i.e., until cytopathic effects were detectable). For harvesting, cells were scraped, and the flask content was pooled and centrifuged first at 300× *g* for 10 min and then at 500× *g* for 10 min to remove cell debris. Using the same harvesting protocol, supernatants of non-infected cells were collected and used as the non-infectious control (referred to from here on as Mock). Viral titers of supernatants were determined by immunocytochemistry in 96-well plates with Hep-2 cells, as described previously [29,30].

2.3. *Mycobacterium Bovis-BCG-Culture, and Storage*

The BCG Danish 1331 strain was grown in medium 7H9 (Sigma-Aldrich, Saint Louis, MO, USA, M0178-500G), a specific mycobacteria broth [30], supplemented with 10% Middlebrook oleic acid, albumin, dextrose, and catalase (OADC) Growth Supplement (Sigma-Aldrich, M0678-1VL), with constant stirring at 120 rpm until reaching an OD600 nm equal to 0.8. At this point, the mycobacteria culture was washed three times with 1X PBS-0.05% Tween 80, resuspended with 1X PBS-glycerol 50% at a final concentration of 1×10^6 colony-forming units (CFU)) per vial and frozen at −80 °C until their use. For the infection, BCG vials were centrifuged at 14,000× *g* and resuspended in PBS for intranasal administration.

2.4. Mouse Immunization and Viral Infection

The effect of a recombinant BCG (rBCG) strain in the modulation of HO-1 was evaluated by immunization with rBCG expressing the nucleoprotein of hRSV (rBCG-N-hRSV) as described next. Six to eight-week-old BALB/cJ mice were immunized by sub-cutaneous 1×10^8 CFU of BCG WT or rBCG-N-hRSV in a final volume of 100 µL per dose at days 0 and 14. Twenty-one days after immunization, mice were intraperitoneally anesthetized and challenged by intranasal infection with ~1×10^7 PFU of hRSV A2, strain 13018-8. On days 7 and 14 post-infection, mice were euthanized. rBCG production was performed as described previously [30].

2.5. Mouse Viral and Mycobacterial Infections

Two infection schemes were conducted to determine the consequences of hRSV infection and the effect of a subsequent infection. The short scheme was performed to evaluate if the inoculation with mycobacteria could infect and damage the pulmonary tissue a few days after the clearance of hRSV (day 10 post-infection with hRSV). Euthanasia was performed 11 days after the inoculation with mycobacteria since the alveolar tissue repair was not complete on this day. For this reason, it would be expected to find an effect induced by the administration of mycobacteria. The choice of this day was mainly due to the slow replicative cycle of this bacterium, which makes it difficult to carry out these tests at earlier times. The long scheme was performed to evaluate if, a significant number of days after the clearance of hRSV (day 21 post-infection with hRSV), the inoculation with mycobacteria could infect and damage the pulmonary tissue. During the day of the inoculation, the cell target of mycobacteria alveolar macrophages was replaced, allowing the mycobacteria more time to proliferate and cause an effect on the lung. Euthanasia was performed 21 days after the inoculation with mycobacteria since, on this day, the alveolar tissue repair is not complete. Six to eight-week-old C57BL/6J mice received an intranasal infection with 1×10^7 PFU of hRSV A2 strain 13018-8 of 100 µL per mouse. After 10- or 21-days post-infection (dpi), mice were intranasally instilled with 1×10^6 CFU/mice of BCG. After 11- and 21-days post inoculation with BCG, mice were euthanized for collection of lung samples, bronchoalveolar lavage (BAL), and mediastinal lymph nodes. The controls were mock and vehicle (Sauton diluent) for the inoculation with hRSV and BCG, respectively. The administration of the first and second inoculation and their description are described in Table 1.

Table 1. Schemes of infection and their times.

	Day of the First Inoculation (Mock or Human Respiratory Syncytial Virus)	Day of the Second Inoculation (Vehicle or *Bacillus* Calmette-Guerin)	Day of the Euthanasia (After the Second Inoculation)
Short scheme	0	10	11
Long scheme	0	21	21

2.6. Evaluation of hRSV-Associated Disease Parameters

To determine the infiltration of polymorphonuclear cells, BAL was collected as previously described [29] and stained with anti–CD11b PerCP-Cy5.5 (clone M1/70, BD Pharmingen, San José, CA, USA), anti-CD11c APC (clone HL3, BD Pharmingen), anti-IA/IE APCcy7 (clone M5/114.15.2, Biolengend), anti-Singlec F PE CF594 (clone E50-2240, BD Bioscience, San José, CA, USA), anti-Ly6C BV605 (clone HK1.4, Biolegend, San José, CA, USA) and anti-Ly6G FITC (Clone 1A8, BD Pharmingen) antibodies. As previously described, viral loads were detected in the lungs by real-time quantitative reverse transcription polymerase chain reaction (RT-qPCR) [28,30,31]. In addition, lung samples were stored in 4% paraformaldehyde solution (PFA), maintained at 4 °C, embedded in paraffin, cut, and stained with H&E as previously described [29]. The BCG count after 21 dpi was performed by seeding 1000 µL of lung homogenate in 7H10 plates and incubation for 14 to 21 days at 37 °C with 5% CO2. To evaluate the presence of HO-I in the lungs by flow cytometry, the lungs were incubated with collagenase IV for 30 min at 37 °C with agitation (120 rpm). Then the cells were homogenized using a 70 µm cell strainer. The cells were incubated with ammonium-chloride-potassium (ACK) lysis buffer for 5 min and centrifugated 300 g for 5 min at 4 °C, and then stained with α-CD45-BV510 (clone 30-FL1, BD Horizon, San José, CA, USA), α-CD11c APC (clone HL3, BD Pharmingen), α-IA/IE-V500 (clone M5/114.15.2, BD Pharmigen), α- CD326(Ep-CAM) PE (clone G8.8, Biolegend). Then, for HO-1 intracellular staining, fixed cells were incubated with anti-mouse HO-1 monoclonal antibody (mAb) (Abcam, UK) in permeabilization buffer (1% saponin, 10% FBS in PBS) for 45 min at 4 °C. For all antibody dilutions, 1 μL of antibody was diluted in 500 μL of a buffer PEB (1X Phosphate Buffered Saline (PBS)-0.5% Bovine Serum Albumin (BSA)-0.2mM Ethylenediaminetetraacetic acid (EDTA)). Data were acquired in an LSRFortessa X2-0 cytometer (BD Biosciences) and analyzed using FlowJo v10.0.7 software (BD Biosciences). The gating strategy for detecting immune cells is shown in Supplementary Figure S1.

2.7. Lung Histopathology Analyses

Before proceeding with BAL collection, the major bronchus of the left lung was clamped using 10 cm Kelly hemostatic forceps to perform histopathology analyses without significantly altering tissue architecture. After obtaining the BAL from the right lung, the left lung was fixed with 4% paraformaldehyde, then paraffin-embedded using a Leica ASP300S enclosed, automatic tissue processor (Leica Microsystems, Wetzlar, Germany). Then, 5 μm-thick tissue sections were obtained using a Microm HM 325 Rotary Microtome (Thermo Scientific, Waltham, MA, USA) before being mounted and stained for histopathology analyses using H&E stain. A histopathological score was used to measure structural alterations in lung sections of control and infected animals [32,33]. The histopathological score was as follows: 0 = normal tissue morphology, normal alveolar architecture, connective tissue associated with bronchi (slight presence of immune cells); 1 = alveolar spaces are reduced, and there are immune cells present; 2 = bronchoalveolar involvement is defined as reduced alveolar spaces and high infiltration of immune cells (neutrophils and lymphocytes) within and surrounding the airways, including bronchi; 3 = pulmonary consolidation is evidenced by loss of alveolar spaces, bronchial walls thickening or bronchial collapse, and high cellular infiltration. In addition, Ziehl-Neelsen (ZN) staining was performed according to standard protocols [34,35]. In brief, the major bronchus of the left lung section was dewaxed by washing with decreasing alcohol concentrations (from 96% to 70% ethanol), heat-fixed, then stained with carbol-fuchsin (Bacto TB Carbolfuchsin KF, Becton-Dickinson, Sparks, MD, USA) for 4 min, washed, and incubated with 3% hydrochloric acid (HCl) until the stain was completely dissolved. Counterstaining was performed with brilliant green (Bacto TB Brilliant Green K, Becton- Dickinson, Sparks, MD, USA) for 20 s. Sections were air-dried after thorough washing.

2.8. Relative Expression by RT-qPCR

Quantitative real-time RT-qPCR total RNA was isolated from lung tissues collected using the Trizol reagent according to the manufacturer's instructions (Thermo Fisher Scientific). Complementary DNA (cDNA) synthesis from total RNAs was performed using the iScript™ Reverse Transcription Supermix for RT-qPCR (Bio-Rad, Hercules, CA, USA) and random primers. RT-qPCR reactions were carried out using a StepOne plus thermocycler (Applied Biosystems, Waltham, MA, USA). The abundance of *ho-1* and *nrf2* mRNAs were determined by relative expression to the respective housekeeping gene (*β-actin* gene) by the 2-ΔΔ threshold cycle (ΔΔCt) method [36]. The RT-qPCR assays performed had 100% efficacy. For *n-hRSV* gene expression, absolute quantification data were expressed as the number of hRSV N-gene copies for every 5×10^3 copies of the *β-actin* transcript, as previously described [28,37]. The choice of using the *β-actin* gene for a single reference gene is based on present high stability [31,38]. The primers used can be found in Table 2.

2.9. Statistical Analyses

All statistical analyses were performed using GraphPad Prism version 6.0 Software. Statistical significance values and analyses are detailed in each figure legend. For Figure 1, only a normal distribution was observed for the data in Figure 1B. The data from Figure 1C–F showed a non-normal distribution, so non-parametric Mann-Whitney tests were performed. One-way ANOVA tests with a post hoc Tukey test were performed in Figures 2–5 and Supplementary Figures S2–S4 since the data showed a normal distribution. Two-way Anova with a post hoc Dunnett's multiple comparisons test were performed to compare the kinetics of the weight curves of Figure 2A and Supplementary Figure S2A. Figure 1A used a two-way ANOVA with a post hoc Šídák's multiple comparisons test.

Table 2. List of primers used for real-time quantitative reverse transcription polymerase chain reaction (RT-qPCR) analysis.

Gene	Forward Primer	Reverse Primer	Gene Accession Code
n-hRSV	5′-GCTAGTGTGCAAGCAGAAATC-3′	5′-TGGAGAAGTGAGGAAATTGAGTC-3′	Gene ID: 1489820
Mouse *ho-1*	5′-CCTCTGACGAAGTGACGCC-3′	5′-CAGCCCCACCAAGTTCAAA-3′	Gene ID: 15368
Mouse *nrf2*	5′-TTCTTTCAGCAGCATCCTCTCCAG-3′	5′-ACAGCCTTCAATAGTCCCGTCCAG-3′	Gene ID: 18024
Mouse *ifn-β*	5′-AGCTCCAAGAAAGGACGAACA-3′	5′-GCCCTGTAGGTGAGGTTGAT-3′	Gene ID: 15977
Mouse *il-6*	5′-TAGTCCTTCCTACCC CAATTTCC-3′	5′-TAGTCCTTCCTACCCCAATTTCC-3′	Gene ID: 16193
Mouse *cd200*	5′-CTCTCCACCTACAGCCTGATT-3′	5′-AGAACATCGTAAGGATGCAGTTG-3′	Gene ID: 17470
Mouse *β-actin*	5′-ACCTTCTACAATGAGCTGCG-3′	5′-CTGGATGGCTACGTACATGG-3′	Gene ID: 11461

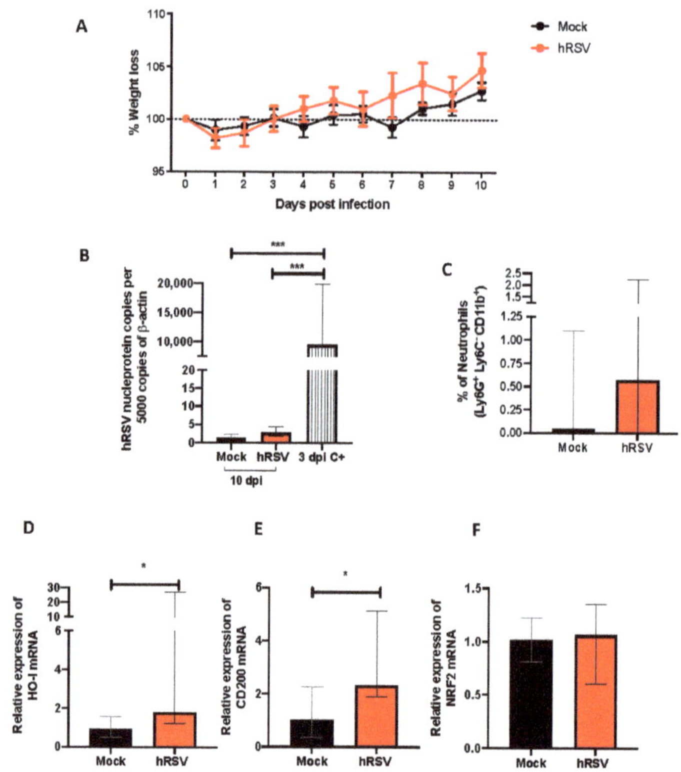

Figure 1. Evaluation of infection, inflammation, and immunomodulatory parameters from human respiratory syncytial virus (hRSV)-infected mice. (**A**) Body mass loss of C57BL/6 mice infected with 1×10^7 plaque formation units (PFU) of hRSV A2 for ten days. All the following parameters were measured at day 10 post hRSV infection. (**B**) Determination of viral load through specific real-time quantitative reverse transcription polymerase chain reaction (RT-qPCR) for hRSV. (**C**) Neutrophils (**D**) Heme Oxygenase (HO)-1, (**E**) OX-2 glycoprotein membrane (CD200), and (**F**) nuclear factor erythroid 2-related factor (NRF2). Data are shown as median ± interquartile range of at least two independent experiments with three animals per group. (**B**) One-way ANOVA was performed with a post-hoc Tukey test. (**C–F**) t-student was performed with the Mann-Whitney U test (* $p < 0.05$; *** $p \leq 0.001$). Created with BioRender.com.

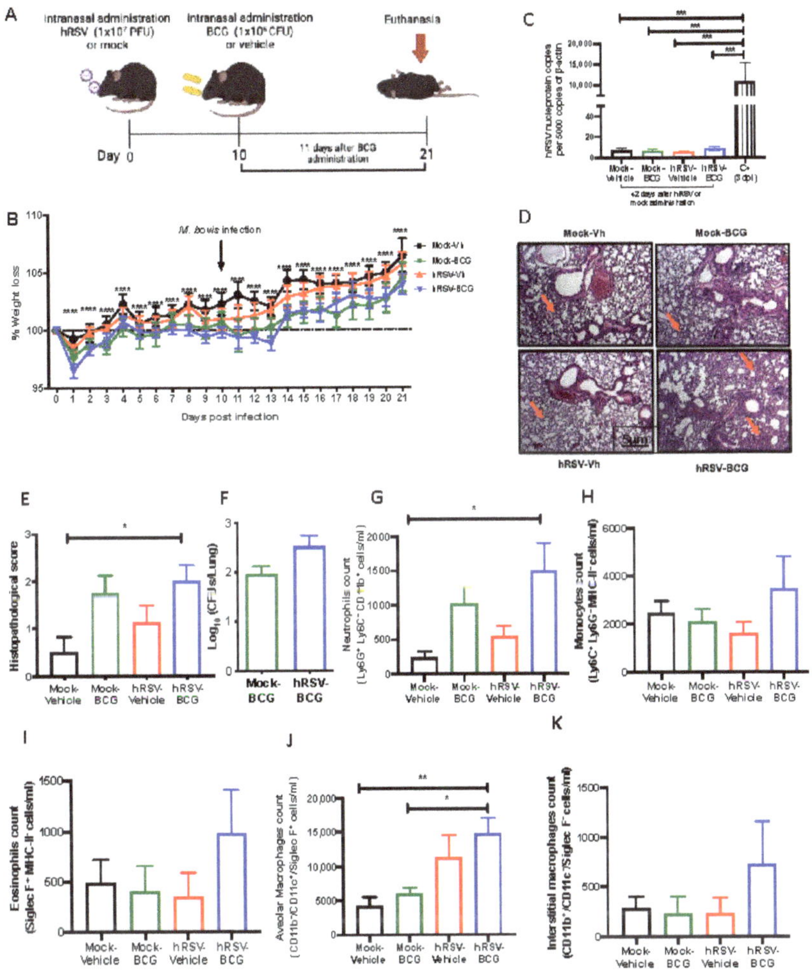

Figure 2. Evaluation of infection parameters from primary infection with a human respiratory syncytial virus (hRSV) and 11 days post-infection with Bacillus Calmette-Guerin (BCG). (**A**) Scheme of infection in mice. (**B**) Body mass loss of C57BL/6 mice infected with 1×10^7 plaque formation units (PFU) of hRSV A2 and a subsequent challenge with BCG intranasal at day 10 post-infection with hRSV. The significant difference corresponds to a two-way ANOVA of multiple comparisons between *hRSV-BCG* versus *mock-vehicle* and hRSV-BCG versus hRSV-vehicle. (**C**) Determination of viral load through specific real-time quantitative reverse transcription polymerase chain reaction (RT-qPCR) for hRSV. (**D**) Lung tissue sections were stained with Hematoxylin and Eosin (10× magnification). (**E**) Histopathological score. (**F**) Bacterial load in the BCG-groups. (**G–K**) Flow cytometry analyses of bronchoalveolar lavages (BAL) from mice infected with *Mycobacterium bovis (M. bovis)*. The figure shows the absolute cell numbers for neutrophils (**G**), monocytes (**H**), eosinophils (**I**), alveolar macrophages (**J**), and interstitial macrophages (**K**) in BAL of *M. bovis*-infected mice. Data are shown as means ± SEM of three independent experiments with 3–4 animals per group. One-way ANOVA was performed with a post-hoc Tukey test. (* p <0.05; ** $p < 0.01$; *** $p \leq 0.001$; **** $p \leq 0.0001$).

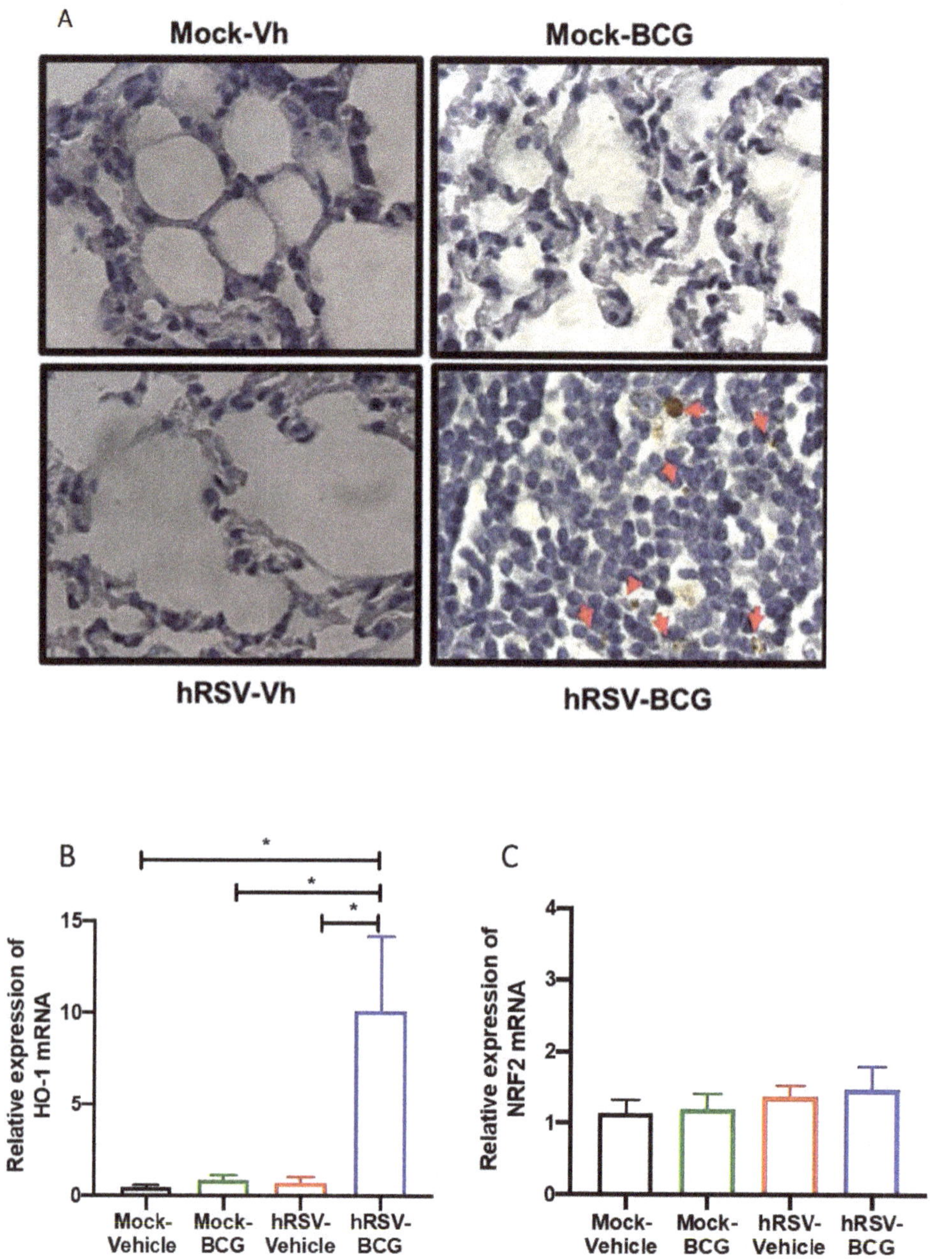

Figure 3. Determination of mycobacteria and Heme Oxygenase (HO)-1 activity during a short infection scheme with the human respiratory syncytial virus (hRSV) and Bacillus Calmette-Guerin (BCG). (**A**) Acid-fast staining of *Mycobacterium bovis*-infected mouse lungs collected on day 11 post-inoculation with BCG (100× magnification). Quantification by real-time quantitative reverse transcription polymerase chain reaction (RT-qPCR) of *ho-1* (**B**) and the nuclear factor E2-related factor 2 (*nrf2*) (**C**). Data are shown as means ± SEM of three independent experiments with 3–4 animals per group. One-way ANOVA was performed with a post-hoc Tukey test (* $p < 0.05$).

3. Results

3.1. HRSV Infection Induces de Expression of Immunomodulatory Molecules

Our study model evaluated different disease and cellular infiltration parameters to characterize the inflammatory immune response induced by hRSV. A significant reduction in viral loads was observed during previous studies using a mouse infection model by day 7 post hRSV instillation [39]. Therefore, we measured weight loss and viral load by quantitative RT-PCR from the lungs of mice at 10 dpi and used as positive controls (C+) the viral load at 3 dpi (Figure 1A,B). No significant differences were found in the weight loss after the infection compared to mock-treated mice (Figure 1A). As shown in Figure 1B, animals showed levels of *n-hRSV* RNA close to non-detectable, indicating an adequate viral clearance by day 10 post-challenge, making it an appropriate inoculation time point for subsequent BCG challenge. As expected, only a significant difference was found in the positive control with respect to the groups evaluated (p = 0.0001). These results correlated with a low percentage of neutrophils recruited in the lungs (Figure 1C). Additionally, to evaluate whether the resolution of the infection was affected by anti-inflammatory modulators, the expression of *ho-1*, nrf2 related to oxidative stress, and *cd200* genes were measured by RT-qPCR. At 10 dpi, we observed a significant increase in *ho-1* (p= 0.0152) and *cd200* (p = 0.0244) relative expression in convalescent mice lungs compared to control animals. However, no differences were observed in the transcription factor controlling *ho-1* gene expression *nrf2* between the groups (Figure 1F). We have recently shown that HO-1 upregulation has an antiviral protective effect in the lungs by limiting viral replication in the tissue and epithelial cells, as well as by modulating the immunogenicity of antigen-presenting cells (APC), such as dendritic cells (DC) and alveolar macrophages [20,40]. Given these data, it could be suggested that upon viral clearance, the high levels of *ho-1* expression in the lower respiratory tract might play a role in modulating the immunity to secondary infections.

3.2. Previous hRSV Infection Causes Further Long-Term Susceptibility to Mycobacterium Bovis-Driven Pneumonia

To further evaluate whether animals that have cleared hRSV from the lungs remain immunocompetent to clear secondary infections, control (mock-instilled) and hRSV-convalescent animals were challenged with *M. bovis* BCG after 10 dpi. BCG inoculation was given intranasally to mice using a saline solution as a control (Figure 2A). As shown in Figure 2B, hRSV-convalescent mice were inoculated with BCG (*hRSV-BCG*), but not their relevant controls (*Mock-BCG*), showed increased weight loss at day 13 post-infection. Statistical differences were found along the kinetics for *hRSV-BCG* vs. *Mock-Vehicle* or *Mock-BCG* groups ($p \leq 0.0001$). This weight loss can be associated with the infection with BCG rather than hRSV persistence, as evidenced by the absence of detectable viral loads at day 21 post-infection (Figure 2C). As expected, only a significant difference was found in the positive control with respect to the groups evaluated (p = 0.0002). Additionally, this was associated with more severe histopathological scores in both groups infected with BCG. Careful histopathological scoring did not show any significant difference in the semiquantitative blinded scoring of both BCG-infected groups despite a more marked thickening of alveolar walls (Figure 2D,E). The more severe lung pathology observed at 21 dpi in the *hRSV-BCG* group was characterized by the thickening of the lung parenchyma due to interstitial inflammation in discrete and well-defined foci developed around bronchi and alveolar sacs (Figure 2E), which showed a significant difference compared to *Mock-Vehicle* (p = 0.00219). Although less severe, a mild inflammation of the lung parenchyma was observed with foci in the bronchi in mice pre-treated with Mock and then inoculated with BCG (*Mock-BCG*) (Figure 2D,E). Additionally, the maintenance of the pulmonary architecture and no significant lung inflammation were observed in both vehicle-inoculated controls (*Mock-Vehicle* and *hRSV-Vehicle* groups). Mycobacteria CFUs were detected in the lungs of animals previously inoculated with either mock or hRSV (Figure 2F). Consistent with the histopathological score, only the *hRSV-BCG* group showed a significant increase of BAL neutrophils (Ly-6C$^-$ CD11b$^+$ Ly-6G^{hi}) compared to the *Mock-Vehicle* group (p = 0.0445) (Figure 2G). Accompanying neutrophils, we observed no significant differences in BAL monocytes (Ly-6C$^+$ CD11b$^+$ Ly-6G$^-$) and eosinophils (CD11b$^+$ Siglec F$^+$) in all groups of mice (Figure 2H,I, respectively). Next, we analyzed other lung immune cell populations by flow cytometry, such as interstitial and alveolar macrophages. Consistent with the infiltration of neutrophils in BAL, alveolar macrophages, defined as CD11b$^-$ CD11c$^+$ Siglec-F$^+$ [41], showed a significant increase only in *hRSV-BCG* compared to both *Mock-Vehicle* (p = 0.0062) and *Mock-BCG* (p = 0.0159) groups (Figure 2J). A slight, non-significant increase was observed in the *hRSV-Vehicle* group (Figure 2J). No significant increase in interstitial macrophages (CD11b$^+$CD11c$^+$Siglec-F$^-$) was detected between groups, but a slight increase was found in the *hRSV-BCG* group. (Figure 2K).

To assess whether the increased susceptibility to *M. bovis* BCG induced by a previous hRSV infection is short-lived, we evaluated a long scheme in which the second inoculation and the euthanasia were performed 21 and 42 days after the first inoculation, respectively (Table 1). Interestingly, the results obtained from these trials showed that there were no significant changes between those reflected in the parameters of disease or inflammation evaluated in the different groups of animals (Supplementary Figure S2), except for more significant damage related to the histological score (Supplementary Figure S2E).

Following our detection of viable bacilli in the lungs of BCG-infected animals, we performed ZN staining of the lung sections shown in Figure 3A to assess whether the significant increase in pathology and infiltration of neutrophils correlated with the increased presence of dormant bacilli, which might not be detected by ex vivo culture of lung homogenates. In agreement with the low numbers of CFU detected in BCG-infected animals, no active *M. bovis* was found in the lungs of both BCG-infected groups of mice (Figure 3A). However, a population of macrophages that reacted with the stain was observed only in the lungs of *hRSV-BCG* treated mice but not in the other treatments (red arrows). Since ZN staining binds to mycolic acids, the observation of weak ZN staining suggests the existence of either myeloid cells processing remainder mycobacterial components, or macrophages containing bacilli with a low metabolic rate (i.e., under dormancy) [42].

Since HO-1 induction has been associated with both impairment of antigen processing by APC [40] and bacilli dormancy [23], we sought to evaluate whether the detection of ZN-stained cells was associated with the modulation of HO-1 expression in the lungs. A significant increase of almost 50% in the relative expression of HO-1 was observed only in *hRSV-BCG*, as compared to *mock-vehicle* (p = 0.0357), *hRSV-vehicle* (p = 0.0263) or *mock-BCG* (p = 0.0239) control groups (Figure 3B). No significant differences were observed in the expression of Nrf2, suggesting that HO-1 up-regulation might imply other transcription factors [43] (Figure 3C).

No significant changes were observed in the determination of dominant BCG in the lungs of the long scheme animals, nor were differences in the expression levels of HO-I or NRF2 (Supplementary Figure S3).

3.3. *Characterization of the Inflammatory Landscape of the Lungs*

As mentioned above, CD200 can induce an inhibitory signal in alveolar macrophages, among other APC, downregulating the inflammation in the pulmonary tissue [25]. No significant differences in the expression of the epithelial anti-inflammatory molecule CD200 were observed between the groups of the short scheme (Figure 4A), suggesting that the absence of CD200 upregulation might support the observed lung pathology [43]. Additionally, the relative expression of pro-inflammatory cytokines IFN-γ, IFN-β, and IL-6 was evaluated, but no significant differences were found for any of these molecules (Figure 4B–D). Only a slight increase in the relative expression of IFN-γ was detected in the *hRSV-BCG* compared to the other groups (Figure 4B). For the long scheme, an increase in the relative expression was observed for all molecules evaluated. Still, only significant differences were found for the expression of *ifn-γ* for the *hRSV-BCG* group compared to *mock-vehicle* (p = 0.0094) (Supplementary Figure S4B).

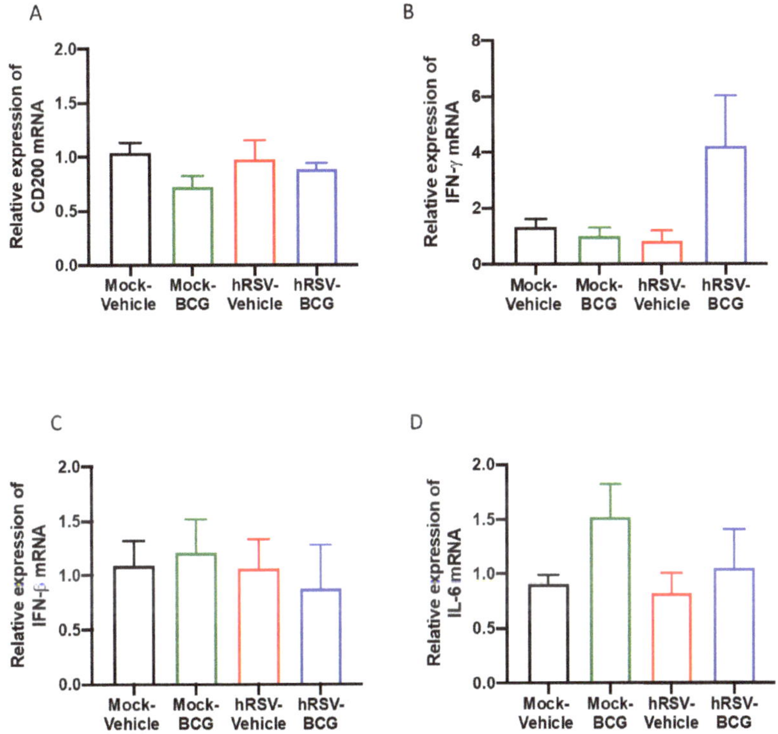

Figure 4. Determination of the relative expression of immunomodulatory molecules and cytokines in infection with Bacillus Calmette-Guerin (BCG) at 10 days post-human respiratory syncytial virus (hRSV)-infection. Quantification by real-time quantitative reverse transcription polymerase chain reaction (RT-qPCR) of OX-2 glycoprotein membrane (*cd200*) at day 10 post-infection (**A**), interferon-gamma (*ifn-γ*) (**B**), interferon beta (*ifn-β*) (**C**), and interleukin (IL)-6 (**D**). Data are shown as means ± SEM of three independent experiments with 3–4 animals per group. One-way ANOVA was performed with a post-hoc Tukey test.

3.4. rBCG-N-hRSV Immunization Promotes a Higher Induction of HO-1 on Epithelial Cells than in DCs at Late Infection Times

BCG is commonly used as a vaccine to prevent infections by *M. tuberculosis* and is often used as a vector for generating new recombinant vaccines [14]. Therefore, we decided to evaluate if the immunization with an rBCG-N-hRSV can modulate the expression of HO-1 in hRSV-infected mice. After 7 and 14-dpi with hRSV, mice were euthanized, and both epithelial and DCs from the lungs were stained to evaluate the presence of HO-1 (Figure 5). The HO-1 expression observed on day 7 post-infection in DCs showed a non-significant decrease in the *rBCG-N-hRSV* immunized mice compared to mock-treated mice. A similar result was observed in the rBCG-N-hRSV group in *BCG-WT* and hRSV-infected mice (Figure 5B). In contrast, when HO-1 was evaluated in epithelial cells, the mock-treated group showed the lowest non-significant levels of expression as compared to the rest of the infected groups (*mock-treated* compared to *hRSV-infected* (p = 0.9680), *BCG-WT* (p = 0.2232), *rBCG-N-hRSV* (p = 0.733)) (Figure 5C). When the HO-1 expression was evaluated on day 14 post-infection in DCs, similar results were obtained compared to day 7 post-infection. Interestingly, although the behavior of the groups was similar to that previously reported, the magnitude order was 4 to 5-fold less (Figure 5D). On the other hand, epithelial cells showed a different effect in the HO-1 expression on day 14 post-infection, demonstrating a different behavior distinct from the HO-1 expression in epithelial cells on day 7 post-infection. In this line, the HO-1 expression was higher in both *mock-treated*, and *rBCG-N-hRSV* immunized mice than in hRSV-infected or *BCG-WT* groups (Figure 5E), suggesting that the immunization with rBCG-N-hRSV promotes a lasting induction of HO-1 through time.

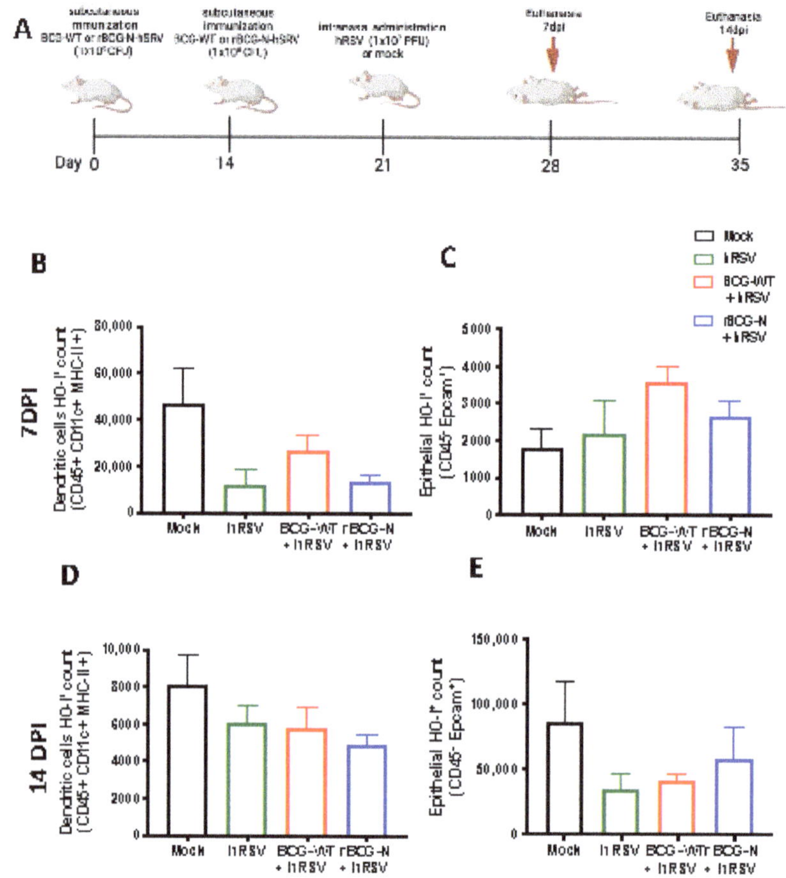

Figure 5. Detection of Heme Oxygenase (HO)-1 in cell populations during an immunization scheme with recombinant BCG expressing the nucleoprotein of human respiratory syncytial virus (rBCG-N-hRSV). Immunization scheme of BALB/c mice (**A**). Analysis by flow cytometry of dendritic cells (DC) (**B**) and epithelial cells (**C**) at 7 days post-challenge, and DCs (**D**) and epithelial cells (**E**) at 14 days post-challenge. Data are shown as means ± SEM of one independent experiment with three animals per group. One-way ANOVA was performed with a post-hoc Tukey test.

4. Discussion

The infection with hRSV is the principal cause of ALRTI, and it has been demonstrated that it can predispose to secondary lung infections [1]. The effects of primary viral infections on the outcome of concomitant secondary acute bacterial infections have only been studied in mice [13], such as *S. pneumoniae,* which induces a more severe pulmonary pathology [11,44]. Some studies have demonstrated that the immunopathology generated during coinfections of viruses, with either other viruses or bacteria, depends on the primary viral infection stage [14,42]. Therefore, it is important to establish the stage of the viral infection that will be evaluated. It has been reported that the administration of *S. pneumoniae* following the peak of hRSV replication in the lungs increases bacterial burden (analyzed as CFU and infiltration of inflammatory cells [13]. However, the link between hyperresponsiveness of the airways and long-term pulmonary sequels is not clear, although two possible explanations have been suggested. First, it was proposed that hRSV produces long-term damage in the airway epithelium, which may favor the infection by other pathogens [13,45]. In this line, studies with other respiratory viruses, such as influenza A virus, have shown a detrimental effect of the viral infection on the survival rate and clearance of *M. tuberculosis* in mice after viral clearance was demonstrated in the lungs [46]. As a second hypothesis, it was proposed that hRSV-infection exerts an immunomodulatory effect on the lungs, which predisposes to allergy and asthma [47]. To the best of our knowledge, no studies address whether hRSV infection exerts long-lasting effects on pulmonary immune responses, especially after the resolution of the initial viral insult.

As mentioned previously, the infection with BCG is a valuable model for studying anti-mycobacterial immunity, including forming granulomatous lung lesions and acquiring a paucibacillary state [15,16]. The mouse model of intranasal BCG infection is suitable for addressing whether hRSV blunts the desired type 1 T helper (Th1)-driven anti-mycobacterial immune responses, which correlate with protection to other mycobacterial species [48,49]. Furthermore, it has been described that hRSV-effects can last up to 42 dpi in the BALB/c mouse model [50] and up to 20 dpi in C57BL/6 mice [51].

This work aimed to unravel whether a previous infection with hRSV can elicit an immune response that facilitates the mycobacterial infection in a murine model. Therefore, we used an intranasal administration of attenuated *M. bovis* strain, BCG, to establish our subsequent infection model. Since BCG has been used as an infection model of TB [49,52,53], here we used an intranasal administration of BCG after the clearance of the infection with hRSV in mice to elucidate if this viral infection promotes an immune response that favors a secondary BCG infection. One of the significant discoveries in this work was the detection of brown structures in the lung epithelium in the *hRSV-BCG* group, where it was possible to identify the presence of mycobacteria in the acid-fast staining (Figure 3A). One hypothesis is that these brown spots, present only in the *hRSV-BCG* group, are lipid droplets that lose the acid-fast staining, as the color is brown and even considered orange or red [42] (Figure 3A and Supplementary Figure S3A). These drops of lipids can result from a mechanism of *Mycobacterium* when it remains dormant, considering that the *Mycobacterium* is surrounded by a monolayer of phospholipids and uses these structures as a primary source of carbon [42]. This stage occurs when the bacteria are in a context that promotes hypoxia, such as when the pulmonary tissue is not yet fully recovered during respiratory infections, such as hRSV [54,55]. The absence of complete recovery in the pulmonary tissue can be suggested due to the significant increase of CD200 (Figure 1F) and HO-1 relative expression (Figure 1E), which promote an anti-inflammatory environment [55]. Additionally, we can suggest that both CD200 and HO-1 might play a role in restructuring the lung's epithelial environment after 10 dpi with hRSV [55]. After inoculating the mice with BCG, further damage was observed, and the cell infiltration remained increased, as shown in Figure 2. Similar results are observed in Supplementary Figure S2. The mycobacteria acquire a dormant state that increases the expression of HO-1 to produce carbon monoxide and promote an anti-inflammatory state in the host cells [23,56]. Accordingly, the increased detection of HO-1 and the possibility of macrophages containing bacilli in a dormant state (Figure 3) can be related to the presence of mycobacteria in the epithelium [57,58]. In this context, tissue oxygenation is impeded, favoring the conditions in which the *Mycobacterium* is put into a dormant state through triglyceride metabolism [59–61]. Unlike hRSV, which produces ALRTI, BCG infection produces a less diffuse pulmonary pathology and, therefore, a less pronounced body weight loss (Figure 2B,E).

One of the limitations of this study was that the observation of mycobacterial infection was performed over an unprolonged period. With an even longer-term scheme, it would be possible to observe how the mycobacteria behave at cellular levels and if the cellular populations change on this day. This is because the airway epithelium has a cell replacement every twenty days but not in the alveolar macrophages, which can stay in the lung for years [62–64]. However, it must be considered that the alveolar macrophages will remain in the lungs for long periods if they are not destroyed by pathogenic infections [65]. Even more, some respiratory viruses, such as influenza virus, can promote changes in the alveolar macrophages to induce an extended antibacterial response [66]. In addition, more days allow for identifying the mycobacteria by acid-fast staining, which is correlated with the CFUs counts (Figures 2F and 3A). This study also used BCG mycobacteria as a model, making it possible to use it in a murine model and address scientific questions with an attenuated pathogen. However, this work proves that it serves as a model for research on secondary *Mycobacterium* infections despite not having the same effect and clinical symptoms as an infection with *M. tuberculosis*.

In addition, relative cytokine expression was evaluated for *ifn-*γ, *ifn-*β, and *il-6* (Figure 4B–D), which participate in the control of mycobacterial infections and other bacterial infections [67,68]. Thus, *ifn-*γ, *ifn-*β, and *il-6* are related to mycobacteria colonization in these schemes. In the case of *ifn-*γ in the short scheme of infection, the relative expression was more elevated in the *hRSV-BCG* group than in *Mock-BCG* and the other groups (Figure 4B). Interestingly, in the short infection scheme, the hRSV-BCG group had a lower relative expression of *ifn-*β than the *hRSV-Vehicle* (Figure 4C). Still, in the long infection scheme, only the *hRSV-BCG* group had a higher relative expression of this gene than the *hRSV-Vehicle* (Supplementary Figure S4C). *ifn-*β is a molecule that helps delay the beginning of mycobacterial infection [67], and as such, the data observed on the long scheme might be explained due to viral clearance. However, in the long scheme, we can see a similar phenomenon to *ifn-*γ, where the group with the highest expression was in the groups with mycobacteria inoculation (Supplementary Figure S4B), which could be attributed to the presence of mycobacteria. This increase in the expression of *ifn-*γ could be associated with the presence of alveolar macrophages, which are sentinels in the pulmonary epithelium that promote a favorable environment [69,70]. Both the short (Figure 4D) and long (Supplementary Figure S4D) schemes presented similar relative expressions of *ifn-*γ. Lastly, the increased relative expression of *il-6* does not inhibit the growth of mycobacteria, even though it has been reported that an increased secretion helps to protect the host against mycobacteria [68]. These data would imply that the presence of mycobacteria activates the expression of these cytokines.

We have previously shown the anti-inflammatory effect of HO-1 induction in vivo and in vitro [20]. Here we extend these observations by showing that at 7 dpi with hRSV, mice pre-vaccinated with BCG showed an elevated expression of HO-1 in DCs, but the group treated with mock had the highest HO-1 within hRSV and BCGs-groups (Figure 5B). A similar effect was found at day 14 dpi, but the activation of HO-1 was four to five-fold less than previously reported (Figure 5D), and this HO-1 in DCs elevated just in the mock group with respect to hRSV-infection was reported [20]. Interestingly, the effect of HO-1 in epithelial cells differed from that observed in DCs. Here, at 7 dpi, all hRSV-infected groups had a similar HO-1 response to the mock-treated group (Figure 5C). However, this effect was different at 14 dpi, where only rBCG-N-hRSV immunized mice had high HO-I levels, similar to the mock-treated group (Figure 5E). This last statement might suggest that hRSV-N-BCG promotes a prolonged state of HO-I activation in epithelial cells that helps eliminate the virus and recover damaged tissue [71]. No changes in the MFI of HO-I were found between the groups at different times (data not shown). Interestingly, it has been reported that the infection with hRSV can modify the methylation profile of immune cells and epithelial cells, promoting the secretion of Th2 cytokines and viral replication [72]. Since the activation of Nrf2 can be regulated epigenetically, it would be interesting to evaluate whether hRSV infection can modulate the activation of Nrf2 in an epigenetically-manner [73]. In this sense, a limitation of our article is that we did not explore whether a primary infection with hRSV can modulate epigenetic changes that may participate in the resolution of the disease against subsequent administration with *M. bovis*.

5. Conclusions

Pulmonary infections modulated by hRSV negatively modulate the respiratory and immune response, which promotes a sub-sequential *M. bovis* exposure, leading to inefficient mycobacterial clearance and increased host inflammatory response. The clearance is impeded by a possible dormancy state established by mycobacteria. Additionally, the mycobacteria promote HO-1 expression activating the dormancy state. In other words, hRSV promotes the development of pulmonary-tuberculosis-like in mice by increasing lung inflammation and the survival of infecting bacilli.

Supplementary Materials: The following are available online at https://www.mdpi.com/article/10.3390/antiox11081453/s1, Figure S1: Gating strategy for identifying pulmonary cells found either in bronchoalveolar lavage (BAL) or in lung homogenates; Figure S2: Evaluation of infection parameters from primary infection with human respiratory syncytial virus (hRSV) and 21 days post-infection (dpi) with Bacillus Calmette-Guerin (BCG). (A) Scheme of infection with a prime infection of hRSV A2 (1×10^7 plaque formation units (PFU) and then a subsequent instillation of BCG at 21 dpi. (B) Percent body weight in C57BL/6 mice background after 21 days of subsequent BCG challenge, the mice have a prime infection with hRSV. (C) Real-time quantitative reverse transcription polymerase chain reaction (RT-qPCR) for determination of viral load. Determination of infiltration in the lungs of mice infected with BCG on day 21 post-infection with hRSV. (D) Lung tissue sections were stained with Hematoxylin and Eosin (10X magnification). (E) Histopathological score. (F) Bacterial load in the BCG-groups. (G-J) Flow cytometry analyses of BAL from mice infected with *Mycobacterium bovis (M. bovis)*. The figure shows the absolute cell numbers for neutrophils (G), monocytes (H), eosinophils (I), and alveolar macrophages (J) in BAL of BCG-infected mice. Data are shown as means ± SEM of three independent experiments with 3–4 animals per group. One-way ANOVA was performed with a post hoc Tukey test. (** $p < 0.01$; **** $p \leq 0.0001$). (F) t-student test was using parametric distribution was used; Figure S3: The long scheme of hRSV-infection favors colonization infection with mycobacteria in the lungs. (A) Acid-fast staining of attenuated *M. bovis*-infected mouse lungs that were collected on day 21 post-infection with hRSV (10X and 100X magnification). Quantification by RT-qPCR of Heme Oxygenase *(ho)-1* (B) and the nuclear factor erythroid 2-related (*nrf2*) (C). Data are shown as means ± SEM of three independent experiments with 3–4 animals per group. One-way ANOVA was performed with a posterior Tukey test; Figure S4: Determination of the relative expression of immunomodulatory molecules and cytokines in infection with BCG at 21 days post-hRSV-infection. Quantification by RT-qPCR of the OX-2 glycoprotein membrane (*cd200*) at day 10 post-infection (A), interferon gamma (*ifn-γ*) (B), interferon beta (*ifn-β*) (C), and interleukin (*il*)*-6* (D). Data are shown as means ± SEM of three independent experiments with 3–4 animals per group. One-way ANOVA was performed with a posterior Tukey test.

Author Contributions: G.C.-M. performed experiments, designed the study, analyzed data, and wrote and revised the manuscript. J.A.S. performed experiments and manuscript revision. C.A.A. participated in manuscript revision and editing. S.M.B. supported in manuscript revision. A.M.K. is the leading investigator and supported in the organization, experimental design, and full manuscript revision. All authors have read and agreed to the published version of the manuscript.

Funding: This work was supported by FONDECYT 1070352; 1190830, FONDEF D061008, FONDEF D11I1080, PAI ANID No. SA77210076, PAI ANID No. SA77210051, ANID/CONICYT scholarship 21210662, and the Millennium Institute on Immunology and Immunotherapy CN09_016/ICN 2021_045; former P09/016-F.

Institutional Review Board Statement: All experimental protocols followed guidelines from the Sanitary Code of Terrestrial Animals of the World Organization for Animal Health (OIE, 24. Edition, 2015) and were reviewed and approved by the Scientific Ethical Committee for Animal and Environment Care of the Pontificia Universidad Católica de Chile (Protocol number 160915010 and 160405005). All mouse experiments were conducted in agreement with international ethical standards and according to the local animal protection law number 20,800.

Informed Consent Statement: Not applicable.

Data Availability Statement: The data generated during this study are readily available in the manuscript.

Acknowledgments: We thank our veterinarian María José Altamirano for her support in handling and caring for the experimental animals and evaluating the histopathological score. We also thank Francisco Salazar for technical assistance. We also thank Pablo Cespedes (CONICYT/FONDECYT POSTDOCTORADO No. 3140455) for technical assistance and experimental design. Finally, we thank Biorender for partial material used for the design of some figures.

Conflicts of Interest: The authors declare no conflict of interest.

References

1. Iwane, M.K.; Farnon, E.C.; Gerber, S.I. Importance of Global Surveillance for Respiratory Syncytial Virus. *J. Infect. Dis.* **2013**, *208*, S165–S166. [CrossRef] [PubMed]
2. Proesmans, M.; Rector, A.; Keyaerts, E.; Vandendijck, Y.; Vermeulen, F.; Sauer, K.; Reynders, M.; Verschelde, A.; Laffut, W.; Garmyn, K.; et al. Risk factors for disease severity and increased medical resource utilization in respiratory syncytial virus (+) hospitalized children: A descriptive study conducted in four Belgian hospitals. *PLoS ONE* **2022**, *17*, e0268532. [CrossRef] [PubMed]
3. Moreira, F.B.; Rosario, C.S.; Santos, J.S.; Avanzi, V.M.; Nogueira, M.B.; Vidal, L.R.; Raboni, S.M. Molecular characterization and clinical epidemiology of human respiratory syncytial virus (HRSV) A and B in hospitalized children, Southern Brazil. *J. Med. Virol.* **2017**, *89*, 1489–1493. [CrossRef] [PubMed]
4. Bohmwald, K.; Gálvez, N.M.S.; Canedo-Marroquín, G.; Pizarro-Ortega, M.S.; Andrade-Parra, C.; Gómez-Santander, F.; Kalergis, A.M. Contribution of Cytokines to Tissue Damage During Human Respiratory Syncytial Virus Infection. *Front. Immunol.* **2019**, *10*, 452. [CrossRef]
5. Rodriguez-Fernandez, R.; Tapia, L.I.; Yang, C.-F.; Torres, J.P.; Chavez-Bueno, S.; Garcia, C.; Jaramillo, L.M.; Moore-Clingenpeel, M.; Jafri, H.; Peeples, M.E.; et al. Respiratory Syncytial Virus Genotypes, Host Immune Profiles, and Disease Severity in Young Children Hospitalized with Bronchiolitis. *J. Infect. Dis.* **2018**, *217*, 24–34. [CrossRef]
6. González, A.E.; Lay, M.K.; Jara, E.L.; Espinoza, J.A.; Gómez, R.S.; Soto, J.; Rivera, C.A.; Abarca, K.; Bueno, S.M.; Riedel, C.A.; et al. Aberrant T cell immunity triggered by human Respiratory Syncytial Virus and human Metapneumovirus infection. *Virulence* **2017**, *8*, 685–704. [CrossRef]
7. Sigurs, N.; Aljassim, F.; Kjellman, B.; Robinson, P.D.; Sigurbergsson, F.; Bjarnason, R.; Gustafsson, P.M. Asthma and allergy patterns over 18 years after severe RSV bronchiolitis in the first year of life. *Thorax* **2010**, *65*, 1045–1052. [CrossRef]
8. Chatterjee, A.; Mavunda, K.; Krilov, L.R. Current State of Respiratory Syncytial Virus Disease and Management. *Infect. Dis. Ther.* **2021**, *10*, 5–16. [CrossRef]
9. Abramson, J.S.; Wheeler, J.G. Virus-induced neutrophil dysfunction: Role in the pathogenesis of bacterial infections. *Pediatr. Infect. Dis. J.* **1994**, *13*, 643–652. [CrossRef]
10. Aguilera, E.R.; Pfeiffer, J.K. Strength in numbers: Mechanisms of viral co-infection. *Virus Res.* **2019**, *265*, 43–46. [CrossRef]
11. Kumar, N.; Sharma, S.; Barua, S.; Tripathi, B.N.; Rouse, B.T. Virological and immunological outcomes of coinfections. *Clin. Microbiol. Rev.* **2018**, *31*, e00111-17. [CrossRef] [PubMed]
12. Sande, C.J.; Njunge, J.M.; Ngoi, J.M.; Mutunga, M.N.; Chege, T.; Gicheru, E.T.; Gardiner, E.M.; Gwela, A.; Green, C.A.; Drysdale, S.B.; et al. Airway response to respiratory syncytial virus has incidental antibacterial effects. *Nat. Commun.* **2019**, *10*, 2218. [CrossRef] [PubMed]
13. Stark, J.M.; Stark, M.A.; Colasurdo, G.N.; LeVine, A.M. Decreased bacterial clearance from the lungs of mice following primary respiratory syncytial virus infection. *J. Med. Virol.* **2006**, *78*, 829–838. [CrossRef] [PubMed]
14. Soto, J.A.; Gálvez, N.M.; Andrade, C.A.; Ramírez, M.A.; Riedel, C.A.; Kalergis, A.M.; Bueno, S.M. BCG vaccination induces cross-protective immunity against pathogenic microorganisms. *Trends Immunol.* **2022**, *43*, 322–335. [CrossRef]
15. Tree, J.A.; Williams, A.; Clark, S.; Hall, G.; Marsh, P.D.; Ivanyi, J. Intranasal bacille Calmette–Guérin (BCG) vaccine dosage needs balancing between protection and lung pathology. *Clin. Exp. Immunol.* **2004**, *138*, 405–409. [CrossRef]
16. Harris, S.A.; White, A.; Stockdale, L.; Tanner, R.; Sibley, L.; Sarfas, C.; Meyer, J.; Peter, J.; O'Shea, M.K.; Thomas, Z.-R.M.; et al. Development of a non-human primate BCG infection model for the evaluation of candidate tuberculosis vaccines. *Tuberculosis* **2017**, *108*, 99–105. [CrossRef]
17. Leversen, N.A.; Sviland, L.; Wiker, H.G.; Mustafa, T. Long-Term Persistence of BCG Pasteur in Lungs of C57BL/6 Mice Following Intranasal Infection. *Scand. J. Immunol.* **2012**, *75*, 489–499. [CrossRef]
18. Dunn, L.L.; Midwinter, R.G.; Ni, J.; Hamid, H.A.; Parish, C.R.; Stocker, R. New Insights into Intracellular Locations and Functions of Heme Oxygenase-1. *Antioxid. Redox Signal.* **2014**, *20*, 1723–1742. [CrossRef]
19. Espinoza, J.A.; González, P.A.; Kalergis, A.M. Modulation of Antiviral Immunity by Heme Oxygenase-1. *Am. J. Pathol.* **2017**, *187*, 487–493. [CrossRef]
20. Espinoza, J.A.; León, M.A.; Céspedes-Donoso, P.F.; Gómez, R.S.; Canedo-Marroquín, G.; Riquelme, S.A.; Salazar-Echegarai, F.J.; Blancou, P.; Simon, T.; Anegon, I.; et al. Heme Oxygenase-1 Modulates Human Respiratory Syncytial Virus Replication and Lung Pathogenesis during Infection. *J. Immunol.* **2017**, *199*, 212–223. [CrossRef] [PubMed]
21. Hill-Batorski, L.; Halfmann, P.; Neumann, G.; Kawaoka, Y. The Cytoprotective Enzyme Heme Oxygenase-1 Suppresses Ebola Virus Replication. *J. Virol.* **2013**, *87*, 13795–13802. [CrossRef] [PubMed]

22. Scharn, C.R.; Collins, A.C.; Nair, V.R.; Stamm, C.E.; Marciano, D.K.; Graviss, E.A.; Shiloh, M.U. Heme Oxygenase-1 Regulates Inflammation and Mycobacterial Survival in Human Macrophages during *Mycobacterium tuberculosis* Infection. *J. Immunol.* **2016**, *196*, 4641–4649. [CrossRef] [PubMed]
23. Zacharia, V.M.; Shiloh, M.U. Effect of carbon monoxide on Mycobacterium tuberculosis pathogenesis. *Med. Gas Res.* **2012**, *2*, 30. [CrossRef]
24. He, G. Hypoxia increases Nrf2-induced HO-1 expression via the PI3K Akt pathway. *Front. Biosci.* **2016**, *21*, 385–396. [CrossRef] [PubMed]
25. Hussell, T.; Bell, T.J. Alveolar macrophages: Plasticity in a tissue-specific context. *Nat. Rev. Immunol.* **2014**, *14*, 81–93. [CrossRef]
26. Plotkin, S. History of vaccination. *Proc. Natl. Acad. Sci. USA* **2014**, *111*, 12283–12287. [CrossRef]
27. Plotkin, S.A.; Orenstein, W.A.; Offit, P.A.; Edwards, K.M. (Eds.) *Plotkin's Vaccines*; Elsevier: Amsterdam, The Netherlands, 2018. [CrossRef]
28. Espinoza, J.A.; Bohmwald, K.; Céspedes, P.F.; Gómez, R.S.; Riquelme, S.A.; Cortés, C.M.; Valenzuela, J.A.; Sandoval, R.A.; Pancetti, F.C.; Bueno, S.M.; et al. Impaired learning resulting from Respiratory Syncytial Virus infection. *Proc. Natl. Acad. Sci. USA* **2013**, *110*, 9112–9117. [CrossRef]
29. Bueno, S.M.; González, P.A.; Cautivo, K.M.; Mora, J.E.; Leiva, E.D.; Tobar, H.E.; Fennelly, G.J.; Eugenin, E.A.; Jacobs, W.R.; Riedel, C.A.; et al. Protective T cell immunity against respiratory syncytial virus is efficiently induced by recombinant BCG. *Proc. Natl. Acad. Sci. USA* **2008**, *105*, 20822–20827. [CrossRef]
30. Soto, J.A.; Gálvez, N.M.S.; Rivera, C.A.; Palavecino, C.E.; Céspedes, P.F.; Rey-Jurado, E.; Bueno, S.M.; Kalergis, A.M. Recombinant BCG Vaccines Reduce Pneumovirus-Caused Airway Pathology by Inducing Protective Humoral Immunity. *Front. Immunol.* **2018**, *9*, 2875. [CrossRef]
31. Soto, J.A.; Gálvez, N.M.; Pacheco, G.A.; Canedo-Marroquín, G.; Bueno, S.M.; Kalergis, A.M. Induction of Protective Immunity by a Single Low Dose of a Master Cell Bank cGMP-rBCG-P Vaccine Against the Human Metapneumovirus in Mice. *Front. Cell. Infect. Microbiol.* **2021**, *11*, 1. [CrossRef]
32. Klopfleisch, R. Multiparametric and semiquantitative scoring systems for the evaluation of mouse model histopathology—A systematic review. *BMC Veter- Res.* **2013**, *9*, 123. [CrossRef] [PubMed]
33. Scudamore, C.L. (Ed.) *A Practical Guide to the Histology of the Mouse*; John Wiley & Sons: Hoboken, NJ, USA, 2014. [CrossRef]
34. Durek, C.; Rüsch-Gerdes, S.; Jocham, D.; Böhle, A. Sensitivity of BCG to modern antibiotics. *Eur. Urol.* **2000**, *37*, 21–25. [CrossRef] [PubMed]
35. Willemse, S.; Oomens, M.; De Lange, J.; Karssemakers, L. Diagnosing nontuberculous mycobacterial cervicofacial lymphadenitis in children: A systematic review. *Int. J. Pediatr. Otorhinolaryngol.* **2018**, *112*, 48–54. [CrossRef] [PubMed]
36. Livak, K.J.; Schmittgen, T.D. Analysis of relative gene expression data using real-time quantitative PCR and the $2^{-\Delta\Delta CT}$ Method. *Methods* **2001**, *25*, 402–408. [CrossRef]
37. González, P.A.; Prado, C.E.; Leiva, E.D.; Carreño, L.J.; Bueno, S.M.; Riedel, C.A.; Kalergis, A.M. Respiratory syncytial virus impairs T cell activation by preventing synapse assembly with dendritic cells Surface expression of RSV F protein was measured 48 h postinfection at a multiplicity of infection equal to 1, with three. *Proc. Natl. Acad. Sci. USA* **2008**, *30*, 14999–15004. [CrossRef]
38. Li, Z.; Yang, L.; Wang, J.; Shi, W.; Pawar, R.; Liu, Y.; Xu, C.; Cong, W.; Hu, Q.; Lu, T.; et al. β-Actin is a useful internal control for tissue-specific gene expression studies using quantitative real-time PCR in the half-smooth tongue sole Cynoglossus semilaevis challenged with LPS or *Vibrio anguillarum*. *Fish Shellfish Immunol.* **2010**, *29*, 89–93. [CrossRef]
39. Domachowske, J.; Bonville, C.A.; Gao, J.-L.; Murphy, P.M.; Easton, A.J.; Rosenberg, H.F. MIP-1α Is Produced but It Does Not Control Pulmonary Inflammation in Response to Respiratory Syncytial Virus Infection in Mice. *Cell. Immunol.* **2000**, *206*, 1–6. [CrossRef]
40. Riquelme, S.A.; Bueno, S.M.; Kalergis, A.M. Carbon monoxide down-modulates Toll-like receptor 4/MD2 expression on innate immune cells and reduces endotoxic shock susceptibility. *Immunology* **2015**, *144*, 321–332. [CrossRef]
41. Misharin, A.V.; Morales-Nebreda, L.; Mutlu, G.M.; Budinger, G.R.S.; Perlman, H. Flow Cytometric Analysis of Macrophages and Dendritic Cell Subsets in the Mouse Lung. *Am. J. Respir. Cell Mol. Biol.* **2013**, *49*, 503–510. [CrossRef]
42. Daniel, J.; Maamar, H.; Deb, C.; Sirakova, T.D.; Kolattukudy, P.E. Mycobacterium tuberculosis Uses Host Triacylglycerol to Accumulate Lipid Droplets and Acquires a Dormancy-Like Phenotype in Lipid-Loaded Macrophages. *PLoS Pathog.* **2011**, *7*, e1002093. [CrossRef]
43. Snelgrove, R.J.; Goulding, J.; Didierlaurent, A.M.; Lyonga, D.; Vekaria, S.; Edwards, L.; Gwyer, E.; Sedgwick, J.D.; Barclay, A.N.; Hussell, T. A critical function for CD200 in lung immune homeostasis and the severity of influenza infection. *Nat. Immunol.* **2008**, *9*, 1074–1083. [CrossRef] [PubMed]
44. Saade, G.; Deblanc, C.; Bougon, J.; Marois-Créhan, C.; Fablet, C.; Auray, G.; Belloc, C.; Leblanc-Maridor, M.; Gagnon, C.A.; Zhu, J.; et al. Coinfections and their molecular consequences in the porcine respiratory tract. *Veter Res.* **2020**, *51*, 80. [CrossRef] [PubMed]
45. Bohmwald, K.; Espinoza, J.A.; Rey-Jurado, E.; Gómez, R.S.; González, P.A.; Bueno, S.M.; Riedel, C.A.; Kalergis, A.M. Human Respiratory Syncytial Virus: Infection and Pathology. *Semin. Respir. Crit. Care Med.* **2016**, *37*, 522–537. [CrossRef] [PubMed]
46. Redford, P.S.; Mayer-Barber, K.D.; McNab, F.W.; Stavropoulos, E.; Wack, A.; Sher, A.; O'Garra, A. Influenza A Virus Impairs Control of Mycobacterium tuberculosis Coinfection Through a Type I Interferon Receptor–Dependent Pathway. *J. Infect. Dis.* **2013**, *209*, 270–274. [CrossRef] [PubMed]

47. Espinoza, J.A.; Bohmwald, K.; Céspedes, P.F.; Riedel, C.A.; Bueno, S.M.; Kalergis, A.M. Modulation of host adaptive immunity by hRSV proteins. *Virulence* **2014**, *5*, 740–751. [CrossRef] [PubMed]
48. Lagranderie, M.; Nahori, M.-A.; Balazuc, A.-M.; Kiefer-Biasizzo, H.; Silva, J.-R.L.E.; Milon, G.; Marchal, G.; Vargaftig, B.B. Dendritic cells recruited to the lung shortly after intranasal delivery of *Mycobacterium bovis* BCG drive the primary immune response towards a type 1 cytokine production. *Immunology* **2003**, *108*, 352–364. [CrossRef]
49. Derrick, S.C.; Kolibab, K.; Yang, A.; Morris, S.L. Intranasal Administration of *Mycobacterium bovis* BCG Induces Superior Protection against Aerosol Infection with Mycobacterium tuberculosis in Mice. *Clin. Vaccine Immunol.* **2014**, *21*, 1443–1451. [CrossRef]
50. Estripeaut, D.; Torres, J.P.; Somers, C.S.; Tagliabue, C.; Khokhar, S.; Bhoj, V.G.; Grube, S.M.; Wozniakowski, A.; Gomez, A.M.; Ramilo, O.; et al. Respiratory Syncytial Virus Persistence in the Lungs Correlates with Airway Hyperreactivity in the Mouse Model. *J. Infect. Dis.* **2008**, *198*, 1435–1443. [CrossRef]
51. Tekkanat, K.K.; Maassab, H.; Berlin, A.A.; Lincoln, P.M.; Evanoff, H.L.; Kaplan, M.H.; Lukacs, N.W. Role of Interleukin-12 and Stat-4 in the Regulation of Airway Inflammation and Hyperreactivity in Respiratory Syncytial Virus Infection. *Am. J. Pathol.* **2001**, *159*, 631–638. [CrossRef]
52. Chávez-Galán, L.; Vesin, D.; Martinvalet, D.; Garcia, I. Low Dose BCG Infection as a Model for Macrophage Activation Maintaining Cell Viability. *J. Immunol. Res.* **2016**, *2016*, 4048235. [CrossRef]
53. Taniguchi, K.; Miyatake, Y.; Hayashi, D.; Takami, A.; Itoh, S.; Yamamoto, S.; Hida, S.; Onozaki, K.; Takii, T. Early-shared *Mycobacterium bovis* bacillus Calmette-Guérin sub-strains induce Th1 cytokine productionin vivo. *Microbiol. Immunol.* **2015**, *59*, 684–689. [CrossRef] [PubMed]
54. Fröhlich, S.; Boylan, J.; McLoughlin, P. Hypoxia-Induced Inflammation in the Lung. *Am. J. Respir. Cell Mol. Biol.* **2013**, *48*, 271–279. [CrossRef] [PubMed]
55. Vitali, S.H.; Fernandez-Gonzalez, A.; Nadkarni, J.; Kwong, A.; Rose, C.; Mitsialis, S.A.; Kourembanas, S. Heme oxygenase-1 dampens the macrophage sterile inflammasome response and regulates its components in the hypoxic lung. *Am. J. Physiol. Cell. Mol. Physiol.* **2020**, *318*, L125–L134. [CrossRef] [PubMed]
56. Dutta, N.; Illei, P.B.; Jain, S.; Karakousis, P.C. Characterization of a Novel Necrotic Granuloma Model of Latent Tuberculosis Infection and Reactivation in Mice. *Am. J. Pathol.* **2014**, *184*, 2045–2055. [CrossRef] [PubMed]
57. Lovey, A.; Verma, S.; Kaipilyawar, V.; Ribeiro-Rodrigues, R.; Husain, S.; Palaci, M.; Dietze, R.; Ma, S.; Morrison, R.D.; Sherman, D.R.; et al. Early alveolar macrophage response and IL-1R-dependent T cell priming determine transmissibility of Mycobacterium tuberculosis strains. *Nat. Commun.* **2022**, *13*, 884. [CrossRef]
58. Gengenbacher, M.; Kaufmann, S.H. *Mycobacterium tuberculosis*: Success through dormancy. *FEMS Microbiol. Rev.* **2012**, *36*, 514–532. [CrossRef] [PubMed]
59. Thirunavukkarasu, S.; de Silva, K.; Plain, K.M.; Whittington, R.J. Role of host- and pathogen-associated lipids in directing the immune response in mycobacterial infections, with emphasis on *Mycobacterium avium* subsp. *paratuberculosis*. *Crit. Rev. Microbiol.* **2014**, *42*, 262–275. [CrossRef]
60. Stehr, M.; Elamin, A.A.; Singh, M. *Lipid Inclusions in Mycobacterial Infections. Tuberculosis: Current Issues in Diagnosis and Management*; BoD Books on Demand: Norderstedt, Germany, 2013. [CrossRef]
61. Barisch, C.; Soldati, T. Breaking fat! How mycobacteria and other intracellular pathogens manipulate host lipid droplets. *Biochimie* **2017**, *141*, 54–61. [CrossRef] [PubMed]
62. Janssen, W.J.; Barthel, L.; Muldrow, A.; Oberley-Deegan, R.E.; Kearns, M.T.; Jakubzick, C.; Henson, P.M. Fas Determines Differential Fates of Resident and Recruited Macrophages during Resolution of Acute Lung Injury. *Am. J. Respir. Crit. Care Med.* **2011**, *184*, 547–560. [CrossRef] [PubMed]
63. Eguíluz-Gracia, I.; Schultz, H.H.L.; Sikkeland, L.I.B.; Danilova, E.; Holm, A.M.; Pronk, C.J.H.; Agace, W.W.; Iversen, M.; Andersen, C.; Jahnsen, F.L.; et al. Long-term persistence of human donor alveolar macrophages in lung transplant recipients. *Thorax* **2016**, *71*, 1006–1011. [CrossRef]
64. Hu, G.; Christman, J.W. Editorial: Alveolar Macrophages in Lung Inflammation and Resolution. *Front. Immunol.* **2019**, *10*, 2275. [CrossRef] [PubMed]
65. Santos, L.D.; Antunes, K.H.; Muraro, S.P.; de Souza, G.F.; da Silva, A.G.; de Souza Felipe, J.; Zanetti, L.C.; Czepielewski, R.S.; Magnus, K.; Scotta, M.; et al. TNF-mediated alveolar macrophage necroptosis drives disease pathogenesis during respiratory syncytial virus infection. *Eur. Respir. J.* **2020**, *57*, 2003764. [CrossRef] [PubMed]
66. Aegerter, H.; Kulikauskaite, J.; Crotta, S.; Patel, H.; Kelly, G.; Hessel, E.M.; Mack, M.; Beinke, S.; Wack, A. Influenza-induced monocyte-derived alveolar macrophages confer prolonged antibacterial protection. *Nat. Immunol.* **2020**, *21*, 145–157. [CrossRef] [PubMed]
67. Kuchtey, J.; Fulton, S.A.; Reba, S.M.; Harding, C.V.; Boom, W.H. Interferon-alphabeta mediates partial control of early pulmonary *Mycobacterium bovis* bacillus Calmette-Guerin infection. *Immunology* **2006**, *118*, 39–49. [CrossRef]
68. Saunders, B.M.; Frank, A.A.; Orme, I.M.; Cooper, A.M. Interleukin-6 Induces Early Gamma Interferon Production in the Infected Lung but Is Not Required for Generation of Specific Immunity to *Mycobacterium tuberculosis* Infection. *Infect. Immun.* **2000**, *68*, 3322–3326. [CrossRef] [PubMed]
69. Darwich, L.; Coma, G.; Peña, R.; Bellido, R.; Blanco, E.J.J.; Este, J.A.; Borras, F.E.; Clotet, B.; Ruiz, L.; Rosell, A.; et al. Secretion of interferon-γ by human macrophages demonstrated at the single-cell level after costimulation with interleukin (IL)-12 plus IL-18. *Immunology* **2009**, *126*, 386–393. [CrossRef] [PubMed]

70. Gao, D.K.; Salomonis, N.; Henderlight, M.; Woods, C.; Thakkar, K.; Grom, A.A.; Thornton, S.; Jordan, M.B.; Wikenheiser-Brokamp, K.A.; Schulert, G.S. IFN-γ is essential for alveolar macrophage–driven pulmonary inflammation in macrophage activation syndrome. *JCI Insight* **2021**, *6*, e147593. [CrossRef] [PubMed]
71. Vijayan, V.; Wagener, F.A.; Immenschuh, S. The macrophage heme-heme oxygenase-1 system and its role in inflammation. *Biochem. Pharmacol.* **2018**, *153*, 159–167. [CrossRef]
72. Fonseca, W.; Lukacs, N.W.; Ptaschinski, C. Factors Affecting the Immunity to Respiratory Syncytial Virus: From Epigenetics to Microbiome. *Front. Immunol.* **2018**, *9*, 226. [CrossRef]
73. Guo, Y.; Yu, S.; Zhang, C.; Kong, A.-N.T. Epigenetic regulation of Keap1-Nrf2 signaling. *Free Radic. Biol. Med.* **2015**, *88*, 337–349. [CrossRef]

antioxidants

Article

Increase of HO-1 Expression in Critically Ill COVID-19 Patients Is Associated with Poor Prognosis and Outcome

Maria G. Detsika [1], Ioanna Nikitopoulou [1], Dimitris Veroutis [2], Alice G. Vassiliou [1], Edison Jahaj [1], Stamatis Tsipilis [1], Nikolaos Athanassiou [1], Hariklia Gakiopoulou [3], Vassilis G. Gorgoulis [2,4,5,6,7], Ioanna Dimopoulou [1], Stylianos E. Orfanos [1] and Anastasia Kotanidou [1,*]

[1] 1st Department of Critical Care Medicine & Pulmonary Services, GP Livanos and M Simou Laboratories, Evangelismos Hospital, National and Kapodistrian University of Athens, 10675 Athens, Greece; mdetsika@med.uoa.gr (M.G.D.); joannaniki@gmail.com (I.N.); alvass75@gmail.com (A.G.V.); edison.jahaj@gmail.com (E.J.); stamostsipil@gmail.com (S.T.); nikolaosathanasiou14@gmail.com (N.A.); idimo@otenet.gr (I.D.); sorfanos@med.uoa.gr (S.E.O.)
[2] Molecular Carcinogenesis Group, Department of Histology and Embryology, Medical School, National and Kapodistrian University of Athens, 10675 Athens, Greece; dimitrisveroutis1@gmail.com (D.V.); vgorg@med.uoa.gr (V.G.G.)
[3] First Department of Pathology, Medical School, National and Kapodistrian University of Athens, 11527 Athens, Greece; charagak28@gmail.com
[4] Biomedical Research Foundation, Academy of Athens, 10675 Athens, Greece
[5] Faculty Institute for Cancer Sciences, Manchester Academic Health Sciences Centre, University of Manchester, Manchester M20 4GJ, UK
[6] Center for New Biotechnologies and Precision Medicine, Medical School, National and Kapodistrian University of Athens, 10675 Athens, Greece
[7] Faculty of Health and Medical Sciences, University of Surrey, Surrey GU2 7YH, UK
* Correspondence: akotanid@med.uoa.gr

Citation: Detsika, M.G.; Nikitopoulou, I.; Veroutis, D.; Vassiliou, A.G.; Jahaj, E.; Tsipilis, S.; Athanassiou, N.; Gakiopoulou, H.; Gorgoulis, V.G.; Dimopoulou, I.; et al. Increase of HO-1 Expression in Critically Ill COVID-19 Patients Is Associated with Poor Prognosis and Outcome. *Antioxidants* **2022**, *11*, 1300. https://doi.org/10.3390/antiox11071300

Academic Editor: Stanley Omaye

Received: 8 June 2022
Accepted: 26 June 2022
Published: 29 June 2022

Abstract: Heme-oxygenase (HO)-1 is a cytoprotective enzyme with strong antioxidant and anti-apoptotic properties and previous reports have also emphasized the antiviral properties of HO-1, either directly or via induction of interferons. To investigate the potential role of HO-1 in patients with coronavirus disease 2019 (COVID-19), the present study assessed changes in HO-1 expression in whole blood and tissue samples. Upregulation of HO-1 protein was observed in lung, liver, and skin tissue independently of severe acute respiratory syndrome coronavirus 2 (SARS-CoV-2) presence. A significant increase of blood HO-1 mRNA levels was observed in critically ill COVID-19 patients compared to those in severe COVID-19 patients and healthy controls. This increase was accompanied by significantly elevated levels of serum ferritin and bilirubin in critically ill compared to patients with severe disease. Further grouping of patients in survivors and non-survivors revealed a significant increase of blood HO-1 mRNA levels in the later. Receiver operating characteristic (ROC) analysis for prediction of ICU admission and mortality yielded an AUC of 0.705 ($p = 0.016$) and 0.789 ($p = 0.007$) respectively indicating that HO-1 increase is associated with poor COVID-19 progression and outcome. The increase in HO-1 expression observed in critically ill COVID-19 patients could serve as a mechanism to counteract increased heme levels driving coagulation and thrombosis or as an induced protective mechanism.

Keywords: heme-oxygenase (HO-1); COVID-19; immune response

1. Introduction

Heme-oxygenase (HO)-1 is a cytoprotective enzyme with strong antioxidant and anti-apoptotic properties attributed to its enzymatic activity, which involves the degradation of heme to biliverdin with simultaneous release of carbon monoxide (CO) and ferrous iron. Heme-oxygenase has a strong anti-inflammatory potential by modulating production of interleukins (IL) [1].

The strong anti-inflammatory characteristic of HO-1 has been confirmed in the condition of HO-1 depletion, both in humans and mice. Lack of HO-1 has been shown to result in hyperinflammation with a significant increase in various factors such as IL-1b, IL-6, interferon (IFN)-g, and tumour necrosis factor-a (TNF-a) both in humans and animals [2,3]. Furthermore, mice lacking HO-1 typically show increased sensitivity in sepsis models induced by endotoxins such as lipopolysaccharide (LPS) [4,5]. The same study also reported an induction of HO-1 mRNA levels following LPS administration [5]. Increased HO-1 mRNA levels have also been reported for critically ill patients irrespective of the cause [6].

HO-1 has also been reported to possess antiviral properties, both through its by-products and directly. A previous study reported that activation of HO-1 and CO release inhibited Enterovirus 71 infection by suppressing formation of reactive oxygen species (ROS) [7]. Furthermore, HO-1 was reported to directly interact with interferon regulatory factor 3 (IRF3) thus triggering INFα/β responses and eliminating influenza virus replication in-vitro [8]. The HO-1 expression has also been reported in a cohort of patients with severe disease from Dengue virus showing a strong suppression of expression with disease severity [9] while expression patterns have also been identified in human immunodeficiency (HIV) patients, hepatitis B (HBV) and hepatitis C (HCV) patients [10]. Specifically, a suppression of HO-1 expression was shown in HIV patients with high viral loads [10], while the opposite was observed in patients with HCV suggesting a variable role of HO-1 depending on the type of viral infection and/or type of virus.

Given the strong anti-inflammatory and antiviral potential of HO-1, therapeutic strategies against severe acute respiratory syndrome coronavirus 2 (SARS-CoV-2) have been proposed by various experts [11,12]. However, few studies have yet investigated the profile of HO-1 expression in coronavirus disease 2019 (COVID-19) patients in either tissue or blood level. The current study assessed HO-1 expression levels in blood (mRNA) and tissues (protein) of COVID-19 patients and correlated these levels with disease outcomes.

2. Materials and Methods

2.1. Patients

A written informed consent was obtained from all patients. Ethical approval was obtained from the Ethics committee of Evangelismos Hospital. All patients had a positive polymerase chain reaction (PCR) test for SARS-CoV-2 performed on a nasopharyngeal sample. A total of 47 patients were recruited. Patients were grouped into those with severe disease (n = 27) and critically ill (n = 20). Patients were further grouped into survivors (n = 38) and non-survivors (n = 9). Healthy volunteers with a mean age of 38 years and equal numbers of female and male individuals (n = 5) were used as controls (n = 10). Blood samples were obtained on admission for all patients and analysed on the same day. Patient demographic and clinical data are shown in Table 1. Formalin-fixed and paraffin-embedded lung, kidney, liver, heart, and skin tissue samples were obtained at autopsy from two non-survivors from the critical patient arm of the study.

Table 1. Patient clinical and demographic data.

	Severe COVID-19	Critical COVID-19	p Value
Age (years)	60.26 ± 15.96	66.65 ± 11.45	0.149
Male	12 (44.44%)	15 (75.00%)	0.111
Comorbidities	19 (70.37%)	15 (75.00%)	0.940
Symptoms			
paO2/FIO2	318.80 ± 63.62	137.90 ± 79.23	<0.0001
Days of illness before admission	5.65± 1.62	5.77 ± 2.98	0.852

Table 1. *Cont.*

	Severe COVID-19	Critical COVID-19	*p* Value
Laboratory baseline			
White blood cells (cells/μL)	6664.00 ± 3841	13897 ± 1231	<0.0001
Neutrophils (cells/μL)	68.34 ± 16.62	79.80 ± 13.01	0.012
Lymphocytes (cells/μL)	27.58 ± 19.17	12.02 ± 13.30	0.001
Platelets (cells/μL)	207333 ± 88056	271737 ± 139869	0.100
C-reactive protein (mg/L)	7.05 ± 7.57	13.67 ± 11.52	0.025
Troponin (ng/mL)	37.58 ± 109.70	311.80 ± 845.90	0.114
Urea (mg/dL)	34.89 ± 24.78	84.16 ± 94.24	<0.001
Creatinine (mg/dL)	0.82 ± 0.19	1.10 ± 0.88	0.368
Aspartate aminotransferase (U/L)	43.33 ± 37.50	150.70 ± 419.20	0.237
Alanine transaminase (U/L)	34.76 ± 22.94	62.17 ± 74.60	0.002
Gamma-Glutamyltransferase (U/L)	46.37 ± 39.22	83.26 ± 96.73	0.062
Lactate Dehydrogenase (U/L)	306.8 ± 128.9	535.6 ± 552.1	0.021
Albumin (g/dL)	3.846 ± 0.409	3.210 ± 0.506	<0.0001
Days of hospital stay	8.00 ± 4.019	18.80 ± 14.10	<0.0001
Survival	25 (92.59%)	12 (60.00%)	0.026

2.2. *Histopathology and Immunohistochemistry*

Tissue sections were heat-treated in citric acid buffer and then incubated with 2% H_2O_2 to inactivate the endogenous peroxidase. Following blocking with 3% normal goat serum (NGS) for 1 h, sections were incubated with anti-HO-1 monoclonal antibody (cat no: ADI-SPA-895, Enzo Life Sciences, Farmingdale, NY 11735, USA) at a 1:100 dilution overnight at 4 °C. After incubation with an HRP-conjugated secondary antibody, binding was detected using Vectastain ABC Kit (Vector Laboratories, Burlingame, CA, USA) according to manufacturer's instructions. The 3,3-diaminobenzidine was used as chromogen and slides were counterstained with haematoxylin and observed under an Olympus (BX50F4) microscope (Centre Valley, PA 18034, USA). For SARS-CoV-2 tissue staining with an anti-SARSCoV-2 (G2) monoclonal antibody [13] was performed at a 1:300 dilution as previously described [13]. Routine procedures for Prussian blue (detection of hemosiderin, a ferritin complex) staining were carried out.

2.3. *RNA Extraction, Reverse Transcription and Real-Time PCR*

For RNA extraction blood samples were collected in Tempus Blood RNA Tubes (Applied Biosystems, Foster City, CA, USA) and stored at −80 °C. RNA was extracted according to the manufacturer's instructions. The RNA concentration was determined for each sample prior to reverse transcription using NanoDrop One (ThermoScientific, Whaltham, MA 02451, USA). Reverse transcription reactions were performed using 4 μL of the 5× FastGene® Scriptase II ReadyMix (Nippon Genetics, Duren, Germany) and 100 ng of RNA for each reaction. Reactions were carried out in a CFX90 cycler (BioRad, Hercules, CA, USA) at the following conditions: 25 °C for 10 min, 42 °C for 60 min and 85 °C for 5 min. Real time PCR reactions were carried out in a CFX90 cycler. Each reaction consisted of 1 μL primer-probe assay mix (IDT, Coralville, IA, USA), 10 μL Luna Master Mix (Biolabs, Waltham, MA, USA) and 1 μL cDNA. The GAPDH was used as a housekeeping gene for data normalization. The sequences for the Real time PCR primers and probes for HO-1 mRNA were as follows: forward primer: 5′-GTT CCT CAT GAA CTC AGC ATT-3′, reverse primer: 5′-GAG CCA GCA CGA ACG AG-3′, probe: 56-FAM/AGC ATG CCC /ZEN/CAG GAT TTG TCA GA/3IABkFQ. Reactions were carried out in triplicate and results were analysed by the ΔΔCT method

2.4. Statistical Analysis

Results are reported as absolute numbers, medians, or means and standard deviations, as appropriate. Statistical analysis was performed using the GraphPad Prism 8.0 software for Windows. Data were tested for normality using the Shapiro–Wilks test. Unpaired *t*-test or Mann–Whitney U was used in the case of data displaying normality or not respectively. One-way Anova was used for analysis of three groups and the Fisher's least significant difference (LSD) test was used for post-hoc analysis. Receiver operating characteristic (ROC) analysis was performed using ICU admission, or survival as the classification variable and HO-1 mRNA levels on admission as prognostic variables. $p < 0.05$ was considered statistically significant.

3. Results

3.1. Tissue HO-1 Induction Is Independent from SARS-CoV-2 Presence

Tissue sections obtained post-mortem from critically ill COVID-19 patients were stained for HO-1 and hemosiderin to assess whether HO-1 induction occurred at the tissue level and whether it was accompanied by Fe deposition. As shown in Figure 1, HO-1 localization was observed in lung, liver, and skin tissue sections. Specifically, HO-1 immunohistochemical expression in the lung was observed in cells lining the cleft-like spaces which are probably endothelial cells, and in cells within the cleft-like spaces most possibly macrophages (point with arrow). In liver, HO-1 immunohistochemical expression was mainly observed in Kupffer cells (arrow) within the liver sinusoids while a weak granular stain was also discerned in the cytoplasm of hepatocytes with a perinuclear localization. In the same locations, Prussian blue histochemistry revealed abundant granular pigments indicative of ferric iron. The HO-1 immunohistochemical expression was also detected in skin tissue. Skin sweat glands with HO-1 expression were observed in the cytoplasm of epithelial cells without Fe deposition. The HO-1 was not detected in kidney or heart tissue sections and these were also negative for Prussian blue staining. To assess whether HO-1 localized in tissues that were positive for presence of SARS-COV-2, tissue sections demonstrating HO-1 induction were further stained for SARS-CoV-2 protein. The SARS-CoV-2 was only detected in the liver and the kidney (Figure 2); SARS-CoV-2 was not detected in any other tissue sample (Figure S1).

3.2. HO-1 Induction in Blood Samples of COVID-19 Patients

We next determined HO-1 induction in whole blood samples of COVID-19 patients obtained on admission in either ward or ICU. The HO-1 mRNA levels were significantly elevated in critically ill COVID-19 patient group compared to those in both healthy and severe disease groups (Figure 3a). There was no significant difference in HO-1 mRNA levels between severe disease and healthy control groups (Figure 3a). In addition to the increase in HO-1 mRNA in the critically ill group, there was also an increase in serum ferritin (Figure 3b), bilirubin (Figure 3c) and iron (Fe) levels (Figure 3d). This increase was statistically significant between the critically and severely ill groups for all parameters apart from Fe levels. Furthermore, there was no difference in haemoglobin, gamma-glutamyl transferase (GGT) and alkaline phosphatase (ALP) levels between the critically and severely ill groups (Figure 4).

3.3. HO-1 Induction in COVID-19 Patients Is Associated with Disease Progression and Mortality

We next performed ROC curve analysis to assess whether HO-1 induction predicts COVID-19 outcomes (progression and mortality). To this end, patients were further grouped into survivors (n = 38) and non-survivors (n = 9). A significant increase of HO-1 whole blood mRNA levels was also observed in non-survivors versus survivors (Figure 5b). The ROC curves were generated for prediction of ICU admission and mortality and revealed an AUC of 0.705 (95% 0.5545–0.8566, p = 0.016) and 0.789 (95% 0.5977–0.9813, p = 0.015), respectively (Figure 5c,d).

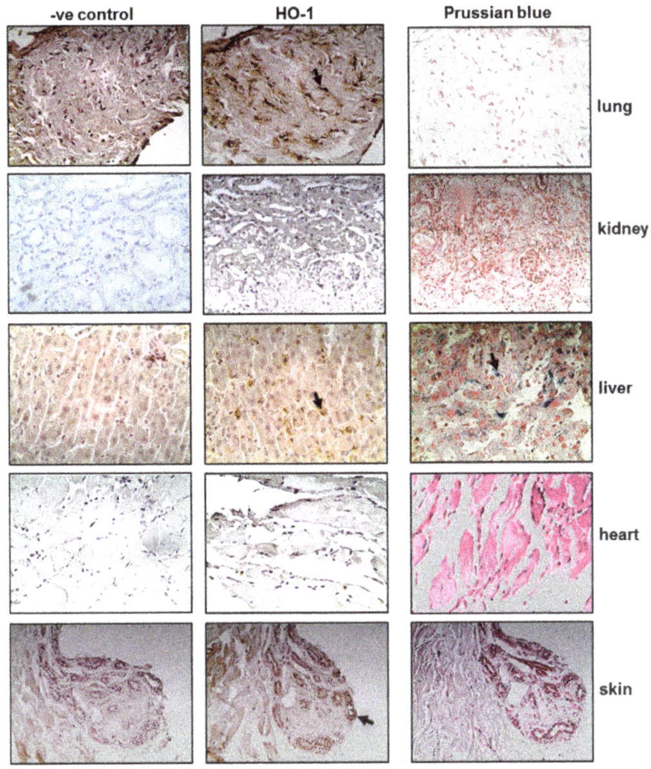

Figure 1. HO-1 upregulation and hemosiderin deposition in tissues of critical COVID-19 patients. Representative images from lung, kidney, liver, heart, and skin tissue sections obtained from critical COVID-19 patients autopsy samples and stained for HO-1 protein (and their respective controls) and Prussian blue. Magnification at ×400.

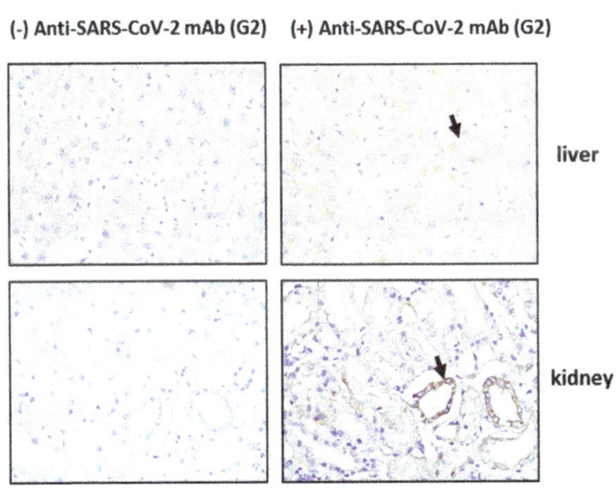

Figure 2. SARS-CoV-2 detection in tissue tissues from critical COVID-19 patients. Post-mortem tissue samples were obtained and directly processed for detection of SARS-CoV-2 by immunohistochemistry. Representative images of SARS-CoV-2 (G2 mab) staining in liver and kidney tissue sections and their respective negative controls. Magnification at ×400.

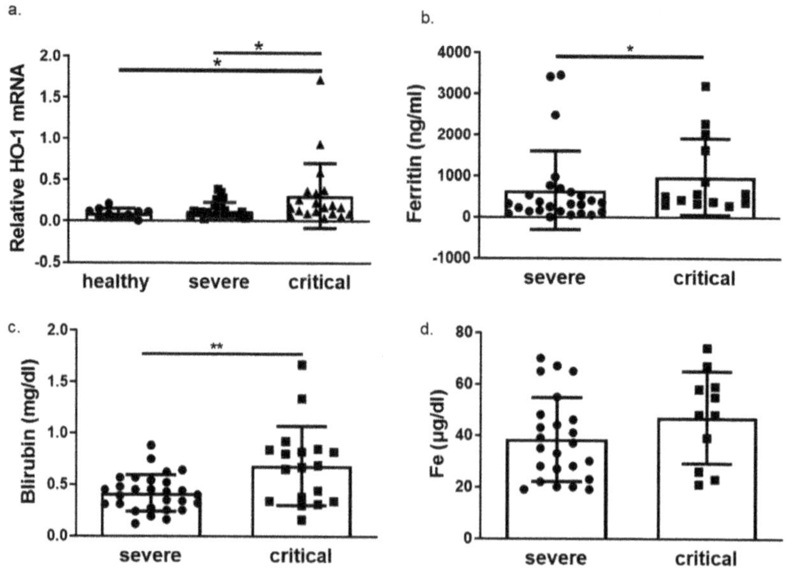

Figure 3. HO-1 is upregulated in critical COVID-19 patients with concomitant increase in ferritin and bilirubin. Whole blood samples were obtained on admission from patients with either severe or critical disease and processed directly. (**a**) HO-1 mRNA upregulation determined in critical patients versus severe and healthy controls from whole blood samples. Increase of ferritin (**b**), bilirubin (**c**) and iron levels (**d**) in critical versus severe COVID-19 patients. Data are expressed as means ± SD. Statistical analysis was performed by one-way ANOVA in three group comparisons and Mann–Whitney U test in two group comparisons. * $p < 0.05$, ** $p < 0.01$. Analysis of variance and post hoc analysis was performed by Uncorrected Fisher's least significant difference test.

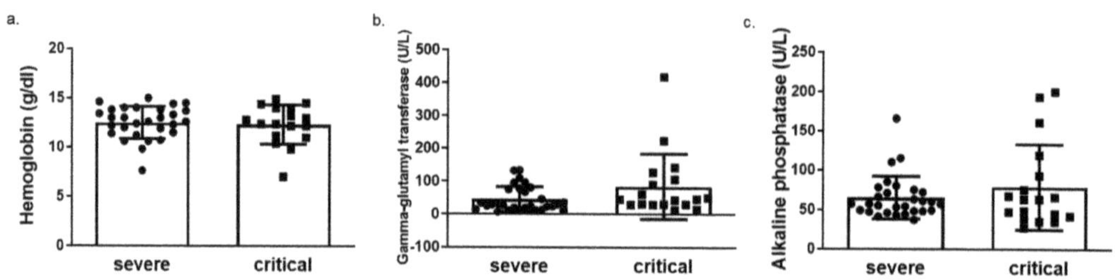

Figure 4. Unaltered levels of haemoglobin, gamma glutamyl transferase (GGT) and alkaline phosphatase (ALP) levels in critical versus severe COVID-19 patients. Levels of (**a**) haemoglobin (**b**) gamma glutamyl transferase (GGT) and (**c**) alkaline phosphatase (ALP) in blood samples obtained on admission from COVID-19 patients by routine in-hospital blood analysis. Data are expressed as means ± SD. Statistical analysis was performed by Mann–Whitney U test. No statistically significant changes were observed.

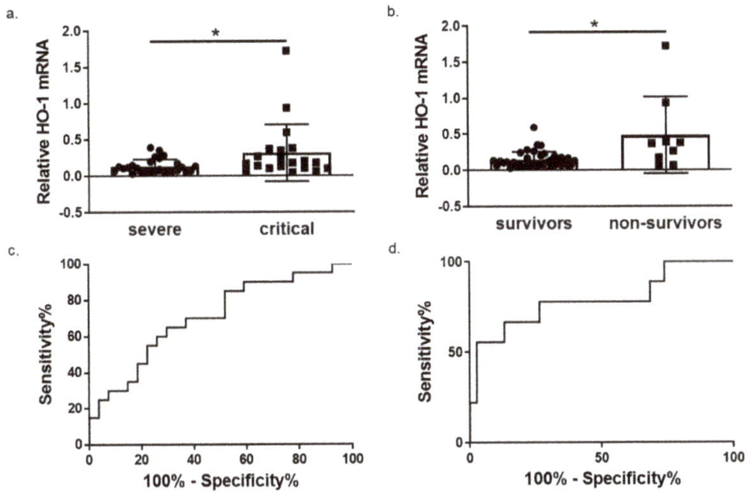

e.

	AUC	p	Sensitivity (%)	Specificity (%)	Odds ratio	Cut off
ICU admission	0.705	0.016	65.00	66.67	1.950	0.126
Mortality	0.789	0.007	77.78	73.68	2.956	0.169

Figure 5. HO-1 mRNA levels are associated with COVID-19 progression and outcome. HO-1 mRNA upregulation in whole blood samples from (**a**) critical versus severe patients (**b**) survivors versus non-survivors. Receiver operating characteristic (ROC) curves of HO-1 mRNA levels for prediction of need for intensive care (**c**) and mortality (**d**). The corresponding areas under the curve (AUC), p, sensitivity and specificity, odd ratio and cut off values are represented in (**e**). Statistical analysis was performed by Mann–Whitney U test. * $p < 0.05$.

4. Discussion

Severe and critical COVID-19 infection largely exhibits characteristics of a multi-inflammatory condition culminating in viral sepsis. The highly pro-inflammatory profile of these patients could be counteracted by the induction of anti-inflammatory molecules such as HO-1. Our results show significantly elevated whole blood HO-1 mRNA levels in critically ill patients indicating that HO-1 induction could be acting as a protective mechanism against the inflammation. The HO-1 has previously been shown to reduce IL-6 and CXCL1 levels in mouse models of ischemia/reperfusion (IRI) models of renal and lung injury [14,15]. To achieve HO-1 induction in these models, hemin was used as a natural HO-1 inducer and was found to reduce IRI cytokine storm. The HO-1 induction in COVID-19 patients could therefore potentially add to reduce the cytokine storm observed. The same studies reported a reduction of lung tissue injury upon hemin induced HO-1 upregulation.

The HO-1 also has an established antiviral effect. Previous studies have reported upregulation of HO-1 by cobalt protoporphyrin which eliminated Zika virus (ZIKV) replication in vitro in HEK-293A cells expressing a ZIKV replicon [16]. Another study described the HO-1 dependent inhibition of Ebola viral replication following HO-1 induction by cobalt protoporphyrin (CoPP) [17]. Furthermore, hemin treatment of monocytes and T cells previously infected with HIV pleiotropic strains was shown to ameliorate viral infection in vitro while tin protoporphyrin, an established inhibitor of HO-1 activity was shown to attenuate HIV infection indicating the antiviral effect of the enzymatic role of HO-1 [18]. A similar study conducted for SARS-CoV-2 recently reported no effect of HO-1 upregulation on SARS-CoV-2 replication in vitro. Specifically, the study utilized monkey kidney and human lung epithelial cell lines, infected with SARS-CoV-2 and assessed the antiviral effect of HO-1 by induction through hemin at 48 and 72 h post infection with no significant reduction in viral replication [19] despite the augmented HO-1 expression.

In our study, HO-1 was detected in lung, liver, and skin tissue sections of critically ill patients. The detection of HO-1 in tissue samples was independent from SARS-CoV-2 presence as HO-1 presence was confirmed in lung and skin tissue sections that were negative for SARS-COV-2, while absence of HO-1 was observed in renal tissue that was positive for SARS-COV-2. This observation indicates that the induction of HO-1 targets COVID-19 independent of a direct antiviral effect possibly because of the systemic inflammatory response triggered during the infection. A limitation of our study regarding staining of tissues from COVID-19 patients is the lack of healthy controls as well as the limited number of patients included in the study ($n = 2$). Instead, we have used a negative control of the same tissue sample. Therefore, although HO-1 was detected in specific tissue samples we could not assess the degree of protein expression. Another study has also described the detection of HO-1 in skin tissue samples from COVID-19 patients [20]. However, the study included patients with mild symptoms and the specimens were obtained from specific skin lesions and reported downregulation of HO-1 protein expression when compared to healthy controls. Further analysis in samples from critically ill patients is needed to confirm our results.

The increase of HO-1 whole blood mRNA levels observed in critical COVID-19 patients was coupled to increased Fe, albeit not significantly, and a significant increase in bilirubin (Figure 3). Whether the increase in Fe and bilirubin levels could be attributed to the HO-1 induction (which releases heme-bound iron and generates bilirubin) is difficult to assess. However, the fact that GGT and ALT levels were no different between severe and critically ill patients (Figure 4) indicates that the increase in bilirubin was not due to liver damage. The elevated levels of Fe observed in these patients, however, could be another mechanism for HO-1 induction. Specifically, Fe is needed for heme synthesis, which is the natural inducer of HO-1. The Fe is usually bound by ferritin, the levels of which were increased in tandem with HO-1 mRNA increase (Figure 3). Any unbound Fe may therefore be captured for heme synthesis leading to HO-1 induction. However, apart from liver tissue, the induction of HO-1 in tissue levels was not always associated with Fe deposition.

The HO-1 mRNA levels were shown to be significantly elevated in non-survivors and ROC analysis showed a high AUC value indicating that this increase of HO-1 expression is associated with mortality. Whether this association of HO-1 with mortality is due to the upregulation of HO-1 as a protective mechanism or whether the overexpression of HO-1 may intensify a 'lethal' effect remains to be unravelled. Specifically, it has been shown that apart from its beneficial properties, HO-1 overexpression may have detrimental effect both at tissue and circulation level and this is mainly attributed to the release of heme-derived catalytically active iron which has been shown to promote a pro-oxidant environment [21,22]. In our study, Fe blood levels were increased in critically ill patients but not significantly. Furthermore, circulating iron levels reflect transferrin bound iron and not heme derived iron. Moreover, iron deposition at tissue levels was only observed in the liver of critically ill patients indicating that the upregulation of HO-1 remains within the 'non-toxic threshold' of expression. Taken together, the observations above, indicate that the association of HO-1 upregulation with mortality of COVID-19 patients is probably due to its induction as a protective mechanism rather than a toxic epiphenomenon of the course of COVID-19 viral infection.

This study showed a strong association of increased HO-1 expression and disease progression as well as outcome, specifically, mortality. The high AUC values shown for the need for ICU treatment and mortality allow for a suggestion of the potential use of HO-1 as a biomarker for COVID-19 disease progression and outcome. However, considering the limited size of the study population, HO-1 may be proposed as a strong indicator rather that a biomarker of COVID-19 progression and mortality. Validation of HO-1 expression levels in larger COVID-19 patient cohorts is therefore necessary to confirm the potential use of HO-1 as a prognostic tool for COVID-19 management and to compare its performance with other established parameters used for COVID-19 disease progression, such as C-reactive protein (CRP) and the neutrophil to lymphocyte ratio (NLR). Another study previously reported elevated levels of HO-1 in COVID-19 patients with low oxygen levels ($SpO2 \leq 95\%$) [23], measured by a standard ELISA method accompanied by increased heme levels. However, the study was also performed on a limited number of patients and stressed the need for further investigation in larger cohorts. Finally, one study reported the use of HO-1 as a marker for disease progression in a patient with exacerbation of idiopathic pulmonary fibrosis (IPF) following SARS-CoV-2 infection and COVID-19, and they found that serum HO-1 which reflected M2 macrophage activation allowed monitoring of the disease progression [24]. However, the study was focused specifically on patients with exacerbation of IPF following COVID-19 and did not suggest HO-1 as a marker for COVID-19 progression in general.

5. Conclusions

This study describes the expression pattern of HO-1 in COVID-19 patients, both at the tissue level and the blood mRNA level. The increase of HO-1 in critically ill and disease COVID-19 patients indicates that its induction serves as a protective mechanism. Further studies are needed to determine the underlying mechanisms for HO-1 induction in COVID-19 to directly assess its potential as a therapeutic strategy against COVID-19.

Supplementary Materials: The following supporting information can be downloaded at https://www.mdpi.com/article/10.3390/antiox11071300/s1. Figure S1: SARS-COV-2 absence in tissue samples from critical COVID-19 patients.

Author Contributions: Conceptualization, M.G.D.; formal analysis, M.G.D.; resources, A.K.; data curation, M.G.D., I.N., D.V., A.G.V., E.J., S.T. and N.A.; writing—original draft preparation, M.G.D.; writing—review and editing, H.G., I.D., S.E.O., V.G.G. and A.K.; supervision, A.K. All authors have read and agreed to the published version of the manuscript.

Funding: This research received no external funding.

Institutional Review Board Statement: The study was conducted according to the guidelines of the Declaration of Helsinki and approved by the Institutional Review Board (Ethics Committee) of Evangelismos Hospital (protocol code 360 and date of approval 17 September 2020).

Informed Consent Statement: Informed consent was obtained from all subjects involved in the study.

Data Availability Statement: The data is contained within the manuscript and the Supplementary Material.

Conflicts of Interest: The authors declare no conflict of interest.

References

1. Maines, M.D. The heme oxygenase system: A regulator of second messenger gases. *Annu. Rev. Pharmacol. Toxicol.* **1997**, *37*, 517–554. [CrossRef] [PubMed]
2. Grochot-Przeczek, A.; Dulak, J.; Jozkowicz, A. Haem oxygenase-1: Non-canonical roles in physiology and pathology. *Clin. Sci.* **2012**, *122*, 93–103. [CrossRef] [PubMed]
3. Poss, K.D.; Tonegawa, S. Heme oxygenase 1 is required for mammalian iron reutilization. *Proc. Natl. Acad. Sci. USA* **1997**, *94*, 10919–10924. [CrossRef] [PubMed]
4. Espinoza, J.A.; Gonzalez, P.A.; Kalergis, A.M. Modulation of Antiviral Immunity by Heme Oxygenase-1. *Am. J. Pathol.* **2017**, *187*, 487–493. [CrossRef]
5. Wiesel, P.; Patel, A.P.; DiFonzo, N.; Marria, P.B.; Sim, C.U.; Pellacani, A.; Maemura, K.; LeBlanc, B.W.; Marino, K.; Doerschuk, C.M.; et al. Endotoxin-induced mortality is related to increased oxidative stress and end-organ dysfunction, not refractory hypotension, in heme oxygenase-1-deficient mice. *Circulation* **2000**, *102*, 3015–3022. [CrossRef]
6. Saukkonen, K.; Lakkisto, P.; Kaunisto, M.A.; Varpula, M.; Voipio-Pulkki, L.M.; Varpula, T.; Pettila, V.; Pulkki, K. Heme oxygenase 1 polymorphisms and plasma concentrations in critically ill patients. *Shock* **2010**, *34*, 558–564. [CrossRef]
7. Tung, W.H.; Hsieh, H.L.; Lee, I.T.; Yang, C.M. Enterovirus 71 induces integrin beta1/EGFR-Rac1-dependent oxidative stress in SK-N-SH cells: Role of HO-1/CO in viral replication. *J. Cell. Physiol.* **2011**, *226*, 3316–3329. [CrossRef]
8. Ma, L.L.; Zhang, P.; Wang, H.Q.; Li, Y.F.; Hu, J.; Jiang, J.D.; Li, Y.H. Heme oxygenase-1 agonist CoPP suppresses influenza virus replication through IRF3-mediated generation of IFN-alpha/beta. *Virology* **2019**, *528*, 80–88. [CrossRef]

9. Wu, Y.H.; Chen, W.C.; Tseng, C.K.; Chen, Y.H.; Lin, C.K.; Lee, J.C. Heme oxygenase-1 inhibits DENV-induced endothelial hyperpermeability and serves as a potential target against dengue hemorrhagic fever. *FASEB J. Off. Publ. Fed. Am. Soc. Exp. Biol.* **2022**, *36*, e22110. [CrossRef]
10. Jablonowska, E.; Wojcik, K.; Szymanska, B.; Omulecka, A.; Cwiklinska, H.; Piekarska, A. Hepatic HMOX1 expression positively correlates with Bach-1 and miR-122 in patients with HCV mono and HIV/HCV coinfection. *PLoS ONE* **2014**, *9*, e95564. [CrossRef]
11. Wagener, F.; Pickkers, P.; Peterson, S.J.; Immenschuh, S.; Abraham, N.G. Targeting the Heme-Heme Oxygenase System to Prevent Severe Complications Following COVID-19 Infections. *Antioxidants* **2020**, *9*, 540. [CrossRef] [PubMed]
12. Rossi, M.; Piagnerelli, M.; Van Meerhaeghe, A.; Zouaoui Boudjeltia, K. Heme oxygenase-1 (HO-1) cytoprotective pathway: A potential treatment strategy against coronavirus disease 2019 (COVID-19)-induced cytokine storm syndrome. *Med. Hypotheses* **2020**, *144*, 110242. [CrossRef] [PubMed]
13. Evangelou, K.; Veroutis, D.; Paschalaki, K.; Foukas, P.G.; Lagopati, N.; Dimitriou, M.; Papaspyropoulos, A.; Konda, B.; Hazapis, O.; Polyzou, A.; et al. Pulmonary infection by SARS-CoV-2 induces senescence accompanied by an inflammatory phenotype in severe COVID-19: Possible implications for viral mutagenesis. *Eur. Respir. J.* **2022**. [CrossRef] [PubMed]
14. Chok, M.K.; Ferlicot, S.; Conti, M.; Almolki, A.; Durrbach, A.; Loric, S.; Benoit, G.; Droupy, S.; Eschwege, P. Renoprotective potency of heme oxygenase-1 induction in rat renal ischemia-reperfusion. *Inflamm. Allergy Drug Targets* **2009**, *8*, 252–259. [CrossRef] [PubMed]
15. Rossi, M.; Delbauve, S.; Roumeguere, T.; Wespes, E.; Leo, O.; Flamand, V.; Le Moine, A.; Hougardy, J.M. HO-1 mitigates acute kidney injury and subsequent kidney-lung cross-talk. *Free. Radic. Res.* **2019**, *53*, 1035–1043. [CrossRef] [PubMed]
16. El Kalamouni, C.; Frumence, E.; Bos, S.; Turpin, J.; Nativel, B.; Harrabi, W.; Wilkinson, D.A.; Meilhac, O.; Gadea, G.; Despres, P.; et al. Subversion of the Heme Oxygenase-1 Antiviral Activity by Zika Virus. *Viruses* **2018**, *11*, 2. [CrossRef] [PubMed]
17. Hill-Batorski, L.; Halfmann, P.; Neumann, G.; Kawaoka, Y. The cytoprotective enzyme heme oxygenase-1 suppresses Ebola virus replication. *J. Virol.* **2013**, *87*, 13795–13802. [CrossRef]
18. Devadas, K.; Dhawan, S. Hemin activation ameliorates HIV-1 infection via heme oxygenase-1 induction. *J. Immunol.* **2006**, *176*, 4252–4257. [CrossRef]
19. Maestro, S.; Cordoba, K.M.; Olague, C.; Argemi, J.; Avila, M.A.; Gonzalez-Aseguinolaza, G.; Smerdou, C.; Fontanellas, A. Heme oxygenase-1 inducer hemin does not inhibit SARS-CoV-2 virus infection. *Biomed. Pharmacother.* **2021**, *137*, 111384. [CrossRef]
20. Cazzato, G.; Colagrande, A.; Cimmino, A.; Cicco, G.; Scarcella, V.S.; Tarantino, P.; Lospalluti, L.; Romita, P.; Foti, C.; Demarco, A.; et al. HMGB1-TIM3-HO1: A New Pathway of Inflammation in Skin of SARS-CoV-2 Patients? A Retrospective Pilot Study. *Biomolecules* **2021**, *11*, 1219. [CrossRef]
21. Suttner, D.M.; Dennery, P.A. Reversal of HO-1 related cytoprotection with increased expression is due to reactive iron. *FASEB J. Off. Publ. Fed. Am. Soc. Exp. Biol.* **1999**, *13*, 1800–1809. [CrossRef] [PubMed]
22. Kvam, E.; Hejmadi, V.; Ryter, S.; Pourzand, C.; Tyrrell, R.M. Heme oxygenase activity causes transient hypersensitivity to oxidative ultraviolet a radiation that depends on release of iron from heme. *Free. Radic. Biol. Med.* **2000**, *28*, 1191–1196. [CrossRef]
23. Su, W.L.; Lin, C.P.; Hang, H.C.; Wu, P.S.; Cheng, C.F.; Chao, Y.C. Desaturation and heme elevation during COVID-19 infection: A potential prognostic factor of heme oxygenase-1. *J. Microbiol. Immunol. Infect.* **2021**, *54*, 113–116. [CrossRef] [PubMed]
24. Hara, Y.; Oshima, Y.; Tagami, Y.; Aoki, A.; Fujii, H.; Izawa, A.; Seki, K.; Kanai, A.; Yabe, A.; Watanabe, K.; et al. Clinical importance of serum heme oxygenase-1 measurement in patients with acute exacerbation of idiopathic pulmonary fibrosis triggered by coronavirus disease 2019. *Respir. Med. Case Rep.* **2022**, *36*, 101615. [CrossRef] [PubMed]

antioxidants

Review

A Journey into the Clinical Relevance of Heme Oxygenase 1 for Human Inflammatory Disease and Viral Clearance: Why Does It Matter on the COVID-19 Scene?

Ayelen Toro [1,2,*], María Sol Ruiz [1,2], Sofia Lage-Vickers [1,2], Pablo Sanchis [1,2], Agustina Sabater [1,2], Gaston Pascual [1,2], Rocio Seniuk [1,2], Florencia Cascardo [1,2], Sabrina Ledesma-Bazan [1,2], Felipe Vilicich [1,2], Elba Vazquez [1,2] and Geraldine Gueron [1,2,*]

1 Laboratorio de Inflamación y Cáncer, Departamento de Química Biológica, Facultad de Ciencias Exactas y Naturales, Universidad de Buenos Aires, Buenos Aires C1428EGA, Argentina; ma.sol.ruiz@qb.fcen.uba.ar (M.S.R.); sofilage@qb.fcen.uba.ar (S.L.-V.); pablosanchis@qb.fcen.uba.ar (P.S.); asabater@qb.fcen.uba.ar (A.S.); gpascual@qb.fcen.uba.ar (G.P.); rseniuk@qb.fcen.uba.ar (R.S.); fcascardo@qb.fcen.uba.ar (F.C.); sledesmabazan@qb.fcen.uba.ar (S.L.-B.); fvilicich@qb.fcen.uba.ar (F.V.); elba@qb.fcen.uba.ar (E.V.)

2 CONICET-Universidad de Buenos Aires, Instituto de Química Biológica de la Facultad de Ciencias Exactas y Naturales (IQUIBICEN), Buenos Aires C1428EGA, Argentina

* Correspondence: ayelentoro@qb.fcen.uba.ar (A.T.); ggueron@gmail.com (G.G.); Tel.: +54-9114-4078-7796 (G.G.); Fax: +54-9114-788-5755 (G.G.)

Abstract: Heme oxygenase 1 (HO-1), the rate-limiting enzyme in heme degradation, is involved in the maintenance of cellular homeostasis, exerting a cytoprotective role by its antioxidative and anti-inflammatory functions. HO-1 and its end products, biliverdin, carbon monoxide and free iron (Fe^{2+}), confer cytoprotection against inflammatory and oxidative injury. Additionally, HO-1 exerts antiviral properties against a diverse range of viral infections by interfering with replication or activating the interferon (IFN) pathway. Severe cases of coronavirus disease 2019 (COVID-19), an infectious disease caused by severe acute respiratory syndrome coronavirus 2 (SARS-CoV-2), are characterized by systemic hyperinflammation, which, in some cases, leads to severe or fatal symptoms as a consequence of respiratory failure, lung and heart damage, kidney failure, and nervous system complications. This review summarizes the current research on the protective role of HO-1 in inflammatory diseases and against a wide range of viral infections, positioning HO-1 as an attractive target to ameliorate clinical manifestations during COVID-19.

Keywords: heme oxygenase 1; COVID-19; influenza A virus; respiratory syncytial virus; human immunodeficiency virus; Ebola virus; Dengue virus; Zika virus; Hepatitis virus; SARS-CoV-2

Citation: Toro, A.; Ruiz, M.S.; Lage-Vickers, S.; Sanchis, P.; Sabater, A.; Pascual, G.; Seniuk, R.; Cascardo, F.; Ledesma-Bazan, S.; Vilicich, F.; et al. A Journey into the Clinical Relevance of Heme Oxygenase 1 for Human Inflammatory Disease and Viral Clearance: Why Does It Matter on the COVID-19 Scene? *Antioxidants* **2022**, *11*, 276. https://doi.org/10.3390/antiox11020276

Academic Editors: Elias Lianos and Maria G. Detsika

Received: 7 January 2022
Accepted: 26 January 2022
Published: 29 January 2022

Publisher's Note: MDPI stays neutral with regard to jurisdictional claims in published maps and institutional affiliations.

1. Introduction

Severe acute respiratory syndrome coronavirus 2 (SARS-CoV-2) emerged in late 2019 in Wuhan, China. The World Health Organization (WHO) declared coronavirus disease 2019 (COVID-19) a pandemic health emergency as of 31 January 2020. The treatment goal in COVID-19 patients is to prevent or to decrease the strong virus induced inflammatory stimuli associated with a wide spectrum of poor prognosis clinical manifestations [1]. Heme oxygenase 1 (HO-1) is a microsomal enzyme with a primary antioxidant and anti-inflammatory role involved in heme degradation, generating carbon monoxide (CO), biliverdin (BV), and free iron (Fe^{2+}) [2]. Hence, HO-1 induction is a useful approach for inflammatory diseases treatment [3–6]. Additionally, HO-1 displays antiviral properties against a wide range of viruses [7]. Hemin, a previously Food and Drug Administration (FDA) and European Medicines Agency (EMA) approved drug for acute intermittent porphyria treatment [8,9], is a well known inducer of HO-1 that increases its plasma concentration in humans. Thus, hemin rises as a promising drug candidate against the replication of different viruses, including SARS-CoV-2. In this review, we summarize the current research on the protective role of HO-1 in inflammatory diseases and in a diverse range of viral infections, positioning this protein as a potential therapeutic target to ameliorate COVID-19's clinical manifestations.

2. Re-Emergence of the Coronavirus Disease

Coronaviruses (CoVs) are a large family of positive sense, single stranded RNA (+ssRNA) viruses that infect humans, other mammals, and birds, causing respiratory, enteric, hepatic, and neurologic diseases [10,11]. CoVs first became renowned in 2002–2003 during an outbreak of a virus with zoonotic origin; the severe acute respiratory syndrome coronavirus (SARS-CoV) originated in China, with 8096 cases and 774 deaths reported between 2002 and 2003 [12], and a case–fatality ratio of 7.2% [13]. In 2012, the Middle East respiratory syndrome coronavirus (MERS-CoV), another virus with zoonotic origin, emerged in Saudi Arabia and caused 927 fatalities among 2581 registered cases [14,15].

By the end of 2019, the Wuhan Health Commission from China reported a number of pneumonia cases of unknown cause and varying severity in the city of Wuhan, China. High throughput sequencing allowed the quick identification of a novel CoV belonging to the beta-coronavirus family, which was named severe acute respiratory syndrome coronavirus 2 (SARS-CoV-2), as the causative agent of Coronavirus disease 2019 (COVID-19) [16]. This pathogen rapidly spread globally via travel related cases; constituting a pandemic and setting an immense challenge for public health [16]. Unlike SARS-CoV and MERS-CoV, SARS-CoV-2 can be transmitted among people before the onset of symptoms or from asymptomatic individuals [17], limiting effective control of the spread. As of 30 December 2021, the WHO reports 281,808,270 confirmed cases and 5,411,759 deaths worldwide [15].

Although COVID-19 is primarily a respiratory disease, SARS-CoV-2 has the capacity to infect a broad range of cell types in different organs and systems, including the central nervous system [18]. SARS-CoV-2 infection begins in the proximal airways and could trigger severe and sometimes fatal symptoms when reaching the distal lung [19]. Among the severe respiratory diseases caused by SARS-CoV-2, acute respiratory distress syndrome (ARDS) and acute lung injury (ALI) [20,21] are complications mainly caused by an exacerbated immune response in elders and patients with comorbidities [22]. Cytokine storm contributes to ALI and the development of ARDS in patients with severe pneumonia caused by SARS-CoV-2, as well as SARS-CoV and MERS-CoV [23,24]. Surprisingly, in addition to the respiratory symptoms, patients may suffer from cardiac, hematological, neurological, hepatic, gastrointestinal and kidney complications [25]. COVID-19 may result in long term sequelae characterized by organ injuries that cannot be completely reversed. Several patients, even those with mild cases, may develop lasting symptoms that can have disabling consequences [26,27].

The COVID-19 era is far from being constrained, and the emergence of new viral variants causing future outbreaks remains a threat. As of December 2021, the WHO has defined five variants of concern (Alpha (B.1.1.7), Beta (B.1.351), Gamma (P.1), Delta (B.1.617.2), and Omicron (B.1.1.529)), as well as two variants of interest (Lambda (C.37) and Mu (B.1.621)) [15]. It has been reported that some variants, such as the recently described Omicron [28], have higher transmission rates and are able to escape neutralizing antibodies generated by natural infection and vaccination, or therapeutic antibodies [29–32]. Incomplete knowledge on the pathogenesis of SARS-CoV-2 and the diverse array of symptoms and manifestations of COVID-19 pose a great challenge for the development of effective treatments that should mainly focus on both, decreasing viral replication and modulating the immune response.

3. Cytokine Storm and Inflammation

Inflammation involves defense mechanisms against infection or injury. It is responsible for activating both innate and adaptive immune responses [33,34]. During infections, innate cells recognize pathogen associated molecular patterns (PAMPs) from the invading agent. In the case of inflammation triggered by tissue damage, trauma or ischemia, innate cells recognize host specific molecules that are released during cell injury or necrotic death, defined as damage associated molecular patterns, such as nucleic acids and adenosine triphosphate (ATP) [33]. During the early stages of inflammation, innate immune cells and endothelial cells (EC) release a diverse set of cytokines: chemotactic cytokines, such as monocyte chemotactic protein-3 (MCP-3) and interferon (IFN) γ-induced protein 10 (IP-10), and recruit other immune cells to the site of infection or inflammation. Proinflammatory cytokines, such as tumor necrosis factor alpha (TNF-α), interleukin-6 (IL-6), and IL-1β [35,36], are also released and trigger the activation of inflammatory pathways, including the mitogen activated protein kinase (MAPK), nuclear factor kappa-B (NF-κB), and Janus kinase (JAK)-signal transducer and activator of transcription (STAT) pathways [34]. Some pathogenic viruses (i.e., highly virulent subtypes of influenza) and bacteria (i.e., *Francisella tularensis*) can induce the acute dysregulated production of inflammatory cytokines, known as "cytokine storm" or hypercytokinemia [37]. The hypercytokinemia and exacerbated secondary events, such as coagulation, eventually result in widespread necrosis, organ failure and death [33,38].

Once SARS-CoV-2 infects target cells, innate immune cells are recruited to the infection site, where they release cytokines and initiate the activation cascade of adaptive B and T cell immune responses [39]. In most cases, the immune system is able to eliminate virus infected cells and resolve the immune response. However, in some patients, this process is dysfunctional, impairing the effective clearance of infected cells, and causing severe damage to the host [40].

4. The Lead Role of Interferons upon Viral Infection

During viral infections, pattern recognition receptors are stimulated to produce IFN by the innate immune cells. IFNs are crucial for the induction of an antiviral state via autocrine and paracrine signaling. There are three types of IFNs: type I, type II and type III. Type I (IFNα, IFNβ, IFNω, IFNτ, IFNε) and Type III (IFNλ1, IFNλ2/3, IFNλ4) share similar dynamics after binding to its receptor, as cross-phosphorylation between JAK1 and tyrosine kinase 2 (TYK2) occurs [41]. Subsequently, a docking site for STAT1 is exposed, STAT1 is phosphorylated, translocates to the nucleus, and induces the transcription of interferon stimulated genes (ISGs). The IFNs biological effects are wide, including immuno-regulation, antiviral, anti-angiogenic, and pro-apoptotic functions [42]. However, many pathogens have evolved to elude the action mechanisms of these powerful cytokines [43–45].

In critically ill COVID-19 patients, a hyperinflammation state prevails. In March 2020, a retrospective study on 150 patients from Wuhan, China, found elevated levels of IL-6 and C-reactive protein (CRP) in SARS-CoV-2 infected patients that died compared with discharged patients [46]. An independent report based on 50 COVID-19 patients with moderate to severe disease, identified IP-10, MCP-3, and IL-1 receptor antagonist (IL-1ra) as independent predictors for disease severity [47]. A longitudinal analysis showcased IL-18 and IFN-α as top biomarkers for predicting mortality. Consequently, higher counts of inflammatory monocytes, plasmablast like neutrophils and eosinophils have been described in patients with severe disease [39].

Blanco-Melo et al. reported an impairment in the response of type I and type III IFNs against SARS-CoV-2 infection [40]. In contrast, a recent study found that severe COVID-19 cases showed an exacerbated expression of type I IFNs, which could lead to augmented inflammation [48]. Several clinical trials evaluating IFNs have been carried out in COVID-19 patients. Two different studies showed a reduction in the mortality rate after IFNβ-1a and IFNβ-1b treatment [49,50]. Another study, using IFNα-2b, reported a decrease in detectable SARS-CoV-2 in the upper respiratory tract associated with lower inflammatory cytokines levels, such as IL-6 and CRP [51]. In addition, peginterferon λ treatment was associated with a reduction in viral RNA [52]. Furthermore, there are several ongoing clinical trials, using recombinant human IFNs or IFNs combined with other drugs [53–55]. These evidences highlight IFNs as potential targets for COVID-19 treatment. In the next section we will focus on the stimulation of IFN pathway by HO-1 induction.

5. Understanding the Protective Role of Heme Oxygenase 1

Heme oxygenases (HO) are metabolic enzymes that partake in the degradation of the heme group [2]. To date, three isoforms of this protein have been found: HO-1, which can be induced by external factors (such as hypoxia, oxidative stress, heat shock, reactive oxygen species (ROS), among others) [56]; HO-2, a constitutively expressed isoform; and HO-3, a nonfunctional isoform in humans [57].

In particular, HO-1, encoded by the *HMOX1* gene, is involved in the maintenance of cellular homeostasis, exerting a cytoprotective role by its anti-inflammatory, anti-oxidative and anti-apoptotic functions, as revealed in a human case of genetic HO-1 deficiency [58]. This enzyme participates not only in normal physiological processes, but also performs a protective role in inflammatory physiopathological conditions, such as kidney disease [59], cancer [60,61], cardiovascular disease [62], asthma [63] and inflammatory bowel diseases [4,64].

HO-1 is expressed in most cell types and tissues; however, its capacity to counteract inflammation seems to be critically dependent on its specific functions in myeloid cells and in EC [65]. In myeloid cells, HO-1 acts as a key regulator of the TLR4/TLR3/IRF3 induced production of IFN-β and primary IRF3 target genes in macrophages [66] and modulates maturation and specific functions of dendritic cells [67,68]. Moreover, HO-1 over-expression in macrophages negatively regulates the expression of diverse proinflammatory molecules and increases the expression of anti-inflammatory cytokines [69–71]. Among HO-1 effects on EC, it is significant to mention its ability to inhibit the expression of pro-inflammatory genes related to EC activation, such as the TNF-α-induced adhesion molecules, E-selectin and VCAM-1, via a mechanism associated with the inhibition of NF-κB activation [72].

HO-1 cleaves the heme group generating BV, CO and Fe^{2+}. Heme is usually bound to a myriad of proteins and it is involved in several homeostatic functions [56]. However, elevated concentrations of heme can cause cell damage because it is a pro-oxidant molecule. It can diffuse through cell membranes and deliver a redox active iron, producing ROS [73]. Excessive amounts of these molecules are toxic and induce oxidative stress that, in turn, generates DNA and protein damage, aggregation and lipid peroxidation, triggering cells permeability and driving cell lysis and death [73].

Several studies highlight heme catabolism end products as potential therapeutic targets in vascular disease, based on their anti-inflammatory and antiproliferative functions [74]. BV and its reduced form, bilirubin (BR), are powerful antioxidants that are able to scavenge ROS and counteract the oxidative stress. BV and BR are critical for the regulation of inflammation by exerting immunosuppressive effects [75], as they have been reported to have potent anti-inflammatory activity against insulin resistance by reducing visceral obesity and adipose tissue inflammation [76].

In addition, CO is considered an anti-apoptotic [77], antiproliferative and anti-inflammatory factor [78]. CO contributes to blood vessel development [79] and promotes angiogenesis, a crucial process involved in tissue reparation after a pathological state [80]. Interestingly, CO also reduces inflammation and inhibits apoptosis by interacting with antigen-presenting cells and suppressing T cell proliferation [81]. Moreover, it has been reported that it downregulates proinflammatory cytokines via the p38/MAPK pathway in RAW 264.7 macrophages and C57BL/6 mice [70], by the c-Jun N-terminal kinase (JNK) pathway in a murine model of sepsis [82] and through the extracellular signal regulated kinase (ERK) signaling pathway in $CD4^+$ T cells [81]. Further, HO-1/CO induced downregulation of the NLRP3 (NOD-, LRR- and pyrin domain-containing protein 3) inflammasome activation has been demonstrated in different models of murine hepatic and lung inflammatory injury [83–85].

Moreover, the HO-1 mediated increase in Fe^{2+} concentration upregulates the expression of ferritin, an iron chelating protein [86]. Ferritin exerts antioxidative and cytoprotective effects [74], as this product scavenges redox active Fe^{2+}, rendering it not harmful for cells and avoiding subsequent production of ROS via Fenton reaction. Fe^{2+} performs its function by inhibiting IL-2 and IgG production, and downregulating the MAPK and NF-κB signaling pathways [56,75,87].

Therapeutic Potential of HO-1 Induction to Treat Chronic Inflammation

As HO-1 and its reaction products exert protective anti-inflammatory effects in different preclinical models [3–5,88–90], the induction of the HO-1 system has emerged as a promising potential therapy for chronic inflammatory diseases. Most of the studied strategies are based on the use of the traditional pharmacological inducers: hemin [91,92], an FDA and EMA approved drug, and cobalt protoporphyrin IX (CoPP) [5,93,94]. In addition, many phytochemicals, such as quercetin, curcumin and resveratrol, are currently under investigation, due to their potential as HO-1 alternative inducers to counteract inflammation processes with lower cytotoxic secondary effects [95–97]. Alternatively, there are also a few approved drugs, such as 5-aminosalicylic acid (5-ASA), dimethyl fumarate (DMF), and 5-aminolevulinic acid (5-ALA), whose beneficial properties in inflammatory conditions are explained, at least in part, by their capacity to induce HO-1 [96,98,99] (Figure 1). Additionally, another effective option is the use of BV/BR based therapies, which have proven to be effective for these chronic pathologies [100–102] and/or the direct administration of CO via inhalation, CO-releasing molecules (CORMs) or hybrid carbon monoxide-releasing molecules (HYCOs). HYCOs are a type of compound where CORMs are combined with DMF, causing a powerful anti-inflammatory action due to its effect on the NRF2/HO-1 pathway [5,100,103–106] (Figure 1).

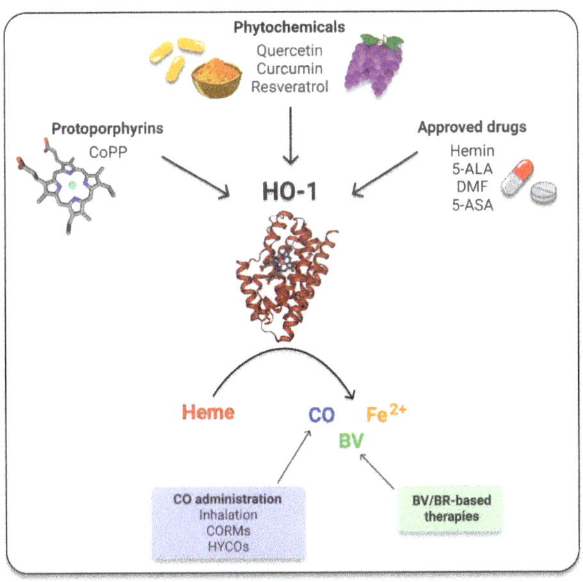

Figure 1. Inducers of HO-1. HO-1 degrades heme producing equimolar amounts of carbon monoxide (CO), biliverdin (BV) and Fe^{2+}. HO-1's inducers are grouped into protoporphyrins, a type of porphyrins that forms heme; phytochemicals, natural antioxidants compounds contained in plants; and approved drugs, compounds that were previously approved by the FDA. CoPP: cobalt protoporphyrin IX; 5-ALA: 5-aminolevulinic acid; DMF: dimethyl fumarate; 5-ASA: 5-aminosalicylic acid; CORMs: CO-releasing molecules; HYCOs: Hybrid carbon monoxide-releasing molecules; BR: bilirubin. The images of HO-1 and CoPP were taken from RCSB PDB (PDB ID: 1N3U) and The National Center for Biotechnology Information [107,108].

6. HO-1 Mechanism of Action against Inflammatory Lung Diseases

There is extensive literature about the role of HO-1 in lung diseases. This protein is expressed in type II pneumocytes and in alveolar macrophages and contributes to the protection of the lung tissue. The main HO-1 inducers in the lungs are pro-inflammatory cytokines, such as TNF-α and IL-6, the heme group and nitric oxide (NO), as well as hypoxia and hyperoxia conditions [109] (Figure 2). There is sound evidence that states that HO-1 induction is a critical defense factor during acute and chronic lung processes [109–111].

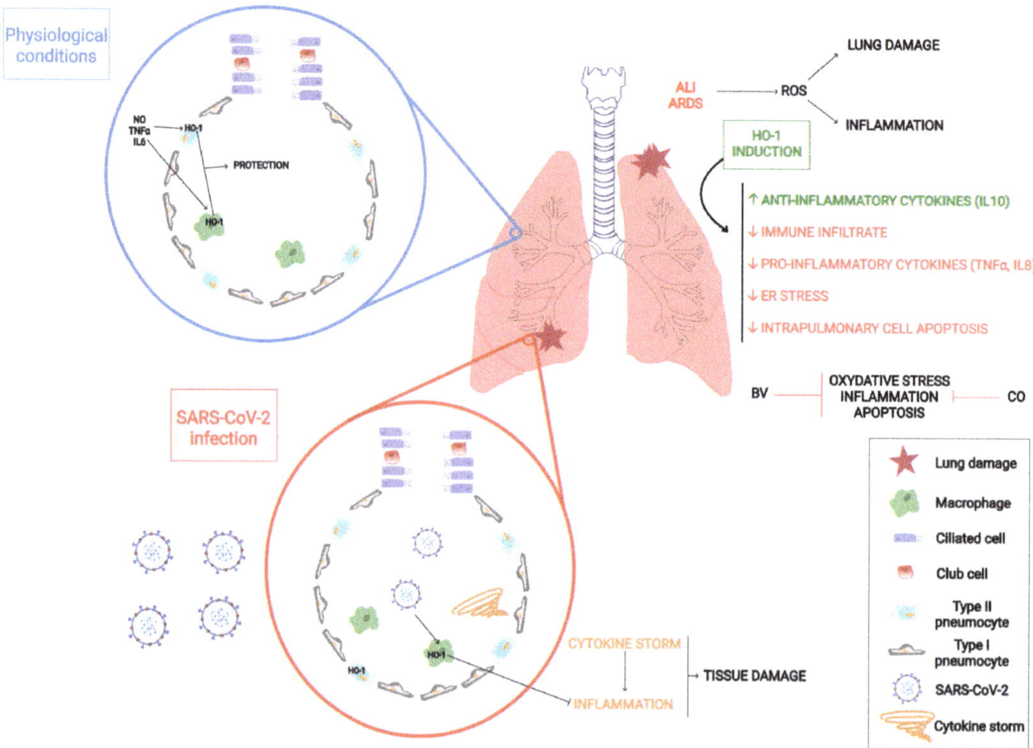

Figure 2. HO-1 and inflammatory lung diseases. HO-1 is expressed in pulmonary cells and confers protection against inflammatory lung diseases such as acute respiratory distress syndrome (ARDS), acute lung injury (ALI), and SARS-CoV-2 infection. Schematic representation displaying HO-1's reaction and its products' protective effects in the lung tissue.

As mentioned earlier, during COVID-19 disease, the number of immune cells infiltrating lung tissues and the pro-inflammatory cytokines levels are augmented [112]. Consequently, anti-inflammatory proteins have a crucial role in halting the cytokine storm and the sequelae generated by viral infection [113].

In particular, ALI and ARDS are the most prevalent diseases emerging from an extended diversity of lung injuries [114,115]. ALI and ARDS are pathophysiologically characterized by lung damage, inflammatory infiltration, and an exacerbation of the host immune response [116]. Several reports indicate that ALI and ARDS might be explained by the presence of high ROS levels, where HO-1 acts as a protective factor against oxidative stress under pharmacological induction [117]. HO-1 induction by hemin shows a protective role against ventilator induced lung injury in rabbits with ALI/ARDS, increasing anti-inflammatory cytokine levels, such as IL-10, as well as decreasing the inflammatory infiltrate of immune system cells and the secretion of inflammatory cytokines, such as TNF-α and IL-8 [118] (Figure 2). Furthermore, it has been found that HO-1 confers protection against ischemia-reperfusion injury (LIRI) [119].

HO-1 regulates diverse signaling pathways that are affected during pulmonary diseases. In rats, HO-1 inhibits the PERK/eIF2-α/ATF4/CHOP pathway, which is involved in the endoplasmic reticulum stress (ERS) characteristic in ALI, and also promotes the decrease in intrapulmonary cell apoptosis [120]. It was also reported that the PI3K/Akt pathway attenuates oxidative damage during ALI/ARDS through HO-1 regulation [121]. In pathologies such as silicosis, characterized by excessive ROS production, lung injury is attenuated by HO-1 induction. The mechanism underlying this cytoprotective effect relies on the ERK pathway inhibition by HO-1, CO and BV [122].

Reaction products derived from the HO-1 mediated heme catalysis have protective roles in lung pathologies as well. CO is known to provide protection against ALI and ARDS by reducing cytokine and chemokine levels [105,117,123]. CO decreases EGR-1 (early growth response protein 1), a proinflammatory protein that regulates the expression of TNF-α and IL-2 [124], in mice lungs [123]. Fujita et al. demonstrated that *Hmox1* deficient mice had increased mortality after lung ischemia, an effect reverted by CO administration [125]. Furthermore, BV exerted antioxidative, anti-inflammatory and anti-apoptotic effects in a rat model of LIRI [126]. BV administration protected against hemorrhagic shock induced ALI through a decrease in the inflammatory infiltrate and proinflammatory cytokines levels [127].

7. Unveiling How HO-1 Promotes Viral Clearance

HO-1 has immunomodulatory properties on the innate immune response and there is compelling evidence suggesting that it also plays a central role in the modulation of adaptive immunity. HO-1 displays antiviral properties against a wide range of viruses [7] (Figure 3). Several reports have demonstrated that HO-1 induction is associated with the activation of the IFN pathway. However, the mechanism underlying the antiviral properties of HO-1 exerted by both its classical and noncanonical activities are yet to be fully elucidated.

Virus	Experimental model	HO-1 inducers	Mechanism	Effect	References
IAV	MDKC cells	None	Decrease of ROS levels Antioxidant activity	Inhibition of cell death	PMID: 29498634
	A549 cells	Hemin	Reduction of lymphocytopenia and inflammatory cytokines	Anti-inflammatory	PMID: 27884416
	MDKC and RAW264.7 cells	CoPP, DMO-CAP	Activation of IFN pathway	Antiviral	PMID: 30580124, 30786886
	MDKC and MRC-5 cells	YZH-106	Activation of IFN pathway	Antiviral	PMID: 27107768
RSV	A549 cell line BALB/cJ mice	CoPP	Decrease of inflammatory cell infiltration and inflammatory cytokines/chemokines	Antiviral and anti-inflammatory state	PMID: 28566367
HIV	Cultured human monocytes and NOD-SCID mice	Hemin	Dependent on HO-1 enzimatic activity	Inhibition of viral replication	PMID: 16547262
	MDM neurotoxicity model	None, CoPP	Dependent on HO-1 enzimatic activity	Reduction of neurotoxin and glutamate prod.	PMID: 21976775, 25202977
EBOV	MDM, Vero E6, HeLa, and HFF1 cells	Hemin, CoPP, overexpression	Dependent on HO-1 enzimatic activity	Inhibition of viral replication	PMID: 24108237, 28133423
DENV	Huh-7 cells and ICR suckling mice	CoPP, hemin, overexpression	BV inhibits NS2B/NS3 protease	Antiviral	PMID: 27553177
	Huh-7 cells and ICR suckling mice	miR-155 overexpression	Activation of IFN pathway	Inhibition of viral protein synthesis and replication	PMID: 27553177
ZIKV	A549 and HEK-293 cells	CoPP and overexpression	Inhibition at the replication level	Limited antiviral effect	PMID: 30577437
HCV	Huh7, Huh-5-15, Huh5-15NS, and Huh7.5FL cells	Hemin, overexpression	Dependent on HO-1 enzimatic activity; iron-dependent inactivation of NS5B	Decreased viral replication, protection against oxidative injury	PMID: 18972446, 15637067
	Huh-5-15, Huh7.5FL, Huh7, and LucUbiNeo-ET cells	None, CoPP	BV-mediated inhibition of NS3/4A, activation of IFN pathway	Decreased viral replication	PMID: 21105106, 20044809
	Huh7 and Ava5 cells	None	Targeting of HO-1 repressor Bach1 by miR-let-7c	Decreased viral replication	PMID: 30706605
HBV	HepG2 cells	Hemin	Interferes with viral replication	Decreased viral replication	PMID: 25633656
	HepG2 and C57BL/6 mice	CoPP, overexpression	Reduces stability of viral core protein and levels of HBV cccDNA	Decreased viral replication	PMID: 17919491
HSV-2	Vero, HeLa and SH-SY5Y cells	CoPP, CORM-2	Dependent on HO-1 enzimatic activity; partly mediated by CO	Decreased viral replication and viral protein expression	PMID: 29163402
EV71	SK-N-SH cells	Overexpression, CORM-2	Downregulation of virus-induced ROS levels	Decreased viral replication	PMID: 21321939
SARS-CoV-2	Vero76 cells	Hemin	Activation of IFN pathway	Antiviral and anti-inflammatory	PMID: 34728736

Figure 3. HO-1's induction and its effect on different viral infections. Table containing previously reported studies about HO-1 involvement in influenza A virus (IAV), respiratory syncytial virus (RSV), human immunodeficiency virus (HIV), ebola virus (EBOV), dengue virus (DENV), zika virus (ZIKV), hepatitis C virus (HCV), hepatitis B virus (HBV), herpes simplex virus 2 (HSV-2), enterovirus 71 (EV71) and severe acute respiratory syndrome coronavirus 2 (SARS-CoV-2) infections. The table includes the experimental model, HO-1 inducers, its mechanism of action, its effect and the study's PMID. CoPP: cobalt protoporphyrin IX, DMO-CAP: 6-demethoxy-4′-O-methylcapillarisin, ROS: Reactive oxygen species, IFN: interferon, MDM: monocyte derived macrophages, BV: Biliverdin, CO: carbon monoxide, CORM-2: CO-releasing molecule-2.

7.1. Respiratory Viruses: IAV and RSV

Influenza A virus (IAV) is a single stranded RNA virus whose infection remains a persistent global health threat with high morbidity and mortality [128]. An estimate of 4.0 to 8.8 deaths per 100,000 individuals with seasonal influenza associated respiratory occur annually (all types of influenza virus) [129]. Considering that the inhibition of virus induced ROS formation impairs IAV replication, proteins such as HO-1 are useful to counteract IAV infections in the host cell [130]. Wang et al. studied the effect of hemin in IAV infections and demonstrated that hemin attenuates the lymphocytopenia caused by IAV infection both in vitro and in vivo [131]. These results suggest that the anti-influenza effect of hemin may be mediated by HO-1's ability to regulate systemic and local inflammatory responses [131]. Furthermore, Hashiba et al. reported that HO-1 gene transfer is a potential strategy to treat lung injury caused by IAV [132] and Cummins et al. suggested that the therapeutic induction of HO-1 expression may represent a novel adjuvant to enhance influenza vaccine effectiveness [133]. It has also been reported that the HO-1 inducers rupestonic acid derivative YZH-106 and the flavonoid 6-demethoxy-4′-O-methylcapillarisin (DMO-CAP) inhibited IAV replication by activating the HO-1 mediated IFN response [134,135]. Additionally, Ma et al. evaluated the effect of CoPP in IAV infection, focusing on the IFN pathways. The authors demonstrated that HO-1 induction attenuates IAV replication, and the most intriguing finding was that the catalytic function of HO-1 was not essential for the anti-IAV effect of CoPP. Interestingly, they found that HO-1 interacts with IFN regulatory factor 3 (IRF3) promoting its phosphorylation and nuclear translocation, thus activating the IFN pathway. Consequently, CoPP treatment increased the expression of *IFITM3*, *PKR* and *OAS1*, three ISGs markedly involved in the anti-IAV response [128].

Respiratory syncytial virus (RSV) is an RNA virus of the Pneumoviridae family and the most common cause of lower respiratory tract infections in children worldwide [136]. It interacts with host cells' toll-like receptors in the primary airway epithelium, and promotes the expression and secretion of inflammatory cytokines [137], under the NF-κB pathway's regulation [138]. Similar to IAV, CoPP HO-1 induction inhibited RSV replication and viral particle production in lung adenocarcinoma (A549) and HEp-2 cells. Most importantly, in vivo assays in BALB/cJ mice treated prophylactically with CoPP also showed a reduction in viral replication and viral particle production, alongside a decrease in inflammatory cell infiltration, and inhibition of proinflammatory cytokine or chemokine secretion during RSV infection [139].

The mentioned reports suggest that HO-1 is involved in host cellular defense mechanisms against IAV and RSV infections. Of note, HO-1's antiviral effects are mediated by its classical and noncanonical functions.

7.2. Retroviridae: HIV

The human immunodeficiency virus-1 (HIV-1) genome consists of two identical single stranded RNA molecules and it is the causative agent of acquired immune deficiency syndrome (AIDS). HIV infection has become a clinically manageable disease since the development of combination antiretroviral therapy (ART). Globally, 1 million people died from HIV/AIDS in 2016; without ART, more than twice as many people would have died from this disease [140]. In 2020, the United Nations Programme on HIV/AIDS (UNAIDS) reported that of all people with HIV worldwide, 66% were virally suppressed [141,142]. However, emerging drug resistance and limitations in access and adherence to ART impose a threat to controlling the spread of the virus [143]. Current drugs do not eradicate the virus, making lifelong treatment necessary [143].

Several reports from in vitro and in vivo models, and HIV infected subjects, have linked HO-1 with HIV replication and its effects on HIV mediated neurodegeneration. In cultured monocytes and HIV infected mice, hemin efficiently inhibited viral replication; the effect observed in vitro was mediated by HO-1 enzymatic activity [144]. Using an in vitro model of HIV mediated neurotoxicity, HIV infection dysregulated the macrophage antioxidant response and reduced levels of HO-1. Restoration of HO-1 expression in HIV-infected monocyte derived macrophages (MDM) reduced neurotoxin release without altering HIV replication [145]. In HIV infected subjects, HO-1 protein levels were reduced in the dorsolateral prefrontal cortex, which correlated with central nervous system (CNS) viral load markers of immune activation. In a model of human astrocytes treated with IFNγ, an HIV associated CNS immune activator, HO-1 was degraded by the immunoproteasome [146]. Additionally, the use of CoPP reduced HIV-MDM glutamate release and neurotoxicity, suggesting a role for HO-1 in HIV associated neurocognitive disorders pathogenesis [147].

Altogether, these reports propose HO-1 induction as a protective mechanism against HIV infection. In particular, the HO-1's classical functions mediate its antiviral properties against HIV.

7.3. *Filoviridae: EBOV*

Ebola virus (EBOV) is a negative-sense RNA virus that belongs to the Filoviridae family [148,149]. From 2013 to 2015 there was an important outbreak in West Africa, which caused >25,000 infections and >10,000 deaths. The average EBOV case fatality rate is ~50% and case fatality rates has ranged from 25% to 90% in past outbreaks [15]. Hemin treatment significantly reduced EBOV replication and delayed pathogenesis in vitro, by stimulating the cellular innate response against the infection [150,151]. In MDM, hemin treatment inhibited EBOV infection in a dose dependent manner. A similar effect was observed in other cell lines, such as HeLa and human foreskin fibroblasts cells, in which hemin also reduced viral replication [150]. Furthermore, it has been reported that the ebola virus protein 35 (VP35) is a critical protein involved in the inhibition of IRF3 phosphorylation as a mechanism that might counteract the antiviral response [152]. Thus, considering that HO-1 promotes IRF3 phosphorylation and activation, its induction may represent a novel therapeutic strategy against EBOV infection.

The cited studies place HO-1 as a novel therapeutic target against EVOB infection. Notably, HO-1's noncanonical functions are involved in the present example of antiviral action.

7.4. *Hepatic Viruses: HCV and HBV*

Hepatitis C virus (HCV), a single stranded positive sense RNA virus, is associated with chronic hepatitis, cirrhosis, steatosis and hepatocellular carcinoma [153]. HCV treatment includes a combination of pegylated IFN-α and ribavirin, which has low efficacy and important side effects. With the development of direct-acting antiviral agents, such as sofosbuvir and simeprevir, patient outcomes have greatly improved; however, the disease remains a concern [154]. Since HCV infection generates oxidative damage to hepatocytes, the modulation of HO-1 expression emerges as an attractive therapeutic approach to reduce chronic liver disease. Abdalla et al. observed lower HO-1 mRNA and protein levels in HCV infected patients' livers, while this alteration was not found in patients with other chronic liver diseases. The authors also reported HO-1 downregulation in hepatocyte cell lines expressing the HCV core protein [155]. Further, the overexpression or hemin-induction of HO-1 in the hepatoma cell line Huh7 decreased HCV replication and conferred protection against oxidative injury [156]. This effect of HO-1 on HCV replication might be explained partly by the iron dependent inactivation of the HCV RNA polymerase NS5B [157], and by the BV mediated inhibition of HCV NS3/4A protease and induction of an antiviral response by IFNα2 and IFNα17 [153,158]. Moreover, overexpression of miR-let-7c, which interferes with the production of proinflammatory cytokines in osteoarthritis and rheumatoid arthritis synovial fibroblasts [159], can reduce HCV replication by targeting HO-1 transcriptional repressor Bach1 [160].

Hepatitis B virus (HBV) is a DNA virus that causes serious liver diseases, representing the most common etiological agent for these pathologies [161]. It has been shown that pharmacological and genetic HO-1 overexpression attenuates viral replication both in vivo and in vitro in HepG2 cells [161–163], while also playing a hepatoprotective role [162]. The effect of HO-1 induction using hemin and CoPP mitigated the effects of HBV replication [6,161,162]. On the other hand, blocking HO-1 by siRNA reversed the inhibition of viral replication [6].

Interestingly, Protzer et al. evaluated the effect of HO-1 on HBV core protein by pulse-chase metabolic labeling experiments finding that HO-1 can destabilize structural proteins to prevent the formation of viral capsids, highlighting a direct HO-1 antiviral mechanism rather than limiting its effect to its anti-inflammatory properties [6].

Hence, the summarized reports established the antiviral effects of HO-1 by impairing HCV's and HBV's replication.

7.5. Arbovirus: DENV and ZIKV

Dengue virus (DENV) is a single stranded positive sense RNA virus [164] that induces oxidative stress by the activation of inflammatory regulators, such as NF-κB, and leads to the progression and pathogenesis of DENV [165]. In this pro-oxidant scenario, Tseng et al. demonstrated that HO-1 promoter activity and protein synthesis gradually increased during the early stages of DENV infection (6 to 12 h), but they were markedly decreased at later stages (24 to 72 h) [166]. Strikingly, pharmacological and genetic HO-1 induction after infection impaired viral protein synthesis and replication, and reduced DENV mortality. This effect was due to BV but not CO nor Fe^{2+} production [166]. The authors demonstrated that BV inhibits NS2B/NS3 DENV protease, thus promoting the antiviral IFN response and impairing its blockage by this protease [166]. Accordingly, Su et al. showed the anti-DENV activity of miR-155, which inhibits Bach1, a protein that negatively regulates the expression of many oxidative stress-response genes, including *HMOX1* [167]. This, in turn, results in the induction of HO-1, boosting the IFN responses against DENV replication by the activation of interferon induced protein kinase R (PKR), 2′-5′-oligoadenylate synthetase 1 (OAS1), OAS2, and OAS3 expression [167]. Interestingly, the summarized studies demonstrate that HO-1's antiviral effects against DENV infection involve both, its canonical, in this case mediated by BV, and noncanonical functions.

Zika virus (ZIKV), a single stranded positive sense RNA, is the causative pathogen of Zika fever in humans [168]. Using A549 and embryonic kidney (HEK-293) cell lines, El Kalamouni et al. demonstrated that ZIKV infection downregulated HO-1 expression by triggering endoplasmic-reticulum-associated protein degradation, thus halting its antiviral effects [168]. This report highlights HO-1's protective role relevance, as it demonstrates that ZIKV infection promotes the decrease in HO-1 expression levels as an evasion mechanism.

7.6. Neurotropic Viruses: HSV-2 and EV71

Herpes simplex virus (HSV) includes HSV-1 and HSV-2, two double stranded DNA viruses that belong to Herpesviridae family. HSV produces recurring lesions in skin and mucosae and can also latently infect neurons of the trigeminal or dorsal root ganglia. HSV infection can result in encephalitis and meningitis [169]. Ibañez et al. demonstrated that pharmacological induction of HO-1 by CoPP hampered HSV-2 propagation in epithelial and neuronal cells. Furthermore, by CORM-2 treatment the authors also showed that the effects of HO-1 were partly mediated by CO [170].

Enterovirus 71 (EV71) is a single stranded positive sense RNA virus that belongs to the Picornaviridae family and is the causative agent of hand foot and mouth disease in children [171]. It was demonstrated that the overexpression of HO-1, as well as CO treatment, decreased viral replication in SK-N-SH cells suggesting that the antiviral effect is mediated by the downregulation of EV71 induced ROS levels [172].

Regarding neurotropic viruses, the summarized reports showcase that HO-1 displays protective effects against HSV-2 and EV71 involving its enzymatic function.

7.7. COVID-19 Causative Agent: SARS-CoV-2

SARS-CoV-2, is the novel beta coronavirus of the Coronaviridae family whose genome is composed of a single stranded RNA molecule [173]. It has been shown that hemin, hemoglobin and protoporphyrin IX bind to SARS-CoV-2 proteins, blocking its adsorption and replication independently from HO-1 induction [174]. However, the current literature regarding HO-1's antiviral effect against SARS-CoV-2 remains unclear. Maestro et al.

showed that hemin does not inhibit SARS-CoV-2 viral replication in vitro [175]. Kidney epithelial Vero-E6 and lung Calu3 cell lines were treated with hemin and results showed that, despite a strong activation of HO-1 in both cell lines, there was no effect on SARS-CoV-2 viral replication, measured by the amplification of the N viral gene by RT-qPCR [175]. However, a more recent report proposed hemin as a potential drug for treating COVID-19 via HO-1 induction [176]. Interestingly, authors observed a reduction in SARS-CoV-2 replication, both when pretreating and after SARS-CoV-2 infection treatment of Vero76 with this drug. Genetic induction or silencing of HO-1 in Vero76 cells demonstrated that the antiviral effect of hemin relies on this protein. Strikingly, this effect was mediated not only by Fe^{2+} and BV, but also by an HO-1 enzymatic independent mechanism. Further, they showed that hemin induced HO-1 boosted ISG15, OAS1 and MX1 protein expression in SARS-CoV-2 infected cells, highlighting the importance of stimulating the host cell's IFN response against this virus [176]. Of note, there are reports from our laboratory showcasing that *MX1* gene expression is increased in COVID-19 patients. However, *MX1* expression is lower in elderly patients, where the disease has been shown to be more severe than in younger people. Additionally, through an in depth proteomics analysis, we described MX1 as a novel HO-1 interactor in prostate cancer (PCa) cell lines [177]. Moreover, genetic and pharmacological HO-1 induction in PCa cells triggered an increase in MX1 at mRNA and protein levels, and altered HO-1 cellular localization, showcasing a clear association between both proteins. Further, *MX1* silencing with a specific siRNA significantly decreased the expression of ERS genes (*HSPA5*, *DDIT3* and *XBP1*), demonstrating the role of MX1 in pro-death events [177].

In summary, the induction of the host infected cells antiviral response appears to be critical for COVID-19 treatment, which could be partly achieved by hemin mediated HO-1 induction, also preventing viral adsorption and replication by binding SARS-CoV-2 proteins. These antiviral effects are mediated by canonical and noncanonical HO-1's functions.

8. HO-1 Induction as a Strategy against COVID-19

There are mainly two different approaches to develop antiviral therapies: (1) therapies directed against viral factors; or (2) therapies targeting the host immune system. To date, the second strategy has received increasing attention due to the fact that targeting viral factors might cause viruses to mutate, increasing the rate of resistance to antiviral drugs [178]. In contrast to the viral genome, host cells' DNA does not have a high mutational frequency. Therefore, overpowering viral infection by targeting host factors involved in the antiviral response is conceivably an effective strategy to counteract the severe consequences, while also fighting the infection [179].

During the last two years, several reports have focused on the understanding of the virus–host interaction underlying COVID-19 disease. The worrying situation of the SARS-CoV-2 pandemic and the threat of new variants, such as Omicron, which is spreading across the globe at an unprecedented rate, drive the interest of scientists to seek for new anti-SARS-CoV-2 strategies. Its enhanced transmissibility compared to the Delta variant could be explained in part by its increased rate of replication in human primary airway cultures, higher binding to ACE2, and ability to efficiently enter cells in a TMPRSS2-independent manner [180]. Fortunately, preliminary data of the Omicron variant suggest a lower virus load in both lower and upper respiratory tract, associated to less inflammatory processes in the lungs, using a mouse model of severe disease [181]. However, exceptionally high transmissibility could result in a great burden on healthcare systems across the globe. In this context, HO-1 emerges as a potential target to boost the host's response to fight the infection and prevent severe COVID-19.

Certainly, HO-1 and its reaction products possess beneficial effects for the host during viral infections: it reduces inflammation and exerts antiviral actions. The most serious COVID-19 complications are: sepsis like inflammation, coagulopathy, and cardiovascular or respiratory complications. Furthermore, respiratory failure triggers hypoxia which, in combination with neuroinflammation, generates neurological complications [182]. When inflammation is not modulated, it turns into hyperinflammation and results in tissue damage or organ failure [183]. Enhancing HO-1 expression might help avoid the severe consequences of this disease. For example, it has been reported that HO-1 induction decreases inflammation, inhibits platelet aggregation, and increases fibrinolysis and phagocytosis, thus preventing tissue damage, thrombosis and sepsis [184]. Additionally, hemin is an activator of neuroglobin, a protein involved in oxygen transport and storage in neurons that increases oxygen's intracellular partial pressure in neurons, and is crucial to protect neurons from hypoxic injury [185–187]. In addition, as mentioned above, HO-1 has a reported antiviral activity against multiple viruses. This effect depends on its classical activity involving its reaction subproducts (BR, BV, CO and Fe^{2+}) and the activation of the IFN pathway; interestingly, its noncanonical activity is also involved in the antiviral effect of HO-1.

HO-1 expression is essentially regulated at the transcriptional level by NRF2. It has been reported that SARS-CoV-2 infection suppresses the NRF2 antioxidant gene expression pathway, and that NRF2 agonists limit viral replication and repress the proinflammatory response of SARS-CoV-2 [188]. This evidence highlights the relevance of the NRF2 signaling pathway on the antiviral response, suggesting that the activation of NRF2 might be a useful strategy against COVID-19 [189].

As explained before, clinical complications associated with COVID-19 disease have been described in different organs, including vascular, cardiac, renal, hepatic, endocrine and neurological complications [190] (Figure 4). Interestingly, HO-1 has been reported to be associated with a reduction in tissue damage, mainly through its anti-inflammatory and antioxidative functions in different organs [4,97,147,191–207] (Figure 3). It would be interesting to address HO-1's vasoprotective and antithrombotic effects for the prevention of thromboembolic events caused by SARS-CoV-2.

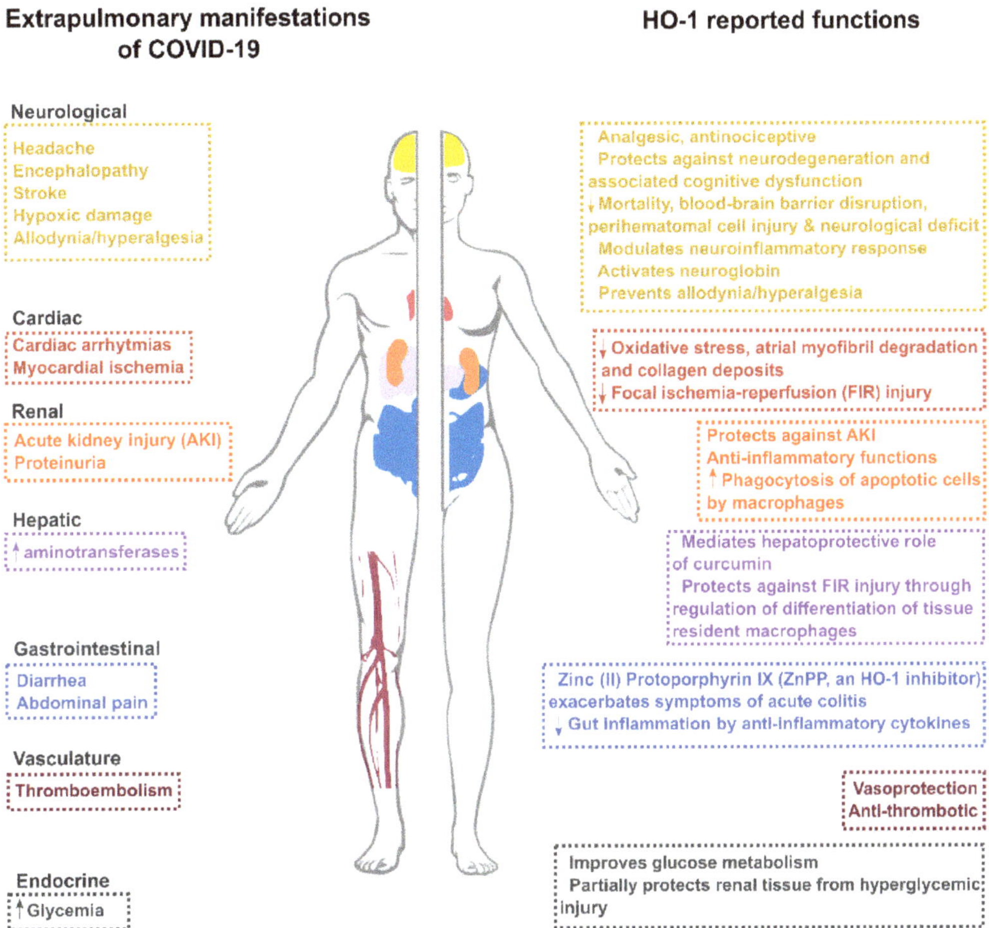

Figure 4. HO-1's role in different sites that can be affected upon SARS-CoV-2 infection. Extra pulmonary manifestations of COVID-19 are grouped according to their site or body system. HO-1's reported functions in different experimental conditions or diseases are grouped according to the model or system in which they are studied. The image of the human body has been adapted from Uhlén et al. (Human Protein Atlas, proteinatlas.org) [208].

9. Closing Remarks

Drug repurposing is an attractive proposition, since it involves the use of derisked and previously approved compounds, with lower development costs and shorter development times [209]. Since the onset of the COVID-19 pandemic, we have witnessed a plethora of alternative drugs as potential therapeutic avenues to fight the disease. Thus, hemin, a previously FDA and EMA approved drug for acute intermittent porphyria treatment, rises as a promising drug candidate, inducing HO-1 plasma concentration in humans, and posing a host defense advantage to fight SARS-CoV-2. Further work on optimal drug concentrations, pharmacokinetics and pharmacodynamics should be performed in order to prove hemin's effectiveness (either alone or in combination with other drugs) to halt infection.

Funding: This work was supported by a grant from AGENCIA-PICT-2019-03215 (ARGENTINA).

Acknowledgments: We thank Javier Cotignola for his helpful comments and critical revision of this article.

Conflicts of Interest: The authors declare no conflict of interest.

References

1. Prete, M.; Favoino, E.; Catacchio, G.; Racanelli, V.; Perosa, F. SARS-CoV-2 Inflammatory Syndrome. Clinical Features and Rationale for Immunological Treatment. *Int. J. Mol. Sci.* **2020**, *21*, 3377. [CrossRef] [PubMed]
2. Dunn, L.L.; Midwinter, R.G.; Ni, J.; Hamid, H.A.; Parish, C.R.; Stocker, R. New insights into intracellular locations and functions of heme oxygenase-1. *Antioxid. Redox Signal.* **2014**, *20*, 1723–1742. [CrossRef] [PubMed]
3. Naito, Y.; Takagi, T.; Yoshikawa, T. Heme oxygenase-1: A new therapeutic target for inflammatory bowel disease. *Aliment. Pharmacol. Ther.* **2004**, *20*, 177–184. [CrossRef] [PubMed]
4. Takagi, T.; Naito, Y.; Mizushima, K.; Hirai, Y.; Harusato, A.; Okayama, T.; Katada, K.; Kamada, K.; Uchiyama, K.; Handa, O.; et al. Heme oxygenase-1 prevents murine intestinal inflammation. *J. Clin. Biochem. Nutr.* **2018**, *63*, 169–174. [CrossRef]
5. Chora, Â.A.; Fontoura, P.; Cunha, A.; Pais, T.F.; Cardoso, S.; Ho, P.P.; Lee, L.Y.; Sobel, R.A.; Steinman, L.; Soares, M.P. Heme oxygenase-1 and carbon monoxide suppress autoimmune neuroinflammation. *J. Clin. Investig.* **2007**, *117*, 438–447. [CrossRef]
6. Protzer, U.; Seyfried, S.; Quasdorff, M.; Sass, G.; Svorcova, M.; Webb, D.; Bohne, F.; Hösel, M.; Schirmacher, P.; Tiegs, G. Antiviral Activity and Hepatoprotection by Heme Oxygenase-1 in Hepatitis B Virus Infection. *Gastroenterology* **2007**, *133*, 1156–1165. [CrossRef]
7. Espinoza, J.A.; González, P.A.; Kalergis, A.M. Modulation of Antiviral Immunity by Heme Oxygenase-1. *Am. J. Pathol.* **2017**, *187*, 487–493. [CrossRef]
8. Anderson, K.E.; Collins, S. Open-label study of hemin for acute porphyria: Clinical practice implications. *Am. J. Med.* **2006**, *119*, 801.e19–801.e24. [CrossRef]
9. European Medicines Agency. Available online: https://www.ema.europa.eu/en/documents/psusa/human-hemin-list-nationally-authorised-medicinal-products-psusa/00001629/202005_en.pdf (accessed on 26 January 2022).
10. Wu, D.; Wu, T.; Liu, Q.; Yang, Z. The SARS-CoV-2 outbreak: What we know. *Int. J. Infect. Dis.* **2020**, *94*, 44–48. [CrossRef]
11. Hu, B.; Guo, H.; Zhou, P.; Shi, Z.L. Characteristics of SARS-CoV-2 and COVID-19. *Nat. Rev. Microbiol.* **2021**, *19*, 141–154. [CrossRef]
12. WHO Summary of Probable SARS Cases with Onset of Illness from 1 November 2002 to 31 July 2003 (Based on Data as of the 31 December 2003). Available online: https://www.who.int/publications/m/item/summary-of-probable-sars-cases-with-onset-of-illness-from-1-november-2002-to-31-july-2003%0Ahttps://www.who.int/publications/m/item/summary-of-probable-sars-cases-with-onset-of-illness-from-1-november-2002-to-31-jul (accessed on 6 January 2022).
13. Walt, G. WHO's World Health Report 2003. *BMJ* **2004**, *328*, 6. [CrossRef] [PubMed]
14. Moh Zaki, A.; Van Boheemen, S.; Bestebroer, T.M.; Osterhaus, A.D.M.E.; Fouchier, R.A.M. Isolation of a Novel Coronavirus from a Man with Pneumonia in Saudi Arabia. *N. Engl. J. Med.* **2012**, *367*, 1814–1834. [CrossRef] [PubMed]
15. World Health Organization. Available online: https://www.who.int/ (accessed on 30 December 2021).
16. Zhu, N.; Zhang, D.; Wang, W.; Li, X.; Yang, B.; Song, J.; Zhao, X.; Huang, B.; Shi, W.; Lu, R.; et al. A Novel Coronavirus from Patients with Pneumonia in China, 2019. *N. Engl. J. Med.* **2020**, *382*, 727–733. [CrossRef] [PubMed]
17. Johansson, M.A.; Quandelacy, T.M.; Kada, S.; Prasad, P.V.; Steele, M.; Brooks, J.T.; Slayton, R.B.; Biggerstaff, M.; Butler, J.C. SARS-CoV-2 Transmission from People Without COVID-19 Symptoms. *JAMA Netw. Open* **2021**, *4*, e2035057. [CrossRef] [PubMed]
18. Song, E.; Zhang, C.; Israelow, B.; Lu-Culligan, A.; Prado, A.V.; Skriabine, S.; Lu, P.; Weizman, O.-E.; Liu, F.; Dai, Y.; et al. Neuroinvasion of SARS-CoV-2 in human and mouse brain. *J. Exp. Med.* **2021**, *218*, e20202135. [CrossRef]
19. Mulay, A.; Konda, B.; Garcia, G.; Yao, C.; Beil, S.; Villalba, J.; Koziol, C.; Sen, C.; Purkayastha, A.; Kolls, J.K.; et al. SARS-CoV-2 infection of primary human lung epithelium for COVID-19 modeling and drug discovery. *Cell Rep.* **2021**, *35*, 109055. [CrossRef] [PubMed]
20. Yaqinuddin, A.; Kashir, J. Novel therapeutic targets for SARS-CoV-2-induced acute lung injury: Targeting a potential IL-1β/neutrophil extracellular traps feedback loop. *Med. Hypotheses* **2020**, *143*, 109906. [CrossRef] [PubMed]
21. Xu, Z.; Shi, L.; Wang, Y.; Zhang, J.; Huang, L.; Zhang, C.; Liu, S.; Zhao, P.; Liu, H.; Zhu, L.; et al. Pathological findings of COVID-19 associated with acute respiratory distress syndrome. *Lancet. Respir. Med.* **2020**, *8*, 420–422. [CrossRef]
22. Soy, M.; Keser, G.; Atagündüz, P.; Tabak, F.; Atagündüz, I.; Kayhan, S. Cytokine storm in COVID-19: Pathogenesis and overview of anti-inflammatory agents used in treatment. *Clin. Rheumatol.* **2020**, *39*, 2085–2094. [CrossRef]
23. Bhaskar, S.; Sinha, A.; Banach, M.; Mittoo, S.; Weissert, R.; Kass, J.S.; Rajagopal, S.; Pai, A.R.; Kutty, S. Cytokine Storm in COVID-19-Immunopathological Mechanisms, Clinical Considerations, and Therapeutic Approaches: The REPROGRAM Consortium Position Paper. *Front. Immunol.* **2020**, *11*, 1648. [CrossRef]
24. Channappanavar, R.; Perlman, S. Pathogenic human coronavirus infections: Causes and consequences of cytokine storm and immunopathology. *Semin. Immunopathol.* **2017**, *39*, 529–539. [CrossRef] [PubMed]
25. Mokhtari, T.; Hassani, F.; Ghaffari, N.; Ebrahimi, B.; Yarahmadi, A.; Hassanzadeh, G. COVID-19 and multiorgan failure: A narrative review on potential mechanisms. *J. Mol. Histol.* **2020**, *51*, 613–628. [CrossRef] [PubMed]
26. Wang, F.; Kream, R.M.; Stefano, G.B. Long-Term Respiratory and Neurological Sequelae of COVID-19. *Med. Sci. Monit. Int. Med. J. Exp. Clin. Res.* **2020**, *26*, e928996. [CrossRef]

27. Al-Aly, Z.; Xie, Y.; Bowe, B. High-dimensional characterization of post-acute sequelae of COVID-19. *Nature* **2021**, *594*, 259–264. [CrossRef]
28. Cao, Y.; Wang, J.; Jian, F.; Xiao, T.; Song, W.; Yisimayi, A.; Huang, W.; Li, Q.; Wang, P.; An, R.; et al. Omicron escapes the majority of existing SARS-CoV-2 neutralizing antobodies. *Res. Briefings* **2021**. [CrossRef]
29. Lu, L.; Mok, B.W.-Y.; Chen, L.-L.; Chan, J.M.-C.; Tsang, O.T.-Y.; Lam, B.H.-S.; Chuang, V.W.-M.; Chu, A.W.-H.; Chan, W.-M.; Ip, J.D.; et al. Neutralization of SARS-CoV-2 Omicron variant by sera from BNT162b2 or Coronavac vaccine recipients. *Clin. Infect. Dis.* **2021**, ciab1041. [CrossRef] [PubMed]
30. Cascella, M.; Rajnik, M.; Aleem, A. *Features, Evaluation, and Treatment of Coronavirus (COVID-19)*; StatPearls: Treasure Island, FL, USA, 2021.
31. Khailany, R.A.; Safdar, M.; Ozaslan, M. Genomic characterization of a novel SARS-CoV-2. *Gene Rep.* **2020**, *19*. [CrossRef]
32. Janik, E.; Niemcewicz, M.; Podogrocki, M.; Majsterek, I.; Bijak, M. The Emerging Concern and Interest SARS-CoV-2 Variants. *Pathogens* **2021**, *10*, 633. [CrossRef]
33. D'Elia, R.V.; Harrison, K.; Oyston, P.C.; Lukaszewski, R.A.; Clark, G.C. Targeting the "cytokine storm" for therapeutic benefit. *Clin. Vaccine Immunol.* **2013**, *20*, 319–327. [CrossRef]
34. Chen, L.; Deng, H.; Cui, H.; Fang, J.; Zuo, Z.; Deng, J.; Li, Y.; Wang, X.; Zhao, L. Inflammatory responses and inflammation-associated diseases in organs. *Oncotarget* **2018**, *9*, 7204–7218. [CrossRef]
35. Medzhitov, R. Recognition of microorganisms and activation of the immune response. *Nature* **2007**, *449*, 819–826. [CrossRef] [PubMed]
36. Speyer, C.L.; Ward, P.A. Role of endothelial chemokines and their receptors during inflammation. *J. Investig. Surg.* **2011**, *24*, 18–27. [CrossRef] [PubMed]
37. Yang, L.; Xie, X.; Tu, Z.; Fu, J.; Xu, D.; Zhou, Y. The signal pathways and treatment of cytokine storm in COVID-19. *Signal Transduct. Target. Ther.* **2021**, *6*, 255. [CrossRef] [PubMed]
38. Tay, M.Z.; Poh, C.M.; Rénia, L.; MacAry, P.A.; Ng, L.F.P. The trinity of COVID-19: Immunity, inflammation and intervention. *Nat. Rev. Immunol.* **2020**, *20*, 363–374. [CrossRef]
39. Lucas, C.; Wong, P.; Klein, J.; Castro, T.B.R.; Silva, J.; Sundaram, M.; Ellingson, M.K.; Mao, T.; Oh, J.E.; Israelow, B.; et al. Longitudinal analyses reveal immunological misfiring in severe COVID-19. *Nature* **2020**, *584*, 463. [CrossRef]
40. Blanco-Melo, D.; Nilsson-Payant, B.E.; Liu, W.C.; Uhl, S.; Hoagland, D.; Møller, R.; Jordan, T.X.; Oishi, K.; Panis, M.; Sachs, D.; et al. Imbalanced Host Response to SARS-CoV-2 Drives Development of COVID-19. *Cell* **2020**, *181*, 1036–1045.e9. [CrossRef]
41. Park, A.; Iwasaki, A. Type I and Type III Interferons—Induction, Signaling, Evasion, and Application to Combat COVID-19. *Cell Host Microbe.* **2020**, *27*, 870–878. [CrossRef]
42. Hervas-Stubbs, S.; Perez-Gracia, J.L.; Rouzaut, A.; Sanmamed, M.F.; Le Bon, A.; Melero, I. Direct Effects of Type I Interferons on Cells of the Immune System. *Clin. Cancer Res.* **2011**, *17*, 2619. [CrossRef]
43. Levy, D.E.; García-Sastre, A. The Virus Battles: IFN Induction of the Antiviral State and Mechanisms of Viral Evasion. *Cytokine Growth Factor Rev.* **2001**, *12*, 143–156. [CrossRef]
44. Goodbourn, S.; Didcock, L.; Randall, R.E. Interferons: Cell signalling, immune modulation, antiviral response and virus countermeasures. *J. Gen. Virol.* **2000**, *81*, 2341–2364. [CrossRef]
45. Cebulla, C.M.; Miller, D.M.; Sedmak, D.D. Viral inhibition of interferon signal transduction. *Intervirology* **1999**, *42*, 325–330. [CrossRef]
46. Ruan, Q.; Yang, K.; Wang, W.; Jiang, L.; Song, J. Clinical predictors of mortality due to COVID-19 based on an analysis of data of 150 patients from Wuhan, China. *Intensive Care Med* **2020**, *46*, 846–848. [CrossRef] [PubMed]
47. Yang, Y.; Shen, C.; Li, J.; Yuan, J.; Wei, J.; Huang, F.; Wang, F.; Li, G.; Li, Y.; Xing, L.; et al. Plasma IP-10 and MCP-3 levels are highly associated with disease severity and predict the progression of COVID-19. *J. Allergy Clin. Immunol.* **2020**, *146*, 119–127.e4. [CrossRef] [PubMed]
48. Lee, J.S.; Shin, E.-C. The type I interferon response in COVID-19: Implications for treatment. *Nat. Rev. Immunol.* **2020**, *20*, 585–586. [CrossRef] [PubMed]
49. Davoudi-Monfared, E.; Rahmani, H.; Khalili, H.; Hajiabdolbaghi, M.; Salehi, M.; Abbasian, L.; Kazemzadeh, H.; Yekaninejad, M.S. A Randomized Clinical Trial of the Efficacy and Safety of Interferon β-1a in Treatment of Severe COVID-19. *Antimicrob. Agents Chemother.* **2020**, *64*, e01061-20. [CrossRef] [PubMed]
50. Rahmani, H.; Davoudi-Monfared, E.; Nourian, A.; Khalili, H.; Hajizadeh, N.; Jalalabadi, N.Z.; Fazeli, M.R.; Ghazaeian, M.; Yekaninejad, M.S. Interferon β-1b in treatment of severe COVID-19: A randomized clinical trial. *Int. Immunopharmacol.* **2020**, *88*, 106903. [CrossRef] [PubMed]
51. Zhou, Q.; Chen, V.; Shannon, C.P.; Wei, X.S.; Xiang, X.; Wang, X.; Wang, Z.H.; Tebbutt, S.J.; Kollmann, T.R.; Fish, E.N. Interferon-α2b Treatment for COVID-19. *Front. Immunol.* **2020**, *11*, 1061. [CrossRef] [PubMed]
52. Feld, J.J.; Kandel, C.; Biondi, M.J.; Kozak, R.A.; Zahoor, M.A.; Lemieux, C.; Borgia, S.M.; Boggild, A.K.; Powis, J.; McCready, J.; et al. Peginterferon lambda for the treatment of outpatients with COVID-19: A phase 2, placebo-controlled randomised trial. *Lancet Respir. Med.* **2021**, *9*, 498–510. [CrossRef]
53. ClinicalTrials.gov [Internet]. IFN-beta 1b and Remdesivir for COVID19. Available online: https://clinicaltrials.gov/ct2/show/NCT04647695 (accessed on 7 January 2022).

54. ClinicalTrials.gov [Internet]. Experimental Trial of rhIFNα Nasal Drops to Prevent 2019-nCOV in Medical Staff; 2021 Nov 30. Available online: https://clinicaltrials.gov/ct2/show/NCT04320238 (accessed on 7 January 2022).
55. ClinicalTrials.gov [Internet]. Efficacy and Safety of IFN-α2β in the Treatment of Novel Coronavirus Patients; 2021 Nov 30. Available online: https://clinicaltrials.gov/ct2/show/NCT04293887 (accessed on 7 January 2022).
56. Kumar, S.; Bandyopadhyay, U. Free heme toxicity and its detoxification systems in human. *Toxicol. Lett.* **2005**, *157*, 175–188. [CrossRef]
57. Hayashi, S.; Omata, Y.; Sakamoto, H.; Higashimoto, Y.; Hara, T.; Sagara, Y.; Noguchi, M. Characterization of rat heme oxygenase-3 gene. Implication of processed pseudogenes derived from heme oxygenase-2 gene. *Gene* **2004**, *336*, 241–250. [CrossRef]
58. Yachie, A.; Niida, Y.; Wada, T.; Igarashi, N.; Kaneda, H.; Toma, T.; Ohta, K.; Kasahara, Y.; Koizumi, S. Oxidative stress causes enhanced endothelial cell injury in human heme oxygenase-1 deficiency. *J. Clin. Investig.* **1999**, *103*, 129–135. [CrossRef] [PubMed]
59. Le Moine, A.; Ryter, S.W.; Jang, H.-S.; Liu, C.; Li, Y.; Ma, K.; Han, Z.; Chi, M.; Sai, X.; Zhu, P.; et al. Immunomodulatory Effects of Heme Oxygenase-1 in Kidney Disease. *Front. Med.* **2021**, *1*, 708453.
60. Chiang, K.-C.; Chang, K.-S.; Hsu, S.-Y.; Sung, H.-C.; Feng, T.-H.; Chao, M.; Juang, H.-H. Human Heme Oxygenase-1 Induced by Interleukin-6 via JAK/STAT3 Pathways Is a Tumor Suppressor Gene in Hepatoma Cells. *Antioxidants* **2020**, *9*, 251. [CrossRef]
61. Gueron, G.; De Siervi, A.; Ferrando, M.; Salierno, M.; De Luca, P.; Elguero, B.; Meiss, R.; Navone, N.; Vazquez, E.S. Critical Role of Endogenous Heme Oxygenase 1 as a Tuner of the Invasive Potential of Prostate Cancer Cells. *Mol. Cancer Res.* **2009**, *7*, 1745–1755. [CrossRef] [PubMed]
62. Immenschuh, S.; Schröder, H. Heme oxygenase-1 and cardiovascular disease. *Histol. Histopathol.* **2006**, *21*, 679–685.
63. Lin, X.; Lv, J.; Ge, D.; Bai, H.; Yang, Y.; Wu, J. Heme oxygenase-1 alleviates eosinophilic inflammation by inhibiting STAT3-SOCS3 signaling. *Pediatr. Pulmonol.* **2020**, *55*, 1440–1447. [CrossRef]
64. Sheikh, S.Z.; Hegazi, R.A.; Kobayashi, T.; Onyiah, J.C.; Russo, S.M.; Matsuoka, K.; Sepulveda, A.R.; Li, F.; Otterbein, L.E.; Plevy, S.E. An anti-inflammatory role for carbon monoxide and heme oxygenase-1 in chronic Th2-mediated murine colitis. *J. Immunol.* **2011**, *186*, 5506–5513. [CrossRef]
65. Paine, A.; Eiz-Vesper, B.; Blasczyk, R.; Immenschuh, S. Signaling to heme oxygenase-1 and its anti-inflammatory therapeutic potential. *Biochem. Pharmacol.* **2010**, *80*, 1895–1903. [CrossRef]
66. Tzima, S.; Victoratos, P.; Kranidioti, K.; Alexiou, M.; Kollias, G. Myeloid heme oxygenase-1 regulates innate immunity and autoimmunity by modulating IFN-β production. *J. Exp. Med.* **2009**, *206*, 1167–1179. [CrossRef]
67. Kotsch, K.; Martins, P.N.A.; Klemz, R.; Janssen, U.; Gerstmayer, B.; Dernier, A.; Reutzel-Selke, A.; Kuckelkorn, U.; Tullius, S.G.; Volk, H.D. Heme oxygenase-1 ameliorates ischemia/reperfusion injury by targeting dendritic cell maturation and migration. *Antioxidants Redox Signal.* **2007**, *9*, 2049–2063. [CrossRef]
68. Chauveau, C.; Rémy, S.; Royer, P.J.; Hill, M.; Tanguy-Royer, S.; Hubert, F.X.; Tesson, L.; Brion, R.; Beriou, G.; Gregoire, M.; et al. Heme oxygenase-1 expression inhibits dendritic cell maturation and proinflammatory function but conserves IL-10 expression. *Blood* **2005**, *106*, 1694–1702. [CrossRef] [PubMed]
69. Kapturczak, M.H.; Wasserfall, C.; Brusko, T.; Campbell-Thompson, M.; Ellis, T.M.; Atkinson, M.A.; Agarwal, A. Heme Oxygenase-1 Modulates Early Inflammatory Responses Evidence from the Heme Oxygenase-1-Deficient Mouse. *Am. J. Pathol.* **2004**, *165*, 1045–1053. [CrossRef]
70. Otterbein, L.E.; Bach, F.H.; Alam, J.; Soares, M.; Tao Lu, H.; Wysk, M.; Davis, R.J.; Flavell, R.A.; Choi, A.M.K. Carbon monoxide has anti-inflammatory effects involving the mitogen-activated protein kinase pathway. *Nat. Med.* **2000**, *6*, 422–428. [CrossRef] [PubMed]
71. Lee, T.-S.; Chau, L.-Y. Heme oxygenase-1 mediates the anti-inflammatory effect of interleukin-10 in mice. *Nat. Med.* **2002**, *8*, 240–246. [CrossRef] [PubMed]
72. Bach Tatiana Vassilevskaia, F.H.; Berberat, P.O.; Yu, J.; Miguel Soares, T.-Y.P.; Seldon, M.P.; Pombo Gregoire, I. Endothelial Cell Activation of Adhesion Molecules Associated with Heme Oxygenase-1 Modulates the Expression. *J. Immunol. Ref.* **2021**, *172*, 3553–3563.
73. Nakamura, T.; Naguro, I.; Ichijo, H. Iron homeostasis and iron-regulated ROS in cell death, senescence and human diseases. *Biochim. Biophys. Acta Gen. Subj.* **2019**, *1863*, 1398–1409. [CrossRef]
74. Ryter, S.W.; Alam, J.; Choi, A.M.K. Heme oxygenase-1/carbon monoxide: From basic science to therapeutic applications. *Physiol. Rev.* **2006**, *86*, 583–650. [CrossRef]
75. Wu, B.; Wu, Y.; Tang, W. Heme Catabolic Pathway in Inflammation and Immune Disorders. *Front. Pharmacol* **2019**, *10*, 825. [CrossRef]
76. Takei, R.; Inoue, T.; Sonoda, N.; Kohjima, M.; Okamoto, M.; Sakamoto, R.; Inoguchi, T.; Ogawa, Y. Bilirubin reduces visceral obesity and insulin resistance by suppression of inflammatory cytokines. *PLoS ONE* **2019**, *14*, e0223302. [CrossRef]
77. Brouard, S.; Otterbein, L.E.; Anrather, J.; Tobiasch, E.; Bach, F.H.; Choi, A.M.K.; Soares, M.P. Carbon Monoxide Generated by Heme Oxygenase 1 Suppresses Endothelial Cell Apoptosis. *J. Exp. Med.* **2000**, *192*, 1015–1026. [CrossRef]
78. Petrache, I.; Otterbein, L.E.; Alam, J.; Wiegand, G.W.; Choi, A.M.K. Heme oxygenase-1 inhibits TNF-α-induced apoptosis in cultured fibroblasts. *Am. J. Physiol. Cell. Mol. Physiol.* **2000**, *278*, L312–L319. [CrossRef] [PubMed]
79. Loboda, A.; Jazwa, A.; Grochot-Przeczek, A.; Rutkowski, A.J.; Cisowski, J.; Agarwal, A.; Jozkowicz, A.; Dulak, J. Heme oxygenase-1 and the vascular bed: From molecular mechanisms to therapeutic opportunities. *Antioxidants Redox Signal.* **2008**, *10*, 1767–1812. [CrossRef] [PubMed]
80. Kliche, S.; Waltenberger, J. VEGF Receptor Signaling and Endothelial Function. *IUBMB Life* **2001**, *52*, 61–66. [CrossRef] [PubMed]

81. Pae, H.-O.; Oh, G.-S.; Choi, B.-M.; Chae, S.-C.; Kim, Y.-M.; Chung, K.-R.; Chung, H.-T. Carbon Monoxide Produced by Heme Oxygenase-1 Suppresses T Cell Proliferation via Inhibition of IL-2 Production. *J. Immunol.* **2004**, *172*, 4744–4751. [CrossRef] [PubMed]
82. Morse, D.; Pischke, S.E.; Zhou, Z.; Davis, R.J.; Flavell, R.A.; Loop, T.; Otterbein, S.L.; Otterbein, L.E.; Choi, A.M.K. Suppression of inflammatory cytokine production by carbon monoxide involves the JNK pathway and AP-1. *J. Biol. Chem.* **2003**, *278*, 36993–36998. [CrossRef] [PubMed]
83. Kim, S.J.; Lee, S.M. NLRP3 inflammasome activation in d-galactosamine and lipopolysaccharide- induced acute liver failure: Role of heme oxygenase-1. *Free Radic. Biol. Med.* **2013**, *65*, 997–1004. [CrossRef]
84. Luo, Y.P.; Jiang, L.; Kang, K.; Fei, D.S.; Meng, X.L.; Nan, C.C.; Pan, S.H.; Zhao, M.R.; Zhao, M.Y. Hemin inhibits NLRP3 inflammasome activation in sepsis-induced acute lung injury, involving heme oxygenase-1. *Int. Immunopharmacol.* **2014**, *20*, 24–32. [CrossRef]
85. Lv, J.; Su, W.; Yu, Q.; Zhang, M.; Di, C.; Lin, X.; Wu, M.; Xia, Z. Heme oxygenase-1 protects airway epithelium against apoptosis by targeting the proinflammatory NLRP3–RXR axis in asthma. *J. Biol. Chem.* **2018**, *293*, 18454–18465. [CrossRef]
86. Abraham, N.G.; Kappas, A. Pharmacological and clinical aspects of heme oxygenase. *Pharmacol. Rev.* **2008**, *60*, 79–127. [CrossRef]
87. Fan, Y.; Zhang, J.; Cai, L.; Wang, S.; Liu, C.; Zhang, Y.; You, L.; Fu, Y.; Shi, Z.; Yin, Z.; et al. The effect of anti-inflammatory properties of ferritin light chain on lipopolysaccharide-induced inflammatory response in murine macrophages. *Biochim. Biophys. Acta Mol. Cell Res.* **2014**, *1843*, 2775–2783. [CrossRef]
88. Sass, G.; Soares, M.C.P.; Yamashita, K.; Seyfried, S.; Zimmermann, W.-H.; Eschenhagen, T.; Kaczmarek, E.; Ritter, T.; Volk, H.-D.; Tiegs, G. Heme oxygenase-1 and its reaction product, carbon monoxide, prevent inflammation-related apoptotic liver damage in mice. *Hepatology* **2003**, *38*, 909–918. [CrossRef] [PubMed]
89. Chen, Z.; Zhong, H.; Wei, J.; Lin, S.; Zong, Z.; Gong, F.; Huang, X.; Sun, J.; Li, P.; Lin, H.; et al. Inhibition of Nrf2/HO-1 signaling leads to increased activation of the NLRP3 inflammasome in osteoarthritis. *Arthritis Res. Ther.* **2019**, *21*, 300. [CrossRef] [PubMed]
90. Zhang, B.; Xie, S.; Su, Z.; Song, S.; Xu, H.; Chen, G.; Cao, W.; Yin, S.; Gao, Q.; Wang, H. Heme oxygenase-1 induction attenuates imiquimod-induced psoriasiform inflammation by negative regulation of Stat3 signaling. *Sci. Rep.* **2016**, *6*, 21132. [CrossRef] [PubMed]
91. Liu, X.; Gao, Y.; Li, M.; Geng, C.; Xu, H.; Yang, Y.; Guo, Y.; Jiao, T.; Fang, F.; Chang, Y. Sirt1 mediates the effect of the heme oxygenase inducer, cobalt protoporphyrin, on ameliorating liver metabolic damage caused by a high-fat diet. *J. Hepatol.* **2015**, *63*, 713–721. [CrossRef]
92. Zhong, W.; Xia, Z.; Hinrichs, D.; Rosenbaum, J.T.; Wegmann, K.W.; Meyrowitz, J.; Zhang, Z. Hemin exerts multiple protective mechanisms and attenuates dextran sulfate sodium-induced colitis. *J. Pediatr. Gastroenterol. Nutr.* **2010**, *50*, 132–139. [CrossRef]
93. Liu, Y.; Zhu, B.; Luo, L.; Li, P.; Paty, D.W.; Cynader, M.S. Heme oxygenase-1 plays an important protective role in experimental autoimmune encephalomyelitis. *Neuroreport* **2001**, *12*, 1841–1845. [CrossRef]
94. Herfarth, H.; Paul, G.; Bataille, F.; Obermeier, F.; Bock, J.; Klebl, F.; Strauch, U.; Lochbaum, D.; Rümmele, P.; Farkas, S.; et al. Analysis of intestinal haem-oxygenase-1 (HO-1) in clinical and experimental colitis. *Clin. Exp. Immunol.* **2005**, *140*, 547–555.
95. Chen, D.; Wu, C.; Qiu, Y.B.; Chu, Q.; Sun, X.Q.; Wang, X.; Chen, J.L.; Lu, M.D.; Chen, D.Z.; Pang, Q. feng Curcumin ameliorates hepatic chronic inflammation induced by bile duct obstruction in mice through the activation of heme oxygenase-1. *Int. Immunopharmacol.* **2020**, *78*, 106054. [CrossRef]
96. Foresti, R.; Bains, S.K.; Pitchumony, T.S.; De Castro Brás, L.E.; Drago, F.; Dubois-Randé, J.L.; Bucolo, C.; Motterlini, R. Small molecule activators of the Nrf2-HO-1 antioxidant axis modulate heme metabolism and inflammation in BV2 microglia cells. *Pharmacol. Res.* **2013**, *76*, 132–148. [CrossRef]
97. Selva Nandakumar, K.; Blank, M.; Kalergis, A.M.; Funes, S.C.; Rios, M.; Fernández-Fierro, A.; Covián, C.; Bueno, S.M.; Riedel, C.A.; Mackern-Oberti, J.P. Naturally Derived Heme-Oxygenase 1 Inducers and Their Therapeutic Application to Immune-Mediated Diseases. *Front. Immunol* **2020**, *11*, 1467.
98. Kang, S.; Kim, W.; Jeong, S.; Lee, Y.; Nam, J.; Lee, S.; Jung, Y. Oxidized 5-aminosalicylic acid activates Nrf2-HO-1 pathway by covalently binding to Keap1: Implication in anti-inflammatory actions of 5-aminosalicylic acid. *Free Radic. Biol. Med.* **2017**, *108*, 715–724. [CrossRef]
99. Nishio, Y.; Fujino, M.; Zhao, M.; Ishii, T.; Ishizuka, M.; Ito, H.; Takahashi, K.; Abe, F.; Nakajima, M.; Tanaka, T.; et al. 5-Aminolevulinic acid combined with ferrous iron enhances the expression of heme oxygenase-1. *Int. Immunopharmacol.* **2014**, *19*, 300–307. [CrossRef]
100. Berberat, P.O.; A-Rahim, Y.I.; Yamashita, K.; Warny, M.M.; Csizmadia, E.; Robson, S.C.; Bach, F.H. Heme Oxygenase-1-Generated Biliverdin Ameliorates Experimental Murine Colitis. *Inflamm. Bowel Dis.* **2005**, *11*, 350–359. [CrossRef]
101. Keum, H.; Kim, T.W.; Kim, Y.; Seo, C.; Son, Y.; Kim, J.; Kim, D.; Jung, W.; Whang, C.H.; Jon, S. Bilirubin nanomedicine alleviates psoriatic skin inflammation by reducing oxidative stress and suppressing pathogenic signaling. *J. Control Release* **2020**, *325*, 359–369. [CrossRef]
102. Rochette, L.; Zeller, M.; Cottin, Y.; Vergely, C. Redox Functions of Heme Oxygenase-1 and Biliverdin Reductase in Diabetes. *Trends Endocrinol. Metab.* **2018**, *29*, 74–85. [CrossRef]
103. Lin, C.-C.; Yang, C.-C.; Hsiao, L.-D.; Chen, S.-Y.; Yang, C.-M. Heme Oxygenase-1 Induction by Carbon Monoxide Releasing Molecule-3 Suppresses Interleukin-1β-Mediated Neuroinflammation. *Front. Mol. Neurosci.* **2017**, *10*, 387. [CrossRef]

104. Motterlini, R.; Mann, B.E.; Foresti, R. Therapeutic applications of carbon monoxide-releasing molecules. *Expert Opin. Investig. Drugs* **2005**, *14*, 1305–1318. [CrossRef]
105. Motterlini, R.; Otterbein, L.E. The therapeutic potential of carbon monoxide. *Nat. Rev. Drug Discov.* **2010**, *9*, 728–743. [CrossRef]
106. Ali, Z.E.; Ollivier, A.; Manin, S.; Rivard, M.; Motterlini, R.; Foresti, R. Therapeutic effects of CO-releaser/Nrf2 activator hybrids (HYCOs) in the treatment of skin wound, psoriasis and multiple sclerosis. *Redox Biol.* **2020**, *34*, 101521. [CrossRef]
107. Lad, L.; Friedman, J.; Li, H.; Bhaskar, B.; Ortiz de Montellano, P.R.; Poulos, T.L. Crystal structure of human heme oxygenase-1 in a complex with biliverdin. *Biochemistry* **2004**, *43*, 3793–3801. [CrossRef]
108. National Center for Biotechnology Information. PubChem Compound Summary for CID 4971, Protoporphyrin IX. 2022. Available online: https://pubchem.ncbi.nlm.nih.gov/compound/protoporphyrin-IX (accessed on 4 January 2022).
109. Fredenburgh, L.E.; Perrella, M.A.; Mitsialis, S.A. The role of heme oxygenase-1 in pulmonary disease. *Am. J. Respir. Cell Mol. Biol.* **2007**, *36*, 158–165. [CrossRef]
110. Slebos, D.-J.; Ryter, S.W.; Choi, A.M.K. Heme oxygenase-1 and carbon monoxide in pulmonary medicine. *Respir. Res.* **2003**, *4*, 7. [CrossRef]
111. Di Pietro, C.; Öz, H.H.; Murray, T.S.; Bruscia, E.M. Targeting the Heme Oxygenase 1/Carbon Monoxide Pathway to Resolve Lung Hyper-Inflammation and Restore a Regulated Immune Response in Cystic Fibrosis. *Front. Pharmacol.* **2020**, *11*, 1059. [CrossRef]
112. Pedersen, S.F.; Ho, Y.-C. SARS-CoV-2: A storm is raging. *J. Clin. Investig.* **2020**, *130*, 2202–2205. [CrossRef]
113. Kritas, S.; Ronconi, G.; Caraffa, A.; Gallenga, C.; Ross, R.; Conti, P. Mast cells contribute to coronavirus-induced inflammation: New anti-inflammatory strategy. *J. Biol. Regul. Homeost. Agents* **2019**, *34*, 9–14.
114. He, Y.Q.; Zhou, C.C.; Yu, L.Y.; Wang, L.; Deng, J.L.; Tao, Y.L.; Zhang, F.; Chen, W.S. Natural product derived phytochemicals in managing acute lung injury by multiple mechanisms. *Pharmacol. Res.* **2021**, *163*, 105224. [CrossRef]
115. Butt, Y.; Kurdowska, A.; Timothy, A.C. Acute Lung Injury A Clinical and Molecular Review. *Arch. Pathol. Lab. Med.* **2016**, *140*, 345–350. [CrossRef]
116. Huang, X.; Xiu, H.; Zhang, S.; Zhang, G. The role of macrophages in the pathogenesis of ali/ards. *Mediat. Inflamm.* **2018**, *2018*, 1–8. [CrossRef]
117. Pereira, M.L.M.; Marinho, C.R.F.; Epiphanio, S. Could heme oxygenase-1 be a new target for therapeutic intervention in malaria-associated acute lung injury/acute respiratory distress syndrome? *Front. Cell. Infect. Microbiol.* **2018**, *8*, 161. [CrossRef]
118. An, L.; Liu, C.T.; Qin, X.B.; Liu, Q.H.; Liu, Y.; Yu, S.Y. Protective effects of hemin in an experimental model of ventilator-induced lung injury. *Eur. J. Pharmacol.* **2011**, *661*, 102–108. [CrossRef]
119. Xia, Z.Y.; Gao, J.; Ancharaz, A.K. Protective effect of ischemic postconditioning on lung ischemia-reperfusion injury in rats and the role of heme oxygenase-1. *Chin. J. Traumatol. Engl. Ed.* **2009**, *12*, 162–166.
120. Chen, X.; Wang, Y.; Xie, X.; Chen, H.; Zhu, Q.; Ge, Z.; Wei, H.; Deng, J.; Xia, Z.; Lian, Q. Heme Oxygenase-1 Reduces Sepsis-Induced Endoplasmic Reticulum Stress and Acute Lung Injury. *Mediators Inflamm.* **2018**, *2018*, 9413876. [CrossRef]
121. Yu, J.; Wang, Y.; Li, Z.; Dong, S.; Wang, D.; Gong, L.; Shi, J.; Zhang, Y.; Liu, D.; Mu, R. Effect of Heme Oxygenase-1 on Mitofusin-1 protein in LPS-induced ALI/ARDS in rats. *Sci. Rep.* **2016**, *6*, 36530. [CrossRef]
122. Nakashima, K.; Sato, T.; Shigemori, S.; Shimosato, T.; Shinkai, M.; Kaneko, T. Regulatory role of heme oxygenase-1 in silica-induced lung injury. *Respir. Res.* **2018**, *19*, 144. [CrossRef]
123. Ryter, S.W. Therapeutic Potential of Heme Oxygenase-1 and Carbon Monoxide in Acute Organ Injury, Critical Illness, and Inflammatory Disorders. *Antioxidants* **2020**, *9*, 1153. [CrossRef]
124. Shen, N.; Gong, T.; Wang, J.D.; Meng, F.L.; Qiao, L.; Yang, R.L.; Xue, B.; Pan, F.Y.; Zhou, X.J.; Chen, H.Q.; et al. Cigarette smoke-induced pulmonary inflammatory responses are mediated by EGR-1/GGPPS/MAPK signaling. *Am. J. Pathol.* **2011**, *178*, 110–118. [CrossRef]
125. Fujita, T.; Toda, K.; Karimova, A.; Yan, S.-F.; Naka, Y.; Yet, S.-F.; Pinsky, D.J. Paradoxical rescue from ischemic lung injury by inhaled carbon monoxide driven by derepression of fibrinolysis. *Nat. Med.* **2001**, *7*, 598–604. [CrossRef]
126. Tian, W.F.; Weng, P.; Sheng, Q.; Chen, J.L.; Zhang, P.; Zhang, J.R.; Du, B.; Wu, M.C.; Pang, Q.F.; Chu, J.J. Biliverdin protects the isolated rat lungs from ischemia-reperfusion injury via antioxidative, anti-inflammatory and anti-apoptotic effects. *Chin. Med. J.* **2017**, *130*, 859–865. [CrossRef]
127. Kosaka, J.; Morimatsu, H.; Takahashi, T.; Shimizu, H.; Kawanishi, S. Effects of Biliverdin Administration on Acute Lung Injury Induced by Hemorrhagic Shock and Resuscitation in Rats. *PLoS ONE* **2013**, *8*, 63606. [CrossRef]
128. Ma, L.L.; Zhang, P.; Wang, H.Q.; Li, Y.F.; Hu, J.; Jiang, J.D.; Li, Y.H. heme oxygenase-1 agonist CoPP suppresses influenza virus replication through IRF3-mediated generation of IFN-α/β. *Virology* **2019**, *528*, 80–88. [CrossRef]
129. Iuliano, A.D.; Roguski, K.M.; Chang, H.H.; Muscatello, D.J.; Palekar, R.; Tempia, S.; Cohen, C.; Gran, J.M.; Schanzer, D.; Cowling, B.J.; et al. Estimates of global seasonal influenza-associated respiratory mortality: A modelling study. *Lancet* **2018**, *391*, 1285–1300. [CrossRef]
130. Kawser Hossain, M.; Kumar, S.; Id, S.; Dayem, A.A.; Kim, J.-H.; Kim, K.; Yang, G.-M.; Choi, H.Y.; Cho, S.-G. Bax Inhibitor-1 Acts as an Anti-Influenza Factor by Inhibiting ROS Mediated Cell Death and Augmenting Heme-Oxygenase 1 Expression in Influenza Virus Infected Cells. *Int. J. Mol. Sci. Artic.* **2018**, *19*, 712. [CrossRef] [PubMed]
131. Wang, C.; Zhang, Y.; Han, L.L.; Guo, L.; Zhong, H.; Wang, J. Hemin ameliorates influenza pneumonia by attenuating lung injury and regulating the immune response. *Int. J. Antimicrob. Agents* **2017**, *49*, 45–52. [CrossRef] [PubMed]

132. Hashiba, T.; Suzuki, M.; Nagashima, Y.; Suzuki, S.; Inoue, S.; Tsuburai, T.; Matsuse, T.; Ishigatubo, Y. Adenovirus-Mediated Transfer of Heme Oxygenase-1 cDNA Attenuates Severe Lung Injury Induced by the Influenza Virus in Mice. *Gene Ther.* **2001**, *8*, 1499–1507. [CrossRef]
133. Cummins, N.W.; Weaver, E.A.; May, S.M.; Croatt, A.J.; Foreman, O.; Kennedy, R.B.; Poland, G.A.; Barry, M.A.; Nath, K.A.; Badley, A.D. Heme oxygenase-1 regulates the immune response to influenza virus infection and vaccination in aged mice. *FASEB J.* **2012**, *26*, 2911–2918. [CrossRef]
134. Zhong, M.; Wang, H.; Ma, L.; Yan, H.; Wu, S.; Gu, Z.; Li, Y. DMO-CAP inhibits influenza virus replication by activating heme oxygenase-1-mediated IFN response. *Virol. J.* **2019**, *16*, 21. [CrossRef]
135. Ma, L.-L.; Wang, H.-Q.; Wu, P.; Hu, J.; Yin, J.-Q.; Wu, S.; Ge, M.; Sun, W.-F.; Zhao, J.-Y.; Aisa, H.A.; et al. Rupestonic acid derivative YZH-106 suppresses influenza virus replication by activation of heme oxygenase-1-mediated interferon response. *Free Radic. Biol. Med.* **2016**, *96*, 347–361. [CrossRef]
136. Nair DNB, H.; Theodoratou, E.; Rudan, I.; Nokes, D.J.; Ngama HND, M.; Munywoki, P.K.; Dherani, M.; Nair, H.; James Nokes, D.; Gessner, B.D.; et al. Global burden of acute lower respiratory infections due to respiratory syncytial virus in young children: A systematic review and meta-analysis. *Lancet* **2010**, *375*, 1545–1555. [CrossRef]
137. Del Vecchio, A.; Ferrara, T.; Maglione, M.; Capasso, L.; Raimondi, F. New perspectives in Respiratory Syncitial Virus infection. *J. Matern. Neonatal Med.* **2013**, *26*, 55–59. [CrossRef]
138. Yoboua, F.; Martel, A.; Duval, A.; Mukawera, E.; Grandvaux, N. Respiratory Syncytial Virus-Mediated NF-B p65 Phosphorylation at Serine 536 Is Dependent on RIG-I, TRAF6, and IKK. *J. Virol.* **2010**, *84*, 7267–7277. [CrossRef]
139. Espinoza, J.A.; León, M.A.; Céspedes, P.F.; Gómez, R.S.; Canedo-Marroquín, G.; Riquelme, S.A.; Salazar-Echegarai, F.J.; Blancou, P.; Simon, T.; Anegon, I.; et al. Heme Oxygenase-1 Modulates Human Respiratory Syncytial Virus Replication and Lung Pathogenesis during Infection. *J. Immunol.* **2017**, *199*, 212. [CrossRef]
140. Antiretroviral Therapy has Saved Millions of Lives from AIDS and Could Save More. Available online: https://ourworldindata.org/art-lives-saved (accessed on 3 January 2022).
141. UNAIDS: Joint United Nations Programme on HIV/AIDS. Available online: https://www.un.org/youthenvoy/2013/08/unaids-joint-united-nations-programme-on-hivaids/ (accessed on 31 December 2021).
142. Global Statistics. Available online: https://www.hiv.gov/hiv-basics/overview/data-and-trends/global-statistics (accessed on 31 December 2021).
143. Collins, D.R.; Gaiha, G.D.; Walker, B.D. CD8+ T cells in HIV control, cure and prevention. *Nat. Rev. Immunol.* **2020**, *20*, 471–482. [CrossRef]
144. Devadas, K.; Dhawan, S. Hemin Activation Ameliorates HIV-1 Infection via Heme Oxygenase-1 Induction. *J. Immunol.* **2006**, *176*, 4252–4257. [CrossRef]
145. Cross, S.A.; Cook, D.R.; Chi, A.W.S.; Vance, P.J.; Kolson, L.L.; Wong, B.J.; Jordan-Sciutto, K.L.; Kolson, D.L. Dimethyl fumarate, an immune modulator and inducer of the antioxidant response, suppresses HIV replication and macrophage-mediated neurotoxicity: A novel candidate for HIV neuroprotection. *J. Immunol.* **2011**, *187*, 5015–5025. [CrossRef]
146. Kovacsics, C.E.; Gill, A.J.; Ambegaokar, S.S.; Gelman, B.B.; Kolson, D.L. Degradation of heme oxygenase-1 by the immunoproteasome in astrocytes: A potential interferon-γ-dependent mechanism contributing to HIV neuropathogenesis. *Glia* **2017**, *65*, 1264–1277. [CrossRef]
147. Gill, A.J.; Kovacsics, C.E.; Cross, S.A.; Vance, P.J.; Kolson, L.L.; Jordan-Sciutto, K.L.; Gelman, B.B.; Kolson, D.L. Heme oxygenase-1 deficiency accompanies neuropathogenesis of HIV-associated neurocognitive disorders. *J. Clin. Investig.* **2014**, *124*, 4459–4472. [CrossRef]
148. Malvy, D.; McElroy, A.K.; de Clerck, H.; Günther, S.; van Griensven, J. Ebola virus disease. *Lancet* **2019**, *393*, 936–948. [CrossRef]
149. Ascenzi, P.; Bocedi, A.; Heptonstall, J.; Capobianchi, M.R.; Di Caro, A.; Mastrangelo, E.; Bolognesi, M.; Ippolito, G. Ebolavirus and Marburgvirus: Insight the Filoviridae family. *Mol. Aspects Med.* **2008**, *29*, 151–185. [CrossRef]
150. Huang, H.; Konduru, K.; Solovena, V.; Zhou, Z.-H.; Kumari, N.; Takeda, K.; Nekhai, S.; Bavari, S.; Kaplan, G.G.; Yamada, K.M.; et al. Therapeutic potential of the heme oxygenase-1 inducer hemin against Ebola virus infection. *Curr. Trends Immunol.* **2016**, *17*, 117–123.
151. Hill-Batorski, L.; Halfmann, P.; Neumann, G.; Kawaoka, Y. The cytoprotective enzyme heme oxygenase-1 suppresses Ebola virus replication. *J. Virol.* **2013**, *87*, 13795–13802. [CrossRef]
152. Basler, C.F.; Mikulasova, A.; Martinez-Sobrido, L.; Paragas, J.; Mühlberger, E.; Bray, M.; Klenk, H.-D.; Palese, P.; García-Sastre, A. The Ebola Virus VP35 Protein Inhibits Activation of Interferon Regulatory Factor 3. *J. Virol.* **2003**, *77*, 7945–7956. [CrossRef] [PubMed]
153. Lehmann, E.; El-Tantawy, W.H.; Ocker, M.; Bartenschlager, R.; Lohmann, V.; Hashemolhosseini, S.; Tiegs, G.; Sass, G. The heme oxygenase 1 product biliverdin interferes with hepatitis C virus replication by increasing antiviral interferon response. *Hepatology* **2010**, *51*, 398–404. [CrossRef] [PubMed]
154. Preciado, M.V.; Valva, P.; Escobar-Gutierrez, A.; Rahal, P.; Ruiz-Tovar, K.; Yamasaki, L.; Vazquez-Chacon, C.; Martinez-Guarneros, A.; Carpio-Pedroza, J.C.; Fonseca-Coronado, S.; et al. Hepatitis C virus molecular evolution: Transmission, disease progression and antiviral therapy. *World J. Gastroenterol.* **2014**, *20*, 15992–16013. [CrossRef]

155. Abdalla, M.Y.; Britigan, B.E.; Wen, F.; Icardi, M.; McCormick, M.L.; LaBrecque, D.R.; Voigt, M.; Brown, K.E.; Schmidt, W.N. Down-Regulation of Heme Oxygenase-1 by Hepatitis C Virus Infection In Vivo and by the In Vitro Expression of Hepatitis C Core Protein. *J. Infect. Dis.* **2004**, *190*, 1109–1118. [CrossRef]
156. Zhu, Z.; Wilson, A.T.; Mathahs, M.M.; Wen, F.; Brown, K.E.; Luxon, B.A.; Schmidt, W.N. Heme oxygenase-1 suppresses hepatitis C virus replication and increases resistance of hepatocytes to oxidant injury. *Hepatology* **2008**, *48*, 1430–1439. [CrossRef] [PubMed]
157. Fillebeen, C.; Rivas-Estilla, A.M.; Bisaillon, M.; Ponka, P.; Muckenthaler, M.; Hentze, M.W.; Koromilas, A.E.; Pantopoulos, K. Iron inactivates the RNA polymerase NS5B and suppresses subgenomic replication of hepatitis C virus. *J. Biol. Chem.* **2005**, *280*, 9049–9057. [CrossRef]
158. Zhu, Z.; Wilson, A.T.; Luxon, B.A.; Brown, K.E.; Mathahs, M.M.; Bandyopadhyay, S.; McCaffrey, A.P.; Schmidt, W.N. Biliverdin inhibits hepatitis C virus nonstructural 3/4A protease activity: Mechanism for the antiviral effects of heme oxygenase? *Hepatology* **2010**, *52*, 1897–1905. [CrossRef]
159. Law, Y.-Y.; Lee, W.-F.; Hsu, C.-J.; Lin, Y.-Y.; Tsai, C.-H.; Huang, C.-C.; Wu, M.-H.; Tang, C.-H.; Liu, J.-F. miR-let-7c-5p and miR-149-5p inhibit proinflammatory cytokine production in osteoarthritis and rheumatoid arthritis synovial fibroblasts. *Aging* **2021**, *13*, 17227–17236. [CrossRef] [PubMed]
160. Chen, W.C.; Wei, C.K.; Lee, J.C. MicroRNA-let-7c suppresses hepatitis C virus replication by targeting Bach1 for induction of haem oxygenase-1 expression. *J. Viral Hepat.* **2019**, *26*, 655–665. [CrossRef] [PubMed]
161. Shen, Y.M.; Zhang, H.L.; Wu, Y.H.; Yu, X.P.; Hu, H.J.; Dai, L.H. Dynamic correlation between induction of the expression of heme oxygenase-1 and hepatitis B viral replication. *Mol. Med. Rep.* **2015**, *11*, 4706–4712. [CrossRef]
162. Schmidt, W.N.; Mathahs, M.M.; Zhu, Z. Heme and HO-1 Inhibition of HCV, HBV, and HIV. *Front. Pharmacol.* **2012**, *3*, 129. [CrossRef] [PubMed]
163. Yeh, C.-N.; Wu, R.-C.; Cheng, C.-T.; Tsai, C.-Y.; Chang, Y.-R.; Yeh, T.-S.; Wu, T.-H.; Lee, W.-C.; Chiang, K.-C. HO-1 is a favorable prognostic factor for HBV-HCC patients who underwent hepatectomy. *Cancer Manag. Res.* **2018**, *10*, 6049–6059. [CrossRef] [PubMed]
164. Murugesan, A.; Manoharan, M. Dengue virus. In *Emerging and Reemerging Viral Pathogens: Volume 1: Fundamental and Basic Virology Aspects of Human, Animal and Plant Pathogens*; Elsevier: Amsterdam, The Netherlands, 2019; pp. 281–359. ISBN 9780128194003.
165. Olagnier, D.; Peri, S.; Steel, C.; Van Montfoort, N.; Chiang, C. Cellular Oxidative Stress Response Controls the Antiviral and Apoptotic Programs in Dengue Virus-Infected Dendritic Cells. *PLoS Pathog* **2014**, *10*, 1004566. [CrossRef] [PubMed]
166. Tseng, C.-K.; Lin, C.-K.; Wu, Y.-H.; Chen, Y.-H.; Chen, W.-C.; Young, K.-C.; Lee, J.-C. Human heme oxygenase 1 is a potential host cell factor against dengue virus replication. *Sci. Rep.* **2016**, *6*, 32176. [CrossRef]
167. Su, Y.-C.; Huang, Y.-F.; Wu, Y.-W.; Chen, H.-F.; Wu, Y.-H.; Hsu, C.-C.; Hsu, Y.-C.; Lee, J.-C. MicroRNA-155 inhibits dengue virus replication by inducing heme oxygenase-1-mediated antiviral interferon responses. *FASEB J.* **2020**, *34*, 7283–7294. [CrossRef]
168. El Kalamouni, C.; Frumence, E.; Bos, S.; Turpin, J.; Nativel, B.; Harrabi, W.; Wilkinson, D.A.; Meilhac, O.; Gadea, G.; Desprès, P.; et al. Subversion of the Heme Oxygenase-1 Antiviral Activity by Zika Virus. *Viruses* **2018**, *11*, 2. [CrossRef]
169. Suazo, P.A.; Tognarelli, E.I.; Kalergis, A.M.; González, P.A. Herpes simplex virus 2 infection: Molecular association with HIV and novel microbicides to prevent disease. *Med. Microbiol. Immunol.* **2015**, *204*, 161–176. [CrossRef]
170. Ibáñez, F.J.; Farías, M.A.; Retamal-Díaz, A.; Espinoza, J.A.; Kalergis, A.M.; González, P.A. Pharmacological Induction of Heme Oxygenase-1 Impairs Nuclear Accumulation of Herpes Simplex Virus Capsids upon Infection. *Front. Microbiol.* **2017**, *8*, 2108. [CrossRef]
171. Palacios, G.; Oberste, M.S. Enteroviruses as agents of emerging infectious diseases. *J. Neurovirol.* **2005**, *11*, 424–433. [CrossRef]
172. Tung, W.-H.; Hsieh, H.-L.; Lee, I.-T.; Yang, C.-M. Enterovirus 71 induces integrin β1/EGFR-Rac1-dependent oxidative stress in SK-N-SH cells: Role of HO-1/CO in viral replication. *J. Cell. Physiol.* **2011**, *226*, 3316–3329. [CrossRef]
173. Chen, Y.; Liu, Q.; Guo, D. Emerging coronaviruses: Genome structure, replication, and pathogenesis. *J. Med. Virol.* **2020**, *92*, 418–423. [CrossRef] [PubMed]
174. Lechuga, G.C.; Souza-Silva, F.; Sacramento, C.Q.; Trugilho, M.R.O.; Valente, R.H.; Napoleão-Pêgo, P.; Dias, S.S.G.; Fintelman-Rodrigues, N.; Temerozo, J.R.; Carels, N.; et al. SARS-CoV-2 Proteins Bind to Hemoglobin and Its Metabolites. *Int. J. Mol. Sci.* **2021**, *22*, 9035. [CrossRef] [PubMed]
175. Maestro, S.; Córdoba, K.M.; Olague, C.; Argemi, J.; Ávila, M.A.; González-Aseguinolaza, G.; Smerdou, C.; Fontanellas, A. Heme oxygenase-1 inducer hemin does not inhibit SARS-CoV-2 virus infection. *Biomed. Pharmacother.* **2021**, *137*, 111384. [CrossRef] [PubMed]
176. Kim, D.-H.; Ahn, H.-S.; Go, H.-J.; Kim, D.-Y.; Kim, J.-H.; Lee, J.-B.; Park, S.-Y.; Song, C.-S.; Lee, S.-W.; Ha, S.-D.; et al. Hemin as a novel candidate for treating COVID-19 via heme oxygenase-1 induction. *Sci. Rep.* **2021**, *11*, 21462. [CrossRef]
177. Ortiz, E.; Sanchis, P.; Bizzotto, J.; Lage-Vickers, S.; Labanca, E.; Navone, N.; Cotignola, J.; Vazquez, E.; Gueron, G. Myxovirus Resistance Protein 1 (MX1), a Novel HO-1 Interactor, Tilts the Balance of Endoplasmic Reticulum Stress towards Pro-Death Events in Prostate Cancer. *Biomolecules* **2020**, *10*, 1005. [CrossRef]
178. Müller, B.; Kräusslich, H.-G. Antiviral strategies. *Handb. Exp. Pharmacol.* **2009**, *189*, 1–24.
179. Baillie, J.K.; Digard, P. Influenza—Time to Target the Host? *N. Engl. J. Med.* **2013**, *369*, 191–193. [CrossRef]
180. Brown, J.; Zhou, J.; Peacock, T.P.; Barclay, W.S. The SARS-CoV-2 Variant, Omicron, Shows Enhanced Replication in Human Primary Nasal Epithelial Cells. Available online: https://assets.publishing.service.gov.uk/government/uploads/system/uploads/attachment_data/file/1043652/S1454_Omicron_report_20.12_Imperial.pdf (accessed on 30 December 2021).

181. Bentley, E.G.; Kirby, A.; Sharma, P.; Kipar, A.; Mega, D.F.; Bramwell, C.; Penrice-Randal, R.; Prince, T.; Brown, J.C.; Zhou, J.; et al. SARS-CoV-2 Omicron-B.1.1.529 Variant leads to less severe disease than Pango B and Delta variants strains in a mouse model of severe COVID-19. *bioRxiv* **2021**. [CrossRef]
182. Sheraton, M.; Deo, N.; Kashyap, R.; Surani, S. A Review of Neurological Complications of COVID-19. *Cureus* **2020**, *12*, e8192. [CrossRef]
183. Manjili Rose Manjili, H.H.; Zarei, M.; Habibi, M. COVID-19 as an Acute Inflammatory Disease. *J. Immunol Ref.* **2021**, *205*, 12–19. [CrossRef]
184. Batra, N.; De Souza, C.; Batra, J.; Raetz, A.G.; Yu, A.-M. The HMOX1 Pathway as a Promising Target for the Treatment and Prevention of SARS-CoV-2 of 2019 (COVID-19). *Int. J. Mol. Sci.* **2020**, *21*, 6412. [CrossRef] [PubMed]
185. Zhu, Y.; Sun, Y.; Jin, K.; Greenberg, D.A. Hemin induces neuroglobin expression in neural cells. *Blood* **2002**, *100*, 2494–2498. [CrossRef] [PubMed]
186. Li, H.M.; Shi, Y.L.; Wen, D.; Luo, H.M.; Lin, X.; Xiao, F. A novel effective chemical hemin for the treatment of acute carbon monoxide poisoning in mice. *Exp. Ther. Med.* **2017**, *14*, 5186–5192. [CrossRef] [PubMed]
187. Lu, X.; Chen-Roetling, J.; Regan, R.F. Systemic hemin therapy attenuates blood-brain barrier disruption after intracerebral hemorrhage. *Neurobiol. Dis.* **2014**, *70*, 245–251. [CrossRef] [PubMed]
188. Olagnier, D.; Farahani, E.; Thyrsted, J.; Blay-Cadanet, J.; Herengt, A.; Idorn, M.; Hait, A.; Hernaez, B.; Knudsen, A.; Iversen, M.B.; et al. SARS-CoV2-mediated suppression of NRF2-signaling reveals potent antiviral and anti-inflammatory activity of 4-octyl-itaconate and dimethyl fumarate. *Nat. Commun.* **2020**, *11*, 4938. [CrossRef] [PubMed]
189. Cuadrado, A.; Pajares, M.; Benito, C.; Jiménez-Villegas, J.; Escoll, M.; Fernández-Ginés, R.; Garcia Yagüe, A.J.; Lastra, D.; Manda, G.; Rojo, A.I.; et al. Can Activation of NRF2 Be a Strategy against COVID-19? *Trends Pharmacol. Sci.* **2020**, *41*, 598–610. [CrossRef] [PubMed]
190. Gupta, A.; Madhavan, M.V.; Sehgal, K.; Nair, N.; Mahajan, S.; Sehrawat, T.S.; Bikdeli, B.; Ahluwalia, N.; Ausiello, J.C.; Wan, E.Y.; et al. Extrapulmonary manifestations of COVID-19. *Nat. Med.* **2020**, *26*, 1017–1032. [CrossRef] [PubMed]
191. Kaiser, S.; Frase, S.; Selzner, L.; Lieberum, J.L.; Wollborn, J.; Niesen, W.D.; Foit, N.A.; Heiland, D.H.; Schallner, N. Neuroprotection after hemorrhagic stroke depends on cerebral heme oxygenase-1. *Antioxidants* **2019**, *8*, 496. [CrossRef]
192. Bijjem, K.R.V.; Padi, S.S.V.; lal Sharma, P. Pharmacological activation of heme oxygenase (HO)-1/carbon monoxide pathway prevents the development of peripheral neuropathic pain in Wistar rats. *Naunyn. Schmiedebergs. Arch. Pharmacol.* **2013**, *386*, 79–90. [CrossRef]
193. Singh, R.D.; Barry, M.A.; Croatt, A.J.; Ackerman, A.W.; Grande, J.P.; Diaz, R.M.; Vile, R.G.; Agarwal, A.; Nath, K.A. The spike protein of SARS-CoV-2 virus induces heme oxygenase-1: Pathophysiologic implications. *Biochim. Biophys. Acta Mol. Basis Dis.* **2021**, *1868*, 166322. [CrossRef]
194. Nath, K.A. Heme oxygenase-1 and acute kidney injury. *Curr. Opin. Nephrol. Hypertens.* **2014**, *23*, 17–24. [CrossRef] [PubMed]
195. Atsaves, V.; Makri, P.; Detsika, M.G.; Tsirogianni, A.; Lianos, E.A. Glomerular epithelial cells-targeted heme oxygenase-1 over expression in the rat: Attenuation of proteinuria in secondary but not primary injury. *Nephron* **2016**, *133*, 270–278. [CrossRef] [PubMed]
196. Yeh, Y.H.; Hsu, L.A.; Chen, Y.H.; Kuo, C.T.; Chang, G.J.; Chen, W.J. Protective role of heme oxygenase-1 in atrial remodeling. *Basic Res. Cardiol.* **2016**, *111*, 58. [CrossRef] [PubMed]
197. Giannini, L.; Vannacci, A.; Fabrizi, F.; Uliva, C.; Bani, D.; Masini, E.; Mannaioni, P.F. Protection from cardiac injury by induction of heme oxygenase-1 and nitric oxide synthase in a focal ischaemia-reperfusion model. *Cell Mol. Biol* **2005**, *51*, 393–401.
198. Farombi, E.O.; Shrotriya, S.; Na, H.-K.; Kim, S.-H.; Surh, Y.-J. Curcumin attenuates dimethylnitrosamine-induced liver injury in rats through Nrf2-mediated induction of heme oxygenase-1. *Food Chem. Toxicol. Int. J. Publ. Br. Ind. Biol. Res. Assoc.* **2008**, *46*, 1279–1287. [CrossRef]
199. Devey, L.; Ferenbach, D.; Mohr, E.; Sangster, K.; Bellamy, C.O.; Hughes, J.; Wigmore, S.J. Tissue-resident macrophages protect the liver from ischemia reperfusion injury via a heme oxygenase-1-dependent mechanism. *Mol. Ther.* **2009**, *17*, 65–72. [CrossRef]
200. Ptilovanciv, E.O.N.; Fernandes, G.S.; Teixeira, L.C.; Reis, L.A.; Pessoa, E.A.; Convento, M.B.; Simões, M.J.; Albertoni, G.A.; Schor, N.; Borges, F.T. Heme oxygenase 1 improves glucoses metabolism and kidney histological alterations in diabetic rats. *Diabetol. Metab. Syndr.* **2013**, *5*, 3–8. [CrossRef]
201. Dolan, M.E.; Hill, D.P.; Mukherjee, G.; McAndrews, M.S.; Chesler, E.J.; Blake, J.A. Investigation of COVID-19 comorbidities reveals genes and pathways coincident with the SARS-CoV-2 viral disease. *Sci. Rep.* **2020**, *10*, 20848. [CrossRef]
202. Durante, W. Targeting heme oxygenase-1 in vascular disease. *Curr. Drug Targets* **2010**, *11*, 1504–1516. [CrossRef]
203. Tracz, M.J.; Juncos, J.P.; Grande, J.P.; Croatt, A.J.; Ackerman, A.W.; Katusic, Z.S.; Nath, K.A. Induction of heme oxygenase-1 is a beneficial response in a murine model of venous thrombosis. *Am. J. Pathol.* **2008**, *173*, 1882–1890. [CrossRef]
204. Arngrim, N.; Schytz, H.W.; Hauge, M.K.; Ashina, M.; Olesen, J. Carbon monoxide may be an important molecule in migraine and other headaches. *Cephalalgia* **2014**, *34*, 1169–1180. [CrossRef] [PubMed]
205. Hervera, A.; Leánez, S.; Negrete, R.; Motterlini, R.; Pol, O. Carbon monoxide reduces neuropathic pain and spinal microglial activation by inhibiting nitric oxide synthesis in mice. *PLoS ONE* **2012**, *7*, e43693. [CrossRef] [PubMed]
206. Chen-Roetling, J.; Song, W.; Schipper, H.M.; Regan, C.S.; Regan, R.F. Astrocyte Overexpression of Heme Oxygenase-1 Improves Outcome after Intracerebral Hemorrhage. *Stroke* **2015**, *46*, 1093–1098. [CrossRef] [PubMed]

207. Lu, X.; Gu, R.; Hu, W.; Sun, Z.; Wang, G.; Wang, L.; Xu, Y. Upregulation of heme oxygenase-1 protected against brain damage induced by transient cerebral ischemia-reperfusion injury in rats. *Exp. Ther. Med.* **2018**, *15*, 4629–4636. [CrossRef] [PubMed]
208. Uhlén, M.; Fagerberg, L.; Hallström, B.M.; Lindskog, C.; Oksvold, P.; Mardinoglu, A.; Sivertsson, Å.; Kampf, C.; Sjöstedt, E.; Asplund, A.; et al. Tissue-based map of the human proteome. *Science* **2015**, *347*, 6220. [CrossRef] [PubMed]
209. Pushpakom, S.; Iorio, F.; Eyers, P.A.; Escott, K.J.; Hopper, S.; Wells, A.; Doig, A.; Guilliams, T.; Latimer, J.; McNamee, C.; et al. Drug repurposing: Progress, challenges and recommendations. *Nat. Rev. Drug Discov.* **2018**, *18*, 41–58. [CrossRef] [PubMed]

antioxidants

MDPI

Article

The *Trypanosoma brucei*-Derived Ketoacids, Indole Pyruvate and Hydroxyphenylpyruvate, Induce HO-1 Expression and Suppress Inflammatory Responses in Human Dendritic Cells

Hannah K. Fitzgerald [1], Sinead A. O'Rourke [1,2], Eva Desmond [1], Nuno G. B. Neto [2], Michael G. Monaghan [2], Miriam Tosetto [3], Jayne Doherty [3], Elizabeth J. Ryan [4], Glen A. Doherty [3], Derek P. Nolan [1], Jean M. Fletcher [1,5,†] and Aisling Dunne [1,5,*,†]

1 School of Biochemistry and Immunology, Trinity Biomedical Sciences Institute, Trinity College Dublin, D02 R590 Dublin, Ireland; hfitzge@tcd.ie (H.K.F.); siorourk@tcd.ie (S.A.O.); evadesmond@gmail.com (E.D.); DENOLAN@tcd.ie (D.P.N.); Fletchj@tcd.ie (J.M.F.)
2 Trinity Centre for Biomedical Engineering, Trinity Biomedical Sciences Institute, Trinity College Dublin, D02 R590 Dublin, Ireland; neton@tcd.ie (N.G.B.N.); monaghmi@tcd.ie (M.G.M.)
3 Centre for Colorectal Disease, St. Vincent's University Hospital, School of Medicine, University College Dublin, D04 YN26 Dublin, Ireland; mtosetto@ucd.ie (M.T.); Jayne.Doherty@svph.ie (J.D.); gdoherty@svhg.ie (G.A.D.)
4 Department of Biological Sciences, Health Research Institute, University of Limerick, V94 T9PX Limerick, Ireland; Elizabeth.Ryan@ul.ie
5 School of Medicine, Trinity Biomedical Sciences Institute, Trinity College Dublin, D02 R590 Dublin, Ireland
* Correspondence: aidunne@tcd.ie
† These authors contributed equally to this work.

Citation: Fitzgerald, H.K.; O'Rourke, S.A.; Desmond, E.; Neto, N.G.B.; Monaghan, M.G.; Tosetto, M.; Doherty, J.; Ryan, E.J.; Doherty, G.A.; Nolan, D.P.; et al. The *Trypanosoma brucei*-Derived Ketoacids, Indole Pyruvate and Hydroxyphenylpyruvate, Induce HO-1 Expression and Suppress Inflammatory Responses in Human Dendritic Cells. *Antioxidants* **2022**, *11*, 164. https://doi.org/10.3390/antiox11010164

Academic Editors: Elias Lianos and Maria G. Detsika

Received: 20 December 2021
Accepted: 12 January 2022
Published: 15 January 2022

Abstract: The extracellular parasite and causative agent of African sleeping sickness *Trypanosoma brucei* (*T. brucei*) has evolved a number of strategies to avoid immune detection in the host. One recently described mechanism involves the conversion of host-derived amino acids to aromatic ketoacids, which are detected at relatively high concentrations in the bloodstream of infected individuals. These ketoacids have been shown to directly suppress inflammatory responses in murine immune cells, as well as acting as potent inducers of the stress response enzyme, heme oxygenase 1 (HO-1), which has proven anti-inflammatory properties. The aim of this study was to investigate the immunomodulatory properties of the *T. brucei*-derived ketoacids in primary human immune cells and further examine their potential as a therapy for inflammatory diseases. We report that the *T. brucei*-derived ketoacids, indole pyruvate (IP) and hydroxyphenylpyruvate (HPP), induce HO-1 expression through Nrf2 activation in human dendritic cells (DC). They also limit DC maturation and suppress the production of pro-inflammatory cytokines, which, in turn, leads to a reduced capacity to differentiate adaptive $CD4^+$ T cells. Furthermore, the ketoacids are capable of modulating DC cellular metabolism and suppressing the inflammatory profile of cells isolated from patients with inflammatory bowel disease. This study therefore not only provides further evidence of the immune-evasion mechanisms employed by *T. brucei*, but also supports further exploration of this new class of HO-1 inducers as potential therapeutics for the treatment of inflammatory conditions.

Keywords: heme oxygenase 1; *Trypanosoma brucei*; inflammatory bowel disease; aromatic ketoacids; dendritic cells; immunomodulation; anti-inflammatory therapies

1. Introduction

Infection of the mammalian vasculature and central nervous system (CNS) with the extracellular protozoan parasite *Trypanosoma brucei* (*T. brucei*) can lead to fatal human sleeping sickness, also known as African trypanosomiasis. Like most parasites, trypanosomes are continuously challenged by the host-immune system, however, they have evolved very effective evasion strategies in order to maintain infection and prolong the host's survival [1].

An infection with *T. brucei* is accompanied by the excretion of high levels of ketoacids into the host's bloodstream [2,3], a phenomenon that was, until recently, largely unexplained. The ketoacids are derived from the conversion of aromatic amino acids (tryptophan, tyrosine, and phenylalanine) to indole pyruvate (IP), hydroxyphenylpyruvate (HPP), and phenyl pyruvate (PP), through the action of a cytoplasmic aspartate aminotransferase (TbcASAT), which is expressed by the parasite.

In recent years, we, and others, have demonstrated that these molecules have potent immunomodulatory properties, which likely contribute to the suppression of host immune responses, but that may also be exploited for potential therapeutic benefit [4–8]. For example, IP and HPP are potent inducers of the immunosuppressive enzyme heme oxygenase-1 (HO-1) in murine glia and macrophages [5]. This occurs in a nuclear factor (erythroid-derived 2)-like 2 (Nrf2) dependent manner and leads to a reduction in LPS-induced pro-inflammatory cytokines and innate immune cell maturation. Indeed, HO-1 has well established anti-inflammatory properties and is induced by a vast number of stimuli during oxidative stress and inflammation. It catalyses the conversion of heme to biliverdin with the liberation of free iron and carbon monoxide (CO). Biliverdin is then converted to bilirubin by biliverdin reductase [9,10] and both biliverdin and bilirubin are considered powerful antioxidants, while many of the anti-inflammatory effects of HO-1 are attributed to CO. The potent anti-inflammatory properties of HO-1 are underlined by rare HO-deficiencies in humans and HO-1 knockout mice, which, along with the expected sensitivity to oxidative stress, show high levels of chronic inflammation [11–13]. Unsurprisingly, induction of HO-1 with known and novel HO-1 inducers is now being explored as a therapy for a number of autoimmune and inflammatory diseases and we have recently reviewed this in detail elsewhere [14].

In addition to HO-1 induction, *T. brucei*-derived ketoacids are also capable of inhibiting HIF-1α–induced pro-IL-1β expression, as well as prostaglandin production, an effect which was shown to be dependent on the activation of the aryl hydrocarbon receptor [4,5,8]. The immunomodulatory effects of IP and HPP have also been confirmed in animal models of disease. In a murine model of skin damage caused by exposure to ultraviolet B radiation, administration of IP resulted in a reduction in damage lesions and expression of the pro-inflammatory cytokines, IL-1β and IL-6 [6]. Furthermore, IP administration reduced disease severity in the DSS colitis model, and this was accompanied by a decrease in the expression of pro-inflammatory cytokines IL-12, TNF, IFNγ, and IL-1β, and an increase in the expression of the anti-inflammatory cytokine IL-10 [7]. A reduction in Th1 cells, as well as a reduced capacity for dendritic cells (DC) to activate T cells, was also observed [7].

Despite the recent interest in these ketoacids as immunomodulators, to date, they have been primarily studied in murine immune cells with very little evidence to support their mechanism of action in human immune cells. In this study, we investigate the immunomodulatory properties of the *T. brucei*-derived ketoacids, IP and HPP, in primary human immune cells and further examine their potential as a therapy for inflammatory diseases.

2. Materials and Methods

2.1. Reagents and Chemicals

4-hydroxyphenylpyruvic acid (HPP) and indole-3-pyruvic acid (IP) were purchased from Merck (Darmstadt, Germany) and dissolved in RPMI to a final concentration of 2 mM before use. Ultrapure lipopolysaccharide (LPS) from *E. Coli* O111:B4 was purchased from Enzo Life Sciences (Bruxelles, Belgium). Complete RPMI or complete IMDM were prepared by supplementing with 10% foetal bovine serum (FBS), 2 mM L-glutamine, 100 U/mL penicillin, and 100 µg/mL streptomycin, which were all purchased from Merck (Darmstadt, Germany). Lymphoprep is manufactured by Axis-Shield poC (Dundee, Scotland). GM-CSF and IL-4 were purchased from Miltenyi Biotec (Bergisch Gladbach, Germany). The protease inhibitor cocktail and High-Pure RNA Isolation Kit were purchased from Roche (Basel, Switzerland). The western blot antibodies for Nrf2, HK2, p-AMPK, t-AMPK, p62 and LC3 were all purchased from Cell Signalling Technology (Danvers, MA, USA), while the antibody for HO-1 was purchased from Enzo Life Sciences (Bruxelles, Belgium), and the β-actin antibody and secondary anti-rabbit were purchased from Merck (Darmstadt, Germany). The Fixable Viability Dye, antibodies for CD40, CD80, CD83, and CD86, and Annexin V & PI staining kit were all purchased from eBioscience (San Diego, CA, USA). The DQ-Ovalbumin, CD3 and IL-17 flow antibodies, FIX & PERM™ Cell Permeabilization Kit, anti-CD3, and all ELISA kits used were purchased from Invitrogen (Waltham, MA, USA). The High-Capacity cDNA reverse transcription kit was purchased from Applied Biosystems (Beverly, MA, USA) and the iTaq Universal SYBR Green mastermix from Bio-Rad (Hercules, CA, USA). The MagniSort Human CD14 Positive Selection kit and MagniSort Human CD4 T cell Positive Selection kit were purchased from Thermo Fisher Scientific (Waltham, MA, USA). The Zombie NIR™ Fixable Viability kit was purchased from BioLegend (San Diego, CA, USA). The antibodies for CD8, ki67 and IFNγ were purchased from BD biosciences (East Rutherford, NJ, USA). The Complete XF assay medium was purchased from Agilent (Santa Clara, CA, USA). The Corning™ Cell-Tak Cell and Tissue Adhesive was purchased from Fisher Scientific (Waltham, MA, USA). Oligomycin was purchased from Cayman Chemicals (Ann Arbor, MI, USA) and carbonyl cyanide-p trifluoromethoxyphenylhydrazone (FCCP) from Santa Cruz biotechnology (Dallas, TX, USA). The HEK-Blue™ hTLR4 assay system was purchased from InvivoGen (San Diego, CA, USA). Dulbecco's phosphate buffered saline (PBS), enhanced chemiluminescent substrate, phosphatase inhibitor cocktail, Antioxidant assay kit, Phorbol 12-myristate 13-acetate (PMA), ionomycin, brefeldin A, rotenone, antimycin A, and 2-deoxy-D-glucose (2-DG) were all purchased from Merck (Darmstadt, Germany).

2.2. Human Blood Samples

The Irish Blood Transfusion Service (IBTS) at St. James's Hospital in Dublin supplied leukocyte-enriched buffy coats for these studies, from donors who provided informed written consent. Ethical approval was obtained from the research ethics committee of the School of Biochemistry and Immunology at Trinity College Dublin, and all experiments were carried out in accordance with the Declaration of Helsinki. Peripheral blood mononuclear cells (PBMC) were isolated by a density gradient centrifugation. The cells were cultured in complete RPMI medium and maintained in humidified incubators at 37 °C with 5% CO_2.

2.3. Dendritic Cell Culture

The MagniSort Human CD14 Positive Selection kit was used according to the manufacturer's protocol to positively select for $CD14^+$ monocytes from PBMC. $CD14^+$ monocytes were then cultured at 1×10^6 cells/mL in complete RPMI and supplemented with GM-CSF (50 ng/mL) and IL-4 (40 ng/mL) to generate monocyte-derived DC. On day three of culture, half of the media was replaced with fresh media supplemented with cytokines at the same starting concentration. On day six, non-adherent and loosely adherent cells were gently removed for use. The purity of $CD14^{lo}$DC-$SIGN^+$ DC was confirmed by flow cytometry and was routinely >95%. DC were cultured at 1×10^6 cells/mL for all further assays.

2.4. Western Blot Experiments

For detection of HO-1 expression, DC were cultured in the presence of HPP or IP (250–1000 μM) for 3, 6, and 24 h, or treated with the Nrf2 inhibitor ML385 (10 μM) for 1 h prior to treatment with HPP or IP at 1000 μM for 24 h. Cell lysates were prepared by washing cells in PBS prior to lysis in RIPA buffer (Tris 50 mM; NaCl 150 mM; SDS 0.1%; Na.Deoxycholate 0.5%; Triton X 100). For detection of Nrf2 expression, DC were cultured in the presence of HPP (1000 μM) or IP (1000 μM) for 6 or 24 h, then washed in PBS and lysed in Laemmli loading buffer. For detection of hexokinase 2 (HK2) expression, DC were cultured in the presence of HPP or IP at 1000 μM for 6 h prior to stimulation with LPS (100 ng/mL) for 12 h, then washed in PBS and lysed in Laemmli loading buffer. For detection of p-AMPK and t-AMPK expression, DC were cultured in the presence of

HPP or IP (both 1000 μM) for 15 min, then washed in PBS and lysed in Laemmli loading buffer. For the detection of p62 and LC3, DC were cultured in the presence of HPP or IP (both 1000 μM) for 6, 12, or 24 h, then washed in PBS, lysed in Laemmli loading buffer, and sonicated. All samples were lysed in the presence of a protease inhibitor cocktail and phosphatase inhibitor cocktail set. Samples were electrophoresed and transferred to PVDF. The membranes were incubated overnight at 4 °C with monoclonal antibodies specific for HO-1, Nrf2, HK2, p-AMPK, t-AMPK, p62, and LC3. The membranes were washed in TBS–Tween prior to incubation with an anti-rabbit streptavidin-conjugated secondary antibody for 2 h at room temperature. A Bio-Rad ChemiDoc MP system was used for developing the blots. The membranes were subsequently re-probed with a loading control, β-actin, in order to normalise the protein of interest to the loading control for densitometric analysis.

2.5. *Antioxidant Assay*

DC were cultured in the presence of IP or HPP (both 1000 μM) for 1 h. The cells were lysed by sonication in ice-cold PBS and centrifuged at 14,000 rpm for 10 min to pellet any debris. The total antioxidant capacity of the cells was analysed using an Antioxidant assay kit according to the manufacturer's protocol. The assay measures the reduction of Cu^{2+} by an antioxidant to Cu^{+}, which can subsequently form a coloured complex with a dye reagent in the kit. The absorbances of the samples were read at 570 nm and compared to the absorbances of a range of known concentrations of Trolox standards. The data is displayed as the total antioxidant capacity of the cells expressed as an equivalent concentration of Trolox (μM).

2.6. *DC Flow Cytometry Experiments*

DC were collected upon completion of the experiment, washed in PBS, and stained accordingly. DC were stained using an Annexin V & PI staining kit according to the manufacturer's protocol for viability assays. For the maturation marker assay, DC were initially stained with Fixable Viability Dye, for dead cells, and subsequently with fluorochrome-conjugated antibodies for CD40, CD80, CD83, and CD86. In order to measure the phagocytic capacity, DC were incubated with RPMI-containing DQ-Ovalbumin (500 ng/mL) for 20 min at 37 °C, before transferring to 4 °C for a further 10 min incubation to stop the uptake of the model antigen. DC were then washed in PBS and immediately acquired. All of the above experiments were acquired on a BD FACS Canto II, and the analysis was performed using FlowJo v.10 software (Tree Star Inc., Ashland, OR, USA).

2.7. *Quantitative Real-Time PCR*

DC were cultured in the presence of HPP or IP (both 1000 μM) for 6 or 24 h. Detection of NAD(P)H dehydrogenase (quinone 1) (NQO1, accession number P15559) expression and glutathione reductase (GSR, accession number P00390) expression were measured using quantitative real-time PCR. RNA was first extracted using the High-Pure RNA Isolation Kit, followed by cDNA synthesis using the High-Capacity cDNA reverse transcription kit. iTaq Universal SYBR Green mastermix was used in the reaction along with relevant primers—the sequences are listed in Table 1. A Bio-Rad CFX96 Real-Time System was used to carry out the reaction. mRNA expression levels for genes of interest were quantified and normalized to the housekeeping (β-actin) mRNA levels.

Table 1. Primer sequences. Table containing the forward and reverse primer sequences for NQO1, GSR, and β-actin.

Gene	Forward Primer	Reverse Primer
NQO1	5′ TGAAGAAGAAAGGATGGGAG 3′	5′ TTTACCTGTGATGTCCTTTC 3′
GSR	5′ GACCTATTCAACGAGCTTTAC 3′	5′ CAACCACCTTTTCTTCCTTG 3′
β-actin	5′ GGACTTCGAGCAAGAGATGG 3′	5′ AGCACTGTGTTGGCGTACAG 3′

2.8. DC ELISA Experiments

For detection of cytokines, DC were cultured in the presence of HPP or IP (500–1000 μM) for 6 h prior to stimulation with LPS (100 ng/mL) for 24 h. Concentrations of IL-12p70, IL-23, TNF, IL-6, and IL-10 were quantified from supernatants by ELISA as per the manufacturers' protocols.

2.9. DC-CD4$^+$ T Cell Co-Cultures

The MagniSort Human CD4 T cell Positive Selection kit was used according to the manufacturer's protocol to isolate CD4$^+$ T cells from PBMC. CD4$^+$ T cells were co-cultured with allogeneic DC that had been pre-treated with IP or HPP and stimulated with LPS as before. The cells were co-cultured at a ratio of 10:1 T cells to DC for five days with no ketoacid present. The supernatants were removed for analysis of IL-10 by ELISA prior to restimulation of the cells in complete RPMI in the presence of 50 ng/mL PMA, 500 ng/mL ionomycin, and 5 μg/mL brefeldin A for 4 h. The cells were washed in PBS and stained for viability (Zombie NIR™ Fixable Viability kit) for 15 min at room temperature. The cells were then washed again in PBS and stained with fluorochrome-conjugated antibodies for surface markers CD3 and CD8 for 15 min at room temperature. The cells were then washed and fixed (FIX & PERM™ Cell Permeabilization Kit) for 15 min at room temperature. Finally, the cells were washed and stained for intracellular markers Ki67 and IFNγ in permeabilization buffer for 15 min at room temperature. A BD LSRFortessa flow cytometer was used to acquire samples, and analysis was performed using FlowJo v.10 software.

2.10. Two-Photon Fluorescence Lifetime Imaging Microscopy (FLIM)

DC were cultured in the presence of HPP (1000 μM) for 6 h prior to stimulation with LPS (100 ng/mL) for 12 h. Two-photon excited NAD(P)H- Fluorescence Lifetime Imaging Microscopy (FLIM) was used to measure the levels of free and protein-bound NADH within these cells, and was performed on a custom multiphoton system (further details regarding experimental setup can be found at the following [15,16]). At least three images for each model were acquired. Afterwards, regions of interest (ROI) were selected, and the NAD(P)H fluorescence decay was analysed.

For the NAD(P)H fluorescence decay analysis, an overall decay curve was generated by the contribution of all pixels in the ROI area. Afterwards, it was fitted with a double exponential decay curve (Equation (1)):

$$I(t) = \alpha_1 e^{-\frac{t}{\tau_1}} + \alpha_2 e^{-\frac{t}{\tau_2}} + c \quad (1)$$

$I(t)$ represents the fluorescence intensity at time (t) after laser excitation. α_1 and α_2 represent the fraction of the overall signal comprised of a short and long lifetime component, respectively. τ_1 and τ_2 are the long and short lifetime components, respectively. C corresponds to background light. X^2 is calculated to evaluate the goodness of multi-exponential fit to the raw fluorescence decay data—the lowest χ^2 values were considered in this study.

For NAD(P)H, a two-component fit was used to differentiate between the free (τ_1) and protein-bound (τ_2) NAD(P)H. The average lifetime (τ_{avg}) of NAD(P)H for each pixel is calculated by a weighted average of both the free and bound lifetime contributions (Equation (2)):

$$\tau_{avg} = \frac{(\alpha_1 \times \tau_1) + (\alpha_2 \times \tau_2)}{(\alpha_1 + \alpha_2)} \quad (2)$$

2.11. Metabolic Profiling Using Seahorse Analysis

DC were cultured in the presence of IP or HPP (both 1000 μM) for 6 h prior to stimulation with LPS (100 ng/mL) for 12 h. The cell culture medium was replaced with complete XF assay medium (pH of 7.4, supplemented with 10 mM glucose, 1 mM sodium pyruvate, 2 mM L-glutamine) and DC were then transferred at a density of 2×10^5

cells/well to a Seahorse 96-well microplate, which was coated with Corning™ Cell-Tak Cell and Tissue Adhesive and incubated in a non-CO_2 incubator. Blank wells were prepared containing XF assay medium only to subtract the background oxygen consumption rate (OCR) and extracellular acidification rate (ECAR) during analysis. Oligomycin (1 mM), FCCP (1 mM), rotenone (500 nM), antimycin A (500 nM), and 2-DG (25 mM) were prepared in XF assay medium. Inhibitors were loaded into the appropriate injection ports on the cartridge plate and incubated for 10 min in a non-CO_2 incubator at 37 °C. Oligomycin, FCCP, rotenone and antimycin A, and 2-DG were sequentially injected while the OCR and ECAR readings were simultaneously measured. Wave software (Agilent Technologies, Santa Clara, CA, USA) was used to analyse the results. The rates of basal glycolysis, max glycolysis, glycolytic reserve, basal respiration, max respiration, and respiratory reserve were calculated as detailed in the manufacturer's protocol and as supplied in Supplementary Table S1.

2.12. IBD Patient PBMC Experiments

This study received ethical approval from St Vincent's University Hospital Ethics and Medical Research Committee to take blood samples from consenting patients (N = 14) attending a specialist outpatient clinic for inflammatory bowel disease (IBD). PBMC were isolated as above and frozen at −80 °C. PBMC were thawed and treated with either IP or HPP (250–1000 μM) for 6 h prior to stimulation with anti-CD3 (1 μg/mL) for 12 h. The media was then removed and replaced with fresh media to circumvent issues that occur as the compounds become fluorescent after incubations over long periods of time, and cells were maintained in the presence of anti-CD3 for a further four days. The supernatants were then removed for analysis of IL-10, IFNγ, and IL-17A by ELISA. PBMC were restimulated in complete IMDM medium in the presence of 50 ng/mL PMA, 500 ng/mL ionomycin, and 5 μg/mL brefeldin A for 4 h. The cells were washed in PBS and stained for viability (Zombie NIR™ Fixable Viability kit) for 15 min at room temperature. The cells were then washed in PBS and stained with fluorochrome-conjugated antibodies for surface markers CD3 and CD8 for 15 min at room temperature. After this, the cells were then washed and fixed (FIX & PERM™ Cell Permeabilization Kit) for 15 min at room temperature. Finally, the cells were washed and stained for intracellular markers Ki67, IFNγ and IL-17 in permeabilization buffer for 15 min at room temperature. The cells were then washed and acquired on a BD LSRFortessa flow cytometer. The analysis was performed with FlowJo v.10. All antibodies used in this experiment were chosen carefully to avoid channels which still had some fluorescence issues, despite the steps taken to overcome this.

2.13. Assessment of Endotoxin Contamination

The HEK-Blue™ hTLR4 assay system was used to test IP and HPP for LPS contamination. HEK-blue cells (5×10^5 cells/mL) expressing TLR4 were stimulated with LPS (0.1–100 ng/mL; positive control), or HPP or IP (both 1000 μM) for 24 h. Supernatants from the HEK-blue cells were incubated with HEK-blue detection medium, to measure SEAP expression, for 30 min at 37 °C and absorbance was read at 650 nm.

2.14. Statistical Analysis

Prism 9 software (GraphPad Software Inc., San Diego, CA, USA) was used to perform the statistical analysis on all datasets. A repeated measures one-way ANOVA, with either Dunnett's or Šídák's post hoc test, as appropriate, was used for analysis of three or more datasets. A Paired Student's t-test was used for the analysis of only two datasets. The analysis of datasets with more than one variable were performed using a two-way ANOVA with Šídák's multiple comparisons post hoc test. Asterisks are used in the figures to denote p values < 0.05, which were considered significant.

3. Results

3.1. HPP and IP Upregulate HO-1 in Primary Human DC

We have previously shown that HPP and IP are capable of inducing HO-1, as well as having immunomodulatory effects, in murine macrophages and glia [5]. Therefore, we sought to investigate if *T. brucei*-derived ketoacids have similar effects in primary human DC, as these are crucial immune cells linking the innate and adaptive immune system. DC were treated with HPP or IP (500–1000 μM) for 24 h (these concentrations are similar to the levels of aromatic ketoacids in circulation close to the peak of parasitaemia during trypanosomiasis and have been used in previously published studies [4,5,8,17]). Both HPP and IP were found to be non-toxic to human DC, having no effect on cell viability at these concentrations, and were confirmed to be endotoxin free (Supplementary Figures S1 and S2). HO-1 expression in HPP- and IP-treated DC was next examined by western blot at a range of timepoints (3, 6, and 24 h) and concentrations (250–1000 μM). Immature DC constitutively expressed HO-1 in line with previous reports [18–20], however, treatment with either HPP or IP resulted in a trend towards increased expression of HO-1 at all concentrations tested with a significance observed at 1000 μM in each case (Figure 1A). At this concentration, significant upregulation of HO-1 occurred within 3 h of HPP treatment, while significant induction of HO-1 occurred following 24 h treatment with IP (Figure 1B). Both IP and HPP also increased the total antioxidant capacity of DC and this was observed within 1 h of incubation with either compound (Figure 1C). The ketoacids also appear to be more potent antioxidants than the established HO-1 inducers, carnosol and curcumin (Figure 1C). These results indicate that IP and HPP rapidly enhance the total antioxidant capacity of human DC, while also upregulating the stress-response protein, HO-1.

3.2. HPP and IP Induce HO-1 through Nrf2 Activation

We have previously reported that HPP and IP activate Nrf2 in murine immune cells, and this is likely the mechanism through which they upregulate HO-1 [5]. We therefore sought to investigate if they have a similar mechanism of action in human DC. To test this, DC were treated with HPP or IP (1000 μM) for 6 and 24 h, and Nrf2 protein expression was measured by western blot. An increase in Nrf2 protein was observed in HPP-treated DC at 6 h (Figure 2A), while IP-treated DC showed increased Nrf2 protein expression (and therefore accumulation) at 24 h (Figure 2A). Both HPP and IP were also found to drive mRNA expression of the additional Nrf2-regulated genes, NQO1 and GSR, further confirming their ability to activate this transcription factor (Figure 2B). Finally, treatment of DC with the Nrf2 inhibitor ML385 (10 μM, a non-toxic dose, Supplementary Figure S3) for 1 h, prior to treatment with either HPP or IP (1000 μM) for 24 h, resulted in a significant decrease in the expression of HO-1 (Figure 2C), further confirming that ketoacid driven HO-1 expression is regulated by Nrf2.

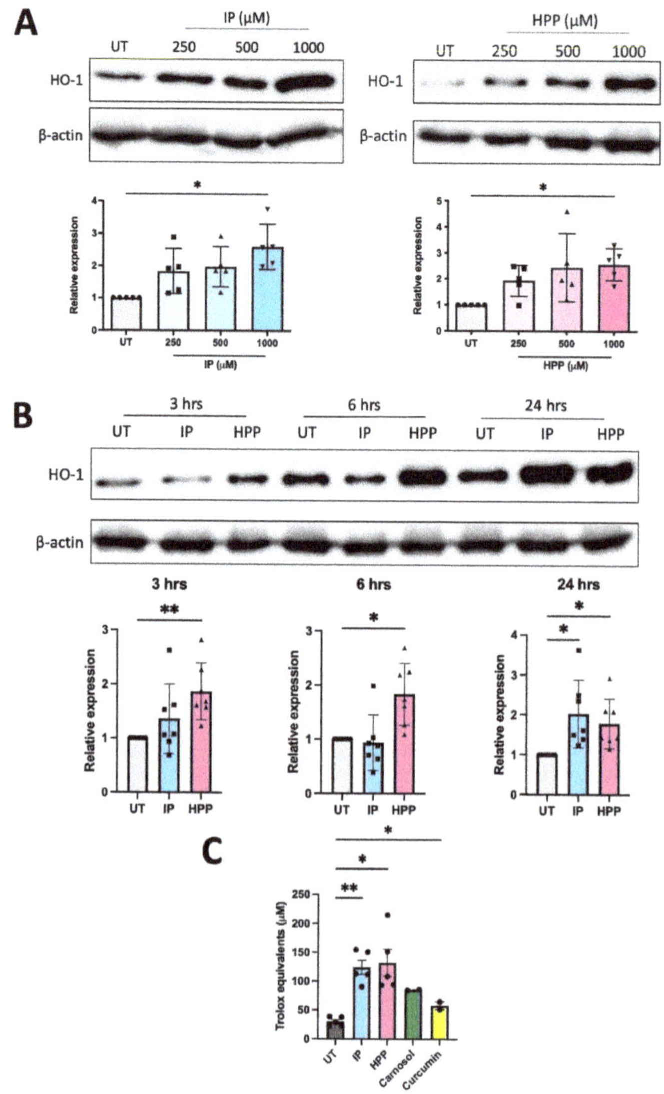

Figure 1. Hydroxyphenylpyruvate (HPP) and indole pyruvate (IP) upregulate HO-1 in primary human dendritic cells (DC). (**A**) Primary human DC were left untreated (UT) or incubated with HPP or IP (250–1000 μM) for 24 h. HO-1 expression was detected by western blot. Densitometry results shown are mean ± SEM of the relative expression of HO-1: β-actin from five healthy donors. (**B**) DC were left UT or incubated with HPP or IP at 1000 μM for 3, 6, or 24 h. HO-1 expression was detected by western blot. Densitometry results shown are mean ± SEM of the relative expression of HO-1: β-actin from seven healthy donors. (**C**) Primary human DC were left UT or incubated with HPP or IP at 1000 μM, or carnosol or curcumin (both 10 μM), for 1 h. Total antioxidant capacity of the cells was determined and expressed as an equivalent concentration of Trolox (μM). Pooled data showing the mean (±SEM) from five healthy donors. Repeated measures one-way ANOVA, with Dunnett's multiple comparisons post hoc test, was used to determine statistical significance by comparing means of treatment groups against the mean of the control group (** $p < 0.01$, * $p < 0.05$). ImageLab (Bio-Rad) software was used to perform densitometric analysis.

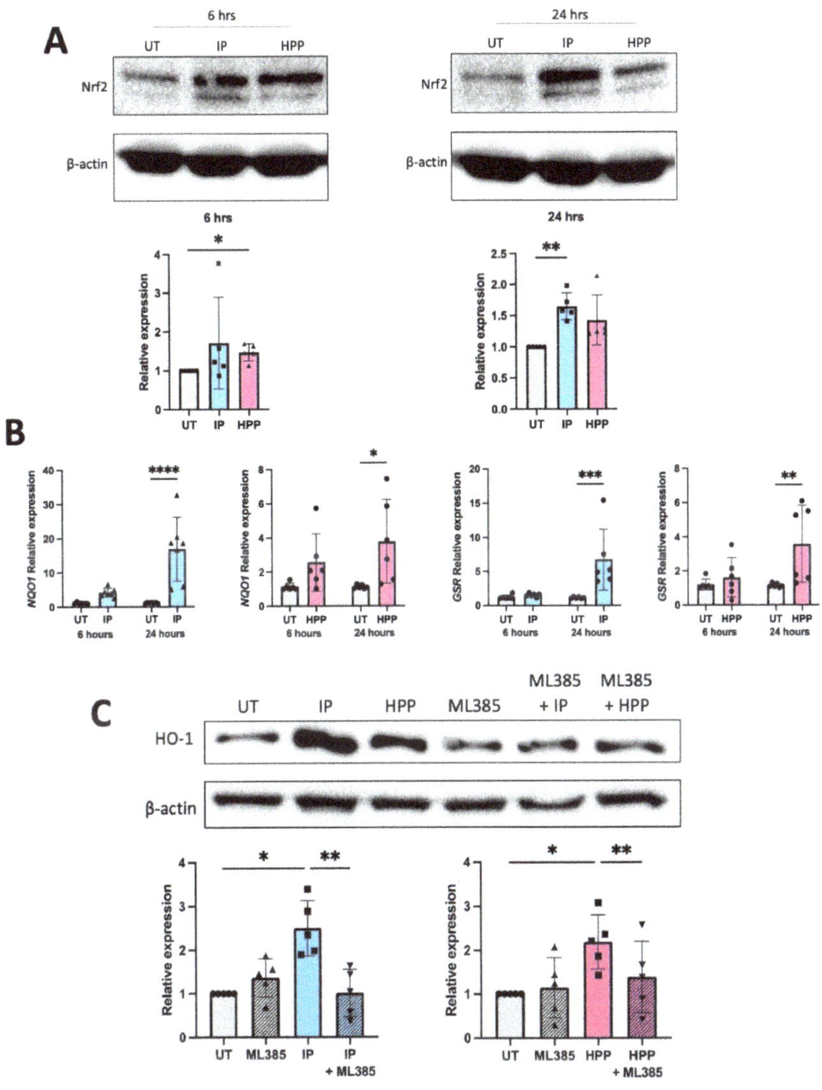

Figure 2. HPP and IP induce HO-1 through Nrf2 activation. (**A**) Primary human DC were left untreated (UT) or incubated with HPP or IP at 1000 μM for 6 or 24 h. Nrf2 expression was measured by western blot. Densitometry results shown are mean ± SEM of the relative expression of Nrf2: β-actin from five healthy donors. (**B**) Primary human DC were left UT or incubated with HPP or IP at 1000 μM for 24 h. mRNA expression of the Nrf2-dependent genes, NQO-1 and GSR, were measured by RT-PCR. Results show mean (±SEM) for six healthy donors. (**C**) Primary human DC were pre-treated either with or without the Nrf2 inhibitor ML385 (10 μM) for 1 h, prior to incubation with HPP or IP at 1000 μM for 24 h. HO-1 expression was measured by western blot. Densitometry results shown are mean ± SEM of the relative expression of HO-1: β-actin from five healthy donors. (**A**) One-way ANOVA, with Dunnett's multiple comparisons post hoc test, was used to determine statistical significance. (**B**) Two-way ANOVA, with Šídák's multiple comparisons post hoc test, was used to determine statistical significance. (**C**) One-way ANOVA, with Šídák's multiple comparisons post hoc test, was used to determine statistical significance (**** $p < 0.0001$, *** $p < 0.001$, ** $p < 0.01$, * $p < 0.05$). ImageLab (Bio-Rad) software was used to perform densitometric analysis.

3.3. HPP and IP Reduce the Production of Pro-Inflammatory Cytokines in LPS-Stimulated Human DC

We recently demonstrated that HPP and IP are capable of reducing the production of pro-inflammatory cytokines in murine glia and macrophages [5]. We next sought to determine if they have similar immune modulating activity in human DC. To test this, DC were treated with either HPP or IP (500–1000 μM) for 6 h prior to stimulation with LPS (100 ng/mL) for 24 h. Cytokine concentrations were measured in cell supernatants by ELISA. Both HPP and IP treatment dose-dependently reduced production of the pro-inflammatory cytokines TNF, IL-6, IL-12p70, and IL-23, and this effect was most potent at the 1000 μM concentration (Figure 3A–H). As well as driving the production of pro-inflammatory cytokines, LPS treatment over time is usually accompanied by production of the anti-inflammatory cytokine IL-10 as a means of regulating inflammatory responses. Interestingly, IP treatment resulted in a significant enhancement of IL-10, while HPP treatment showed a similar, albeit non-significant, trend (Figure 3I,J). These results suggest that both IP and HPP are capable of reducing the production of pro-inflammatory cytokines in LPS-stimulated DC, whilst also promoting a more anti-inflammatory phenotype.

3.4. HPP Treatment Inhibits the Maturation of LPS-Stimulated Human DC, Resulting in Reduced Activation of $CD4^+$ T Cells

In order to determine if the ketoacids impact DC maturation (and in turn T cell activation), human DC were treated with HPP (500–1000 μM) for 6 h prior to stimulation with LPS (100 ng/mL) for 24 h. Surface expression of maturation and co-stimulatory markers (CD40, CD80, CD83 and CD86) were measured by flow cytometry (due to the fluorescent nature of IP it is unsuitable for this flow cytometric analysis and hence was not included in these experiments). As expected, the expression of co-stimulatory markers was increased in LPS-stimulated DC. The average Median Fluorescent Intensity (MFI) of these cells was set to 100% and used as a control for comparison to the HPP-treated cells. There was a significant decrease in MFI at both HPP concentrations when compared to the LPS-stimulated control (Figure 4A). The phagocytic capacity of DC was next measured upon incubation of DC with FITC-conjugated DQ-Ovalbumin (DQ-Ova; 500 ng/mL) and uptake of the model antigen was assessed by flow cytometry. Compared to untreated cells, the ability of LPS-treated DC to phagocytose DQ-OVA was significantly impaired, signifying a heightened maturation status, however, pre-incubation with HPP prior to LPS treatment attenuated this effect and maintained the DC in an immature state (Figure 4B). Finally, and in order to determine if the reduced DC activation status that occurs in the presence of HPP has an impact on T cell activation, DC were treated with HPP (1000 μM) for 6 h prior to stimulation with LPS (100 ng/mL) for 24 h. The cells were then incubated with $CD4^+$ T cells at a ratio of 10:1 ($CD4^+$ T cells:DC) for five days. The supernatants were removed for cytokine analysis by ELISA and cells were stimulated with PMA (50 ng/mL), ionomycin (500 ng/mL) and brefeldin A (5 μg/mL) for 4 h. Expression of ki67, which is a measure of cell proliferation, was assessed by flow cytometry in $CD3^+CD8^-$ cells (gating strategy shown in Supplementary Figure S4), as was production of the pathogenic Th1 cytokine, IFNγ. T cells co-cultured with HPP-treated, LPS-stimulated DC showed a trend towards reduced ki67 and IFNγ expression when compared to T cells co-cultured with LPS-stimulated DC alone (Figure 4C). Interestingly, the T cells exhibited a trend towards enhanced production of the anti-inflammatory cytokine IL-10, when compared to T cells co-cultured with LPS-stimulated DC alone (Figure 4D). Overall, these results indicate that HPP reduces the maturation of innate DC, which in turn impacts adaptive T cell responses and may skew them towards a more anti-inflammatory phenotype.

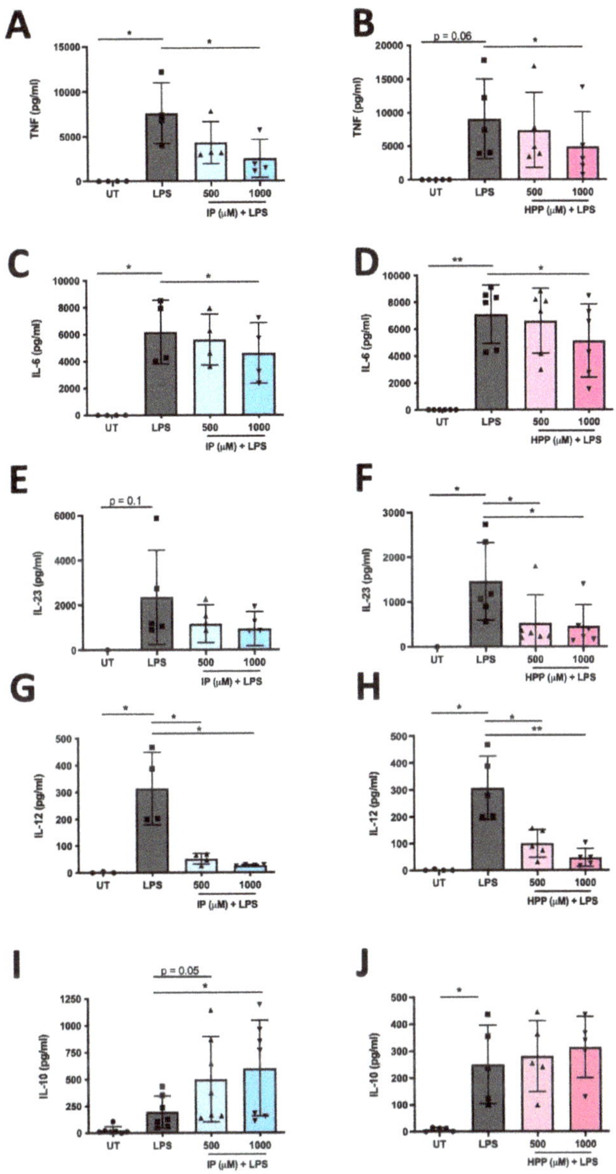

Figure 3. HPP and IP reduce the production of pro-inflammatory cytokines in LPS-stimulated human DC. Primary human DC were left untreated (UT) or incubated with IP (**A**,**C**,**E**,**G**,**I**) or HPP (**B**,**D**,**F**,**H**,**J**) (500–1000 μM) for 6 h prior to stimulation with LPS (100 ng/mL) for 24 h. Cell supernatants were assessed for TNF, IL-6, IL-23, IL-12p70, and IL-10 secretion by ELISA. Pooled data depict mean (±SEM) cytokine concentrations for four to seven healthy donors (means of three technical replicates per donor). Repeated measures one-way ANOVA, with Dunnett's multiple comparisons post hoc test, was used to determine statistical significance, by comparing means of treatment groups against the mean of the control group (** $p < 0.01$, * $p < 0.05$).

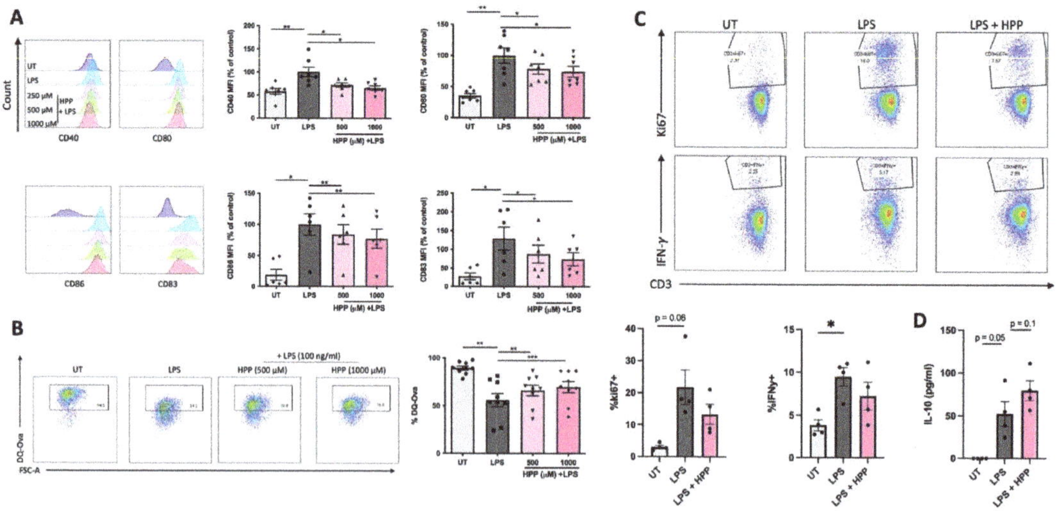

Figure 4. HPP treatment reduces DC maturation and subsequent CD4$^+$ T cell activation. Primary human DC were left untreated (UT) or incubated with HPP (500–1000 μM) for 6 h prior to stimulation with LPS (100 ng/mL) for 24 h. (**A**) Cells were stained for CD40, CD80, CD86, and CD83 and analysed by flow cytometry. Histograms showing the expression of maturation markers for HPP-treated, LPS-stimulated DC compared to unstimulated cells or LPS stimulation alone from one representative experiment. Pooled data showing the mean (±SEM) MFI for each marker expressed as a percentage of control (LPS stimulation alone) from six to seven healthy donors. (**B**) DC were incubated with FITC-conjugated DQ-Ovalbumin (DQ-Ova; 500 ng/mL) for 20 min and were immediately acquired by flow cytometry. Dot plots depicting DQ-Ova uptake from one representative experiment. Pooled data showing the mean (±SEM) DQ-Ova uptake as a percentage of total cells from nine healthy donors. (**C**) DC were pre-treated with HPP prior to stimulation with LPS, and subsequently cultured with CD4$^+$ T cells for five days. Dot plots depicting ki67 expression (as a measure of proliferation) and IFNγ expression from one representative experiment. Pooled data showing the mean (±SEM) of ki67$^+$ and IFNγ$^+$ cells as a percentage of CD3$^+$CD8$^-$ cells from four healthy donors. (**D**) Cell supernatants were assessed for IL-10 secretion by ELISA. Pooled data depict mean (±SEM) cytokine concentrations for four healthy donors (means of three technical replicates per donor). Repeated measures one-way ANOVA, with Dunnett's multiple comparisons post hoc test, was used to determine statistical significance by comparing means of treatment groups against the mean of the control group (*** $p < 0.001$, ** $p < 0.01$, * $p < 0.05$).

3.5. HPP and IP Modulate Metabolic Reprogramming in LPS-Stimulated Human DC

Metabolic reprogramming has been observed in immune cells and numerous recent studies have demonstrated that their activation/maturation is accompanied by a metabolic switch favouring glycolysis over oxidative phosphorylation [21]. In order to determine if *T. brucei*–derived ketoacids have an effect on immune cell metabolism, DC were pre-treated with IP or HPP (1000 μM) for 6 h prior to stimulation with LPS (100 ng/mL) for 12 h. These cells were then analysed in a Seahorse XFe96 analyser following the addition of oligomycin (1 mM), an inhibitor of mitochondrial complex V, FCCP (1 mM), a mitochondrial uncoupler, rotenone (500 nM) and antimycin A (500 nM), which are inhibitors of the mitochondrial complexes I & III, respectively, and 2-DG (25 mM), an inhibitor of glycolysis. The metabolic activity of the cells was then determined by measuring the ECAR, which is a measure of glycolysis, and the OCR, which is a measure of oxidative phosphorylation. IP- and HPP-treated DC showed no change in basal glycolysis when compared to LPS stimulation alone (Figure 5B). LPS-treated cells showed a trend towards increased max glycolysis, and this was significantly decreased in the presence of either HPP or IP (Figure 5C). Both IP and

HPP were also capable of significantly decreasing the glycolytic reserve in LPS-stimulated cells (Figure 5C). There were no significant changes in the basal respiration (Figure 5F), max respiration (Figure 5G), and respiratory reserve (Figure 5H) in IP- or HPP-treated DC when compared to LPS stimulation alone, suggesting that they have no impact on oxidative phosphorylation.

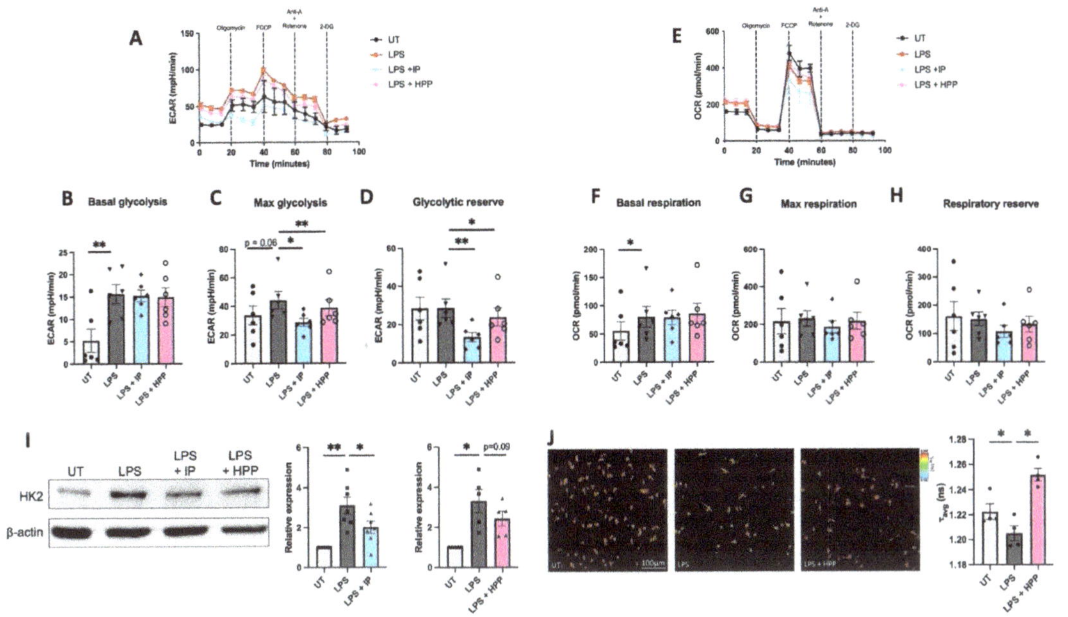

Figure 5. HPP and IP modulate metabolic reprogramming in LPS-stimulated DC. Primary human DC were pre-treated with either HPP or IP at 1000 μM for 6 h before stimulation with LPS (100 ng/mL) for 12 h. The extracellular acidification rate (ECAR) and the oxygen consumption rate (OCR) were measured using a Seahorse XFe96 analyser before and after the injections of oligomycin (1 mM), FCCP (1 mM), antimycin A (500 nM) and rotenone (500 nM), and 2-DG (25 mM). Bioenergetic profiles from one representative experiment depicting (**A**) ECAR and (**E**) OCR measurements over time. Pooled data ($N = 6$) depicts the calculated mean (±SEM) of (**B**) basal glycolytic rate, (**C**) max glycolytic rate, (**D**) glycolytic reserve, (**F**) basal respiratory rate, (**G**) max respiratory rate, and (**H**) respiratory reserve for each treatment group. (**I**) HK2 expression was measured by western blot. Densitometry results shown are mean ± SEM of the relative expression of HK2: β-actin from five to seven healthy donors. (**J**) FLIM images of DC measuring intracellular NADH. Pooled data ($N = 4$) depicts the mean (±SEM) of the ratio of bound:free NADH, represented by the τ average. Repeated measures one-way ANOVA, with Dunnett's multiple comparisons post hoc test, was used to determine statistical significance by comparing means of treatment groups against the mean of the control group (** $p < 0.01$, * $p < 0.05$). ImageLab (Bio-Rad) software was used to perform densitometric analysis.

Expression of HK2, the rate limiting enzyme in the glycolytic pathway, was next assessed by western blotting. HK2 is known to be induced by inflammatory stimuli and, as expected, LPS stimulation induced the upregulation of HK2 in DC. However, there was reduced expression of the enzyme in IP-treated DC, and a trend towards reduced expression (albeit not significant) in HPP-treated DC when compared to LPS-stimulated DC (Figure 5I). The effects of HPP treatment on the metabolism of DC was further investigated using FLIM, which measures the intracellular levels of NADH. Bound NADH, which is associated with oxidative phosphorylation, or free NADH, which is associated with glycolysis, can be distinguished based on their distinct lifetimes upon fluorescence excitation. The ratio of bound to free NADH can be used to measure whether a cell is favouring the engagement of glycolysis (a decrease in the ratio due to increased free NADH) or oxidative phosphorylation (an increase in the ratio due to increased bound NADH). Similar to the Seahorse results reported above, LPS-stimulated DC ramped up glycolysis, as represented by a decrease in the τ average compared to untreated DC (Figure 5J). Cells pre-treated with HPP exhibited a significant increase in the τ average compared to the LPS-stimulated controls, indicating they are favouring oxidative phosphorylation to generate their energy (Figure 5J). Overall, these results indicate that *T. brucei*-derived ketoacids modulate DC metabolism, reducing engagement of glycolysis, which is associated with rapid inflammatory responses.

3.6. HPP and IP Activate Autophagy-Related Proteins

We have previously demonstrated that the HO-1 inducers carnosol and curcumin activate the autophagy regulator AMPK, which incidentally is also known to downmodulate glycolysis in immune activated cells [22]. Furthermore, the key autophagy-related protein, p62, is linked to Nrf2 activation [23–25]. In order to determine if *T. brucei*-derived ketoacids have any impact on autophagy-related proteins, DC were treated with IP or HPP (both 1000 μM) for 15 min and phosphorylation (and therefore activation) of AMPK was assessed by western blotting. IP treatment resulted in a significant increase in AMPK phosphorylation while there was a trend towards increased activation with HPP (Figure 6A). DC treated with IP also showed an increase in both p62 and LC3-II (which is converted from LC3-I during autophagy) over time, and this was most potent after 24 h (Figure 6B,C). HPP treatment significantly increased p62 expression after 6 h and significantly increased LC3-II expression after 24 h (Figure 6B,C). These results indicate both ketoacids are activating autophagy-related proteins in human DC.

3.7. HPP and IP Reduce Proliferation and Cytokine Expression in PBMC Isolated from IBD Patients

It has previously been reported that IP has powerful immune suppressive effects in a murine experimental colitis model [7]. In order to determine if these results translate to a more clinical setting, PBMC were isolated from IBD patients and treated with IP or HPP (250–1000 μM) for 6 h prior to stimulation with anti-CD3 (1 μg/mL) for a further four days. Furthermore, the culture media was replaced with fresh media after 18 h of incubation with the compounds to circumvent issues surrounding the fluorescence of IP over long periods of time. The supernatants were removed for cytokine analysis by ELISA and cells were stimulated with PMA (50 ng/mL), ionomycin (500 ng/mL), and brefeldin A (5 μg/mL) for 4 h. Expression of ki67, and the cytokines IFNγ and IL-17 (both of which are known to play a pathogenic role in IBD), were measured by flow cytometry in $CD3^+CD8^-$ cells (gating strategy shown in Supplementary Figure S5). IP was non-toxic to PBMC at all concentrations tested, however there was some toxicity seen when using the higher concentrations of HPP (Supplementary Figure S6). Despite this being significant, it is unlikely to account for the effects seen, as flow markers were examined in live cells only. Both IP- and HPP-treated cells were capable of dose-dependently reducing the proliferation and expression of IFNγ when compared to anti-CD3 stimulation alone (Figure 7A,B), while having no effect on the intracellular levels of IL-17. However, when cell supernatants were assessed by ELISA, there was a significant reduction in both IFNγ and IL-17 production in IP- and HPP-treated cells (Figure 7C,D). Unlike purified DC, the ketoacids did not enhance the production of IL-10 from PBMC, suggesting a cell-type specific effect (Figure 7C,D).

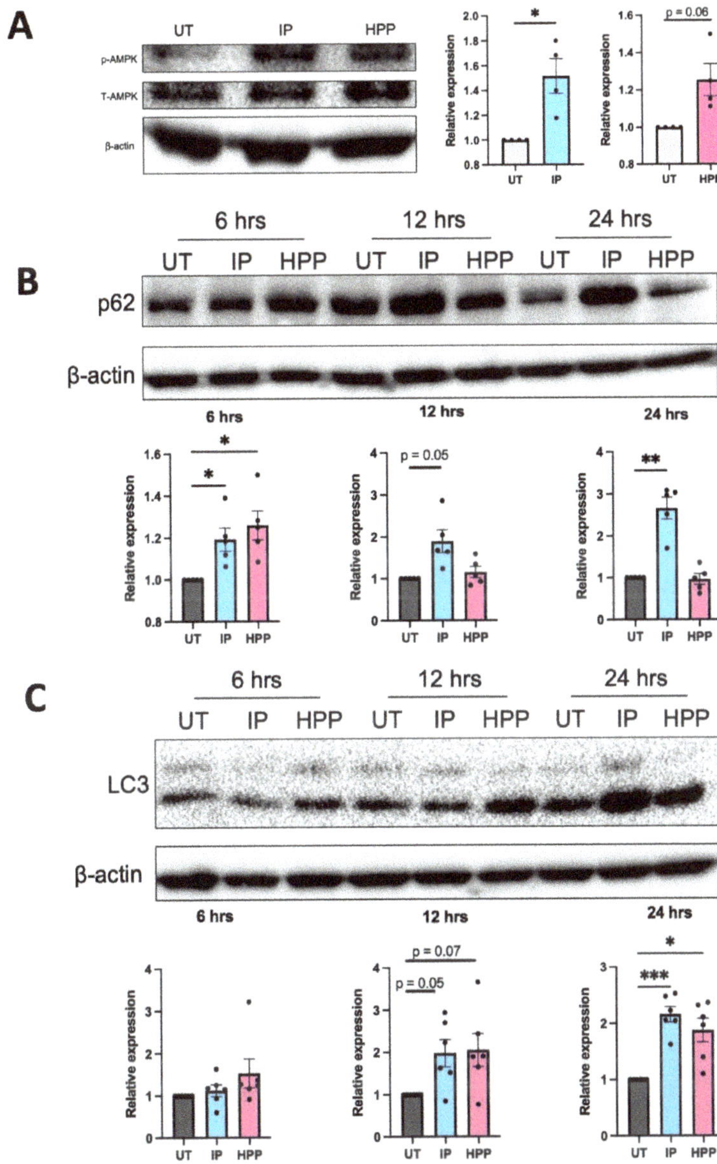

Figure 6. HPP and IP modulate autophagy-related proteins. (**A**) Primary human DC were left untreated (UT) or incubated with IP or HPP at 1000 μM for 15 min. Phosphorylation of AMPK was measured by western blot. Densitometry results shown are mean ± SEM of the relative expression of p-AMPK: β-actin from four healthy donors. (**B**,**C**) Primary human DC were left UT or incubated with IP or HPP at 1000 μM for 6, 12, or 24 h. Expression of (**B**) p62 and (**C**) LC3 were measured by western blot. Densitometry results shown are mean ± SEM of the relative expression of (**B**) p62: β-actin from five healthy donors and (**C**) LC3 II: β-actin from six healthy donors. (**A**) Statistical significance was determined using a Paired *t*-test. (**B**,**C**) Statistical significance was determined by repeated measures one-way ANOVA with Dunnett's multiple comparisons post hoc test to compare means of treatment groups to the control group (*** $p < 0.001$, ** $p < 0.01$, * $p < 0.05$). ImageLab (Bio-Rad) software was used to perform densitometric analysis.

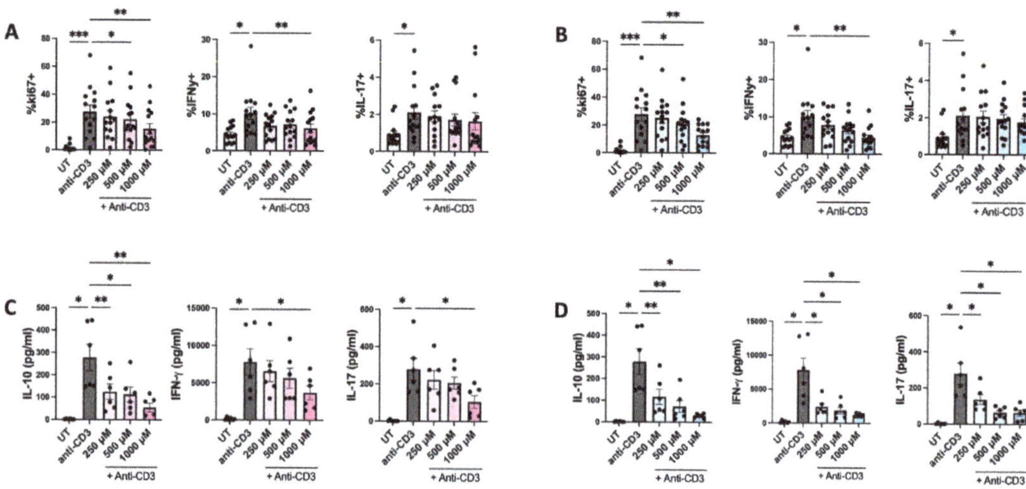

Figure 7. HPP and IP reduce proliferation and cytokine expression in ex vivo stimulated PBMC from patients with Inflammatory Bowel Disease. PBMC isolated from IBD patients were treated with (**A**,**C**) HPP or (**B**,**D**) IP (250 μM–1000 μM) for 6 h prior to stimulation with anti-CD3 for 12 h. After 18 h, culture media was replaced with fresh media and cells were incubated for a further 4 days with anti-CD3 stimulation. Supernatants were removed for analysis of cytokine concentration by ELISA. (**A**,**B**) Proliferation and cytokine production by $CD3^+CD8^-$ cells was analysed by flow cytometry. Pooled data ($N = 14$) depicting the mean ± SEM of ki67 (as a measure of proliferation), IFNγ, and IL-17 in $CD3^+CD8^-$ T cells. (**C**,**D**) Cell supernatants were assessed for concentrations of IL-10, IFNγ, and IL-17 by ELISA. Pooled data depicts mean (±SEM) cytokine concentrations for six IBD patients (means of three technical replicates per donor). Statistical significance was determined by repeated measures one-way ANOVA with Dunnett's multiple comparisons post hoc test to compare means of treatment groups to the control group (*** $p < 0.001$, ** $p < 0.01$, * $p < 0.05$).

4. Discussion

The production of large amounts of immunomodulatory aromatic ketoacids during *T. brucei* infection likely serves to benefit the parasite by prolonging infection, proliferation, and, ultimately, survival in the host. While the secretome of *T. brucei* has been shown to reduce the secretion of IL-12, IL-10, IL-6, and TNF in both murine and human DC [26,27], the ketoacids, IP and HPP, have been shown to directly ameliorate inflammatory cytokine production in murine macrophages and glia [4,5]. This is further supported by studies demonstrating their therapeutic efficacy in murine models of disease [4,6,7]. Given the key role played by DC in shaping both innate and adaptive immune cell responses, we sought to determine if these effects translate to this vital human immune cell population, which not only serves to present antigens during infection, but also plays a key role in determining pathogenic T cell responses during disease. We demonstrate that both IP and HPP are capable of significantly reducing the secretion of a number of pro-inflammatory cytokines, including TNF, IL-6, IL-12, and IL-23 in LPS-stimulated DC. Furthermore, the ketoacids upregulate HO-1 in an Nrf2 dependant manner, which is in line with studies demonstrating that HO-1 induction can promote a more tolerogenic DC phenotype [18,19,28]. In support of this, we also demonstrate that ketoacid-treated DC have a reduced capacity to activate T cells, which in turn limits the production of the pathogenic T cell cytokine IFNγ, which is known to play a deleterious role in a number of inflammatory/autoimmune conditions including IBD [29].

In comparison to the previous data from murine macrophages and glia, a larger repertoire of cytokines is inhibited by the ketoacids in human DC. For example, IP and HPP had no effect on TNF secretion in murine glia and no effect on either TNF or IL-6 secretion in bone marrow derived macrophages [4,5], however, these cytokines were significantly reduced in LPS-activated human DC. Furthermore, IP directly induced the production of the anti-inflammatory cytokine IL-10 in DC. The maturation status of DC is also impacted by ketoacid treatment, with the HPP-treated cells exhibiting a reduction in co-stimulatory and maturation markers, which, in turn, prevents their ability to participate in T cell activation. The upregulation of HO-1 was also accompanied by the expression of additional Nrf2-regulated genes and both IP and HPP appeared to exhibit direct antioxidant activity, which may explain their ability to activate Nrf2, given its well documented capacity to rapidly respond to oxidative stress.

Further analysis of these novel HO-1 inducers also revealed that they can modulate immune cell metabolism. Indeed, recent studies have highlighted an important link between immune cell activation and metabolism. It is now well recognised that, not only do different immune cells engage different metabolic pathways, but that the activation/maturation state of the immune cells is accompanied by metabolic switches. For example, innate immune cells, including macrophages and DC, ramp up glycolysis in order to rapidly generate sufficient energy and the building blocks required to fight infection [21]. This phenomenon is also a feature of pathogenic immune cells, and a significant effort is underway to determine if controlling/preventing dysregulated metabolic reprogramming can serve to ameliorate detrimental immune cell activation during disease. Here we demonstrate that both IP and HPP can decrease the max glycolysis observed in LPS-treated DC while also downregulating the expression of hexokinase 2, the rate-limiting enzyme in glycolysis. These results are similar to our recent observations with the HO-1 inducers, carnosol and curcumin, suggesting that HO-1 may be an important regulator of immune cell metabolism [30]. Further study is required to determine if additional features of metabolism are affected by ketoacids and whether metabolic reprogramming occurs during *T. brucei* infection, but these in vitro results are in line with the notion that metabolism is intricately linked with immune cell activation, and that the downmodulation of glycolysis in immune cells promotes a more tolerogenic phenotype.

From a therapeutic stance, IP in particular has shown potential in a murine model of colitis where administration of the ketoacid not only improved disease outcome, but also decreased expression of pro-inflammatory cytokines, including IL-12, IFNγ, and TNF [7]. We observed similar results in PBMC from IBD patients where both HPP and IP reduced the proliferation of anti-CD3 stimulated T cells, as well as the secretion of both IL-17 and IFNγ. Indeed, induction of HO-1 is being explored as a therapy for IBD and has shown promise in a number of murine models of disease [31–35]. Patients are currently treated largely with anti-inflammatories including 5-aminosalicylic acid (5-ASA), corticosteroids, methotrexate, and anti-TNF therapies [36]. However, many patients are/become refractory to these treatments and will require surgery in their lifetime. While further in vivo study is required to fully elucidate their efficacy and provide further information regarding treatment route, dosing, and long-term effects, these findings provide further impetus to explore aromatic ketoacids as a treatment for IBD (and indeed other inflammatory diseases), either alone or in combination with existing therapies.

Finally, a particularly interesting finding of this study is the observation that IP and HPP can activate key autophagic proteins. Autophagy itself is carried out by a number of autophagy-related (Atg) proteins and initiation is under the control of the protein kinases mTOR and AMPK, which are both intrinsically linked to immune cell metabolism (AMPK inhibits glycolysis while mTOR activation is linked to induction of glycolysis). The autophagic process is complicated and involves many different proteins and has been reviewed in detail elsewhere [37–39]. Briefly, a complex of Atg proteins lipidates LC3-I converting it to LC3-II [37,38]. LC3-II binds to the autophagosome membrane and facilitates the docking of cargos and proteins for degradation through their binding to p62 [40]. Following maturation, the autophagosome fuses with the lysosome to form the autolysosome, the contents of which are then degraded and recycled. In this study, we show that IP and HPP activate AMPK and increase expression of p62 and LC3-II.

Autophagy is of particular importance for DC, as many of their key functions, including antigen uptake and presentation, are strongly associated with autophagy [41]. Despite some conflicting reports, generally it appears that activation of autophagy gives rise to a more tolerogenic DC phenotype, exhibiting reduced antigen presentation and maturation [41], which is similar to the results seen in this study. Furthermore, AMPK activation has been shown to attenuate pro-inflammatory cytokine production and DC maturation [22,42]. Therefore, ketoacid-induced AMPK and subsequent autophagy activation may serve to downmodulate DC maturation, antigen presentation, and glycolysis.

Notably, p62 is not only an important autophagy-related protein, it also plays a crucial role in the activation of Nrf2 [23–25]. Nrf2 is activated upon release from KEAP1, which can occur when p62 sequesters KEAP1, targeting it for degradation and allowing Nrf2 to translocate to the nucleus [23–25]. Therefore, the activation of p62 by both ketoacids, but in particular IP, may be responsible for the subsequent activation of Nrf2, and therefore HO-1, by these ketoacids. While HPP induces HO-1 at an earlier time than IP, the latter appears to have more potent effects overall. In most cases, the in vitro effects of IP are most apparent at 24 h, and we cannot rule out the possibility that it is converting to another form over time. Further study is undoubtedly required to delineate the true impact of aromatic ketoacids (and their potential derivatives) and HO-1 induction on autophagy-related processes, in addition to the noted effects on DC metabolic reprogramming.

5. Conclusions

In conclusion, the data presented here expands our understanding of the mechanism of action of *T. brucei*-derived ketoacids in human immune cells and suggests that HO-1 induction may be useful to regulate the metabolism and, therefore, function of immune cells in inflammatory disease. We firmly believe that these compounds represent novel and exciting HO-1 inducers worthy of further exploration.

Supplementary Materials: The following supporting information can be downloaded at: https://www.mdpi.com/article/10.3390/antiox11010164/s1, Figure S1: HPP and IP are non-toxic to human DC; Figure S2: IP and HPP are not contaminated with endotoxin; Figure S3: ML385 is non-toxic to human DC; Figure S4: Gating strategy used to generate data shown in Figure 4; Figure S5: Gating strategy used to generate data shown in Figure 6; Figure S6: IP is non-toxic to PBMC, while higher concentrations of HPP shows significant, albeit mild, reductions in viability; Table S1: Seahorse calculations.

Author Contributions: H.K.F., J.M.F. and A.D. conceptualized and designed experiments. H.K.F. performed all experiments with laboratory assistance and intellectual input from S.A.O. and E.D. N.G.B.N. performed the FLIM experiments. N.G.B.N. and M.G.M. contributed to the interpretation of the FLIM data. D.P.N. contributed to the interpretation of data. M.T., J.D., E.J.R. and G.A.D. provided access to IBD patient samples. H.K.F., J.M.F. and A.D. wrote, reviewed, and edited the manuscript. All authors have read and agreed to the published version of the manuscript.

Funding: This research was funded by Health Research Board, Ireland, grant number ILP-POR-2017-041.

Institutional Review Board Statement: The study was conducted in accordance with the Declaration of Helsinki, and approved by the Research Ethics Committee of the School of Biochemistry and Immunology, Trinity College, Dublin (BI-HF-300921, approved 18 October 2018).

Informed Consent Statement: Informed consent was obtained from all subjects involved in the study.

Data Availability Statement: The data is contained within the article or supplementary material.

Conflicts of Interest: The authors declare no conflict of interest.

References

1. Stijlemans, B.; Caljon, G.; Van Den Abbeele, J.; Van Ginderachter, J.A.; Magez, S.; De Trez, C. Immune Evasion Strategies of Trypanosoma brucei within the Mammalian Host: Progression to Pathogenicity. *Front. Immunol.* **2016**, *7*, 233. [CrossRef] [PubMed]
2. Berger, B.J.; Dai, W.W.; Wang, H.; Stark, R.E.; Cerami, A. Aromatic amino acid transamination and methionine recycling in trypanosomatids. *Proc. Natl. Acad. Sci. USA* **1996**, *93*, 4126–4130. [CrossRef]
3. Marciano, D.; Llorente, C.; Maugeri, D.A.; de la Fuente, C.; Opperdoes, F.; Cazzulo, J.J.; Nowicki, C. Biochemical characterization of stage-specific isoforms of aspartate aminotransferases from Trypanosoma cruzi and Trypanosoma brucei. *Mol. Biochem. Parasitol.* **2008**, *161*, 12–20. [CrossRef] [PubMed]
4. McGettrick, A.F.; Corcoran, S.E.; Barry, P.J.G.; McFarland, J.; Crès, C.; Curtis, A.M.; Franklin, E.; Corr, S.C.; Mok, K.H.; Cummins, E.P.; et al. Trypanosoma brucei metabolite indolepyruvate decreases HIF-1α and glycolysis in macrophages as a mechanism of innate immune evasion. *Proc. Natl. Acad. Sci. USA* **2016**, *113*, E7778–E7787. [CrossRef] [PubMed]
5. Campbell, N.K.; Williams, D.G.; Fitzgerald, H.K.; Barry, P.J.; Cunningham, C.C.; Nolan, D.P.; Dunne, A. Trypanosoma brucei Secreted Aromatic Ketoacids Activate the Nrf2/HO-1 Pathway and Suppress Pro-inflammatory Responses in Primary Murine Glia and Macrophages. *Front. Immunol.* **2019**, *10*, 2137. [CrossRef]
6. Aoki, R.; Aoki-Yoshida, A.; Suzuki, C.; Takayama, Y. Protective Effect of Indole-3-Pyruvate against Ultraviolet B-Induced Damage to Cultured HaCaT Keratinocytes and the Skin of Hairless Mice. *PLoS ONE* **2014**, *9*, e96804. [CrossRef]
7. Aoki, R.; Aoki-Yoshida, A.; Suzuki, C.; Takayama, Y. Indole-3-Pyruvic Acid, an Aryl Hydrocarbon Receptor Activator, Suppresses Experimental Colitis in Mice. *J. Immunol.* **2018**, *201*, 3683–3693. [CrossRef]
8. Diskin, C.; Corcoran, S.E.; Tyrrell, V.J.; McGettrick, A.F.; Zaslona, Z.; O'Donnell, V.B.; Nolan, D.P.; O'Neill, L.A.J. The Trypanosome-Derived Metabolite Indole-3-Pyruvate Inhibits Prostaglandin Production in Macrophages by Targeting COX2. *J. Immunol.* **2021**, *207*, 2551–2560. [CrossRef]
9. Yamaguchi, T.; Komoda, Y.; Nakajima, H. Biliverdin-IXα reductase and biliverdin-IXβ reductase from human liver. Purification and characterization. *J. Biol. Chem.* **1994**, *269*, 24343–24348. [CrossRef]
10. Singleton, J.W.; Laster, L. Biliverdin reductase of guinea pig liver. *J. Biol. Chem.* **1965**, *240*, 4780–4789. [CrossRef]
11. Radhakrishnan, N.; Yadav, S.P.; Sachdeva, A.; Pruthi, P.K.; Sawhney, S.; Piplani, T.; Wada, T.; Yachie, A. Human heme oxygenase-1 deficiency presenting with hemolysis, nephritis, and asplenia. *J. Pediatr. Hematol. Oncol.* **2011**, *33*, 74–78. [CrossRef] [PubMed]
12. Kapturczak, M.H.; Wasserfall, C.; Brusko, T.; Campbell-Thompson, M.; Ellis, T.M.; Atkinson, M.A.; Agarwal, A. Heme oxygenase-1 modulates early inflammatory responses: Evidence from the heme oxygenase-1-deficient mouse. *Am. J. Pathol.* **2004**, *165*, 1045–1053. [CrossRef]
13. Poss, K.D.; Tonegawa, S. Reduced stress defense in heme oxygenase 1-deficient cells. *Proc. Natl. Acad. Sci. USA* **1997**, *94*, 10925–10930. [CrossRef] [PubMed]
14. Campbell, N.K.; Fitzgerald, H.K.; Dunne, A. Regulation of inflammation by the antioxidant haem oxygenase 1. *Nat. Rev. Immunol.* **2021**, *21*, 411–425. [CrossRef] [PubMed]
15. Neto, N.; Dmitriev, R.I.; Monaghan, M.G. Seeing Is Believing: Noninvasive Microscopic Imaging Modalities for Tissue Engineering and Regenerative Medicine. In *Cell Engineering and Regeneration*; Springer: Cham, Switzerland, 2020; pp. 599–638.
16. Floudas, A.; Neto, N.; Marzaioli, V.; Murray, K.; Moran, B.; Monaghan, M.G.; Low, C.; Mullan, R.H.; Rao, N.; Krishna, V.; et al. Pathogenic, glycolytic PD-1+ B cells accumulate in the hypoxic RA joint. *JCI Insight* **2020**, *5*, e139032. [CrossRef]
17. El Sawalhy, A.; Seed, J.R.; Hall, J.E.; El Attar, H. Increased excretion of aromatic amino acid catabolites in animals infected with Trypanosoma brucei evansi. *J. Parasitol.* **1998**, *84*, 469–473. [CrossRef]
18. Campbell, N.K.; Fitzgerald, H.K.; Malara, A.; Hambly, R.; Sweeney, C.M.; Kirby, B.; Fletcher, J.M.; Dunne, A. Naturally derived Heme-Oxygenase 1 inducers attenuate inflammatory responses in human dendritic cells and T cells: Relevance for psoriasis treatment. *Sci. Rep.* **2018**, *8*, 10287. [CrossRef]
19. Chauveau, C.; Rémy, S.; Royer, P.J.; Hill, M.; Tanguy-Royer, S.S.; Hubert, F.-X.F.X.; Tesson, L.; Brion, R.R.; Beriou, G.G.; Gregoire, M.; et al. Heme oxygenase-1 expression inhibits dendritic cell maturation and proinflammatory function but conserves IL-10 expression. *Blood* **2005**, *106*, 1694–1702. [CrossRef]
20. Hull, T.D.; Agarwal, A.; George, J.F. The mononuclear phagocyte system in homeostasis and disease: A role for heme oxygenase-1. *Antioxid. Redox Signal.* **2014**, *20*, 1770–1788. [CrossRef]
21. O'Neill, L.A.J.; Kishton, R.J.; Rathmell, J. A guide to immunometabolism for immunologists. *Nat. Rev. Immunol.* **2016**, *16*, 553–565. [CrossRef]
22. Krawczyk, C.M.; Holowka, T.; Sun, J.; Blagih, J.; Amiel, E.; DeBerardinis, R.J.; Cross, J.R.; Jung, E.; Thompson, C.B.; Jones, R.G.; et al. Toll-like receptor-induced changes in glycolytic metabolism regulate dendritic cell activation. *Blood* **2010**, *115*, 4742–4749. [CrossRef] [PubMed]
23. Komatsu, M.; Kurokawa, H.; Waguri, S.; Taguchi, K.; Kobayashi, A.; Ichimura, Y.; Sou, Y.S.; Ueno, I.; Sakamoto, A.; Tong, K.I.; et al. The selective autophagy substrate p62 activates the stress responsive transcription factor Nrf2 through inactivation of Keap1. *Nat. Cell Biol.* **2010**, *12*, 213–223. [CrossRef] [PubMed]
24. Katsuragi, Y.; Ichimura, Y.; Komatsu, M. Regulation of the Keap1–Nrf2 pathway by p62/SQSTM1. *Curr. Opin. Toxicol.* **2016**, *1*, 54–61. [CrossRef]

25. Jiang, T.; Harder, B.; Rojo De La Vega, M.; Wong, P.K.; Chapman, E.; Zhang, D.D. p62 links autophagy and Nrf2 signaling. *Free Radic. Biol. Med.* **2015**, *88*, 199. [CrossRef]
26. Dauchy, F.A.; Contin-Bordes, C.; Nzoumbou-Boko, R.; Bonhivers, M.; Landrein, N.; Robinson, D.R.; Rambert, J.; Courtois, P.; Daulouède, S.; Vincendeau, P. Trypanosoma brucei gambiense excreted/secreted factors impair lipopolysaccharide-induced maturation and activation of human monocyte-derived dendritic cells. *Parasite Immunol.* **2019**, *41*, e12632. [CrossRef] [PubMed]
27. Garzón, E.; Holzmuller, P.; Bras-Gonçalves, R.; Vincendeau, P.; Cuny, G.; Lemesre, J.L.; Geiger, A. The Trypanosoma brucei gambiense Secretome Impairs Lipopolysaccharide-Induced Maturation, Cytokine Production, and Allostimulatory Capacity of Dendritic Cells. *Infect. Immun.* **2013**, *81*, 3300. [CrossRef] [PubMed]
28. Moreau, A.; Hill, M.; Thébault, P.; Deschamps, J.Y.; Chiffoleau, E.; Chauveau, C.; Moullier, P.; Anegon, I.; Alliot-Licht, B.; Cuturi, M.C. Tolerogenic dendritic cells actively inhibit T cells through heme oxygenase-1 in rodents and in nonhuman primates. *FASEB J.* **2009**, *23*, 3070–3077. [CrossRef] [PubMed]
29. Andreou, N.P.; Legaki, E.; Gazouli, M. Inflammatory bowel disease pathobiology: The role of the interferon signature. *Ann. Gastroenterol.* **2020**, *33*, 125. [CrossRef] [PubMed]
30. Campbell, N.K.; Fitzgerald, H.K.; Fletcher, J.M.; Dunne, A. Plant-derived polyphenols modulate human dendritic cell metabolism and immune function via AMPK-dependent induction of heme oxygenase-1. *Front. Immunol.* **2019**, *10*, 345. [CrossRef]
31. Sheikh, S.Z.; Hegazi, R.A.; Kobayashi, T.; Onyiah, J.C.; Russo, S.M.; Matsuoka, K.; Sepulveda, A.R.; Li, F.; Otterbein, L.E.; Plevy, S.E. An Anti-Inflammatory Role for Carbon Monoxide and Heme Oxygenase-1 in Chronic Th2-Mediated Murine Colitis. *J. Immunol.* **2011**, *186*, 5506–5513. [CrossRef]
32. Hegazi, R.A.F.; Rao, K.N.; Mayle, A.; Sepulveda, A.R.; Otterbein, L.E.; Plevy, S.E. Carbon monoxide ameliorates chronic murine colitis through a heme oxygenase 1–dependent pathway. *J. Exp. Med.* **2005**, *202*, 1703–1713. [CrossRef] [PubMed]
33. Zhang, L.; Zhang, Y.; Zhong, W.; Di, C.; Lin, X.; Xia, Z. Heme oxygenase-1 ameliorates dextran sulfate sodiuminduced acute murine colitis by regulating Th17/Treg cell balance. *J. Biol. Chem.* **2014**, *289*, 26847–26858. [CrossRef]
34. Paul, G.; Bataille, F.; Obermeier, F.; Bock, J.; Klebl, F.; Strauch, U.; Lochbaum, D.; Rümmele, P.; Farkas, S.; Schölmerich, J.; et al. Analysis of intestinal haem-oxygenase-1 (HO-1) in clinical and experimental colitis. *Clin. Exp. Immunol.* **2005**, *140*, 547–555. [CrossRef] [PubMed]
35. Longhi, M.S.; Vuerich, M.; Kalbasi, A.; Kenison, J.E.; Yeste, A.; Csizmadia, E.; Vaughn, B.; Feldbrugge, L.; Mitsuhashi, S.; Wegiel, B.; et al. Bilirubin suppresses Th17 immunity in colitis by upregulating CD39. *JCI Insight* **2017**, *2*, e92791. [CrossRef] [PubMed]
36. Mowat, C.; Cole, A.; Windsor, A.; Ahmad, T.; Arnott, I.; Driscoll, R.; Mitton, S.; Orchard, T.; Rutter, M.; Younge, L.; et al. Guidelines for the management of inflammatory bowel disease in adults. *Gut* **2011**, *60*, 571–607. [CrossRef] [PubMed]
37. Yang, Z.; Klionsky, D.J. Mammalian autophagy: Core molecular machinery and signaling regulation. *Curr. Opin. Cell Biol.* **2010**, *22*, 124–131. [CrossRef]
38. Glick, D.; Barth, S.; Macleod, K.F. Autophagy: Cellular and molecular mechanisms. *J. Pathol.* **2010**, *221*, 3–12. [CrossRef] [PubMed]
39. Feng, Y.; He, D.; Yao, Z.; Klionsky, D.J. The machinery of macroautophagy. *Cell Res.* **2014**, *24*, 24–41. [CrossRef]
40. Pankiv, S.; Clausen, T.H.; Lamark, T.; Brech, A.; Bruun, J.A.; Outzen, H.; Øvervatn, A.; Bjørkøy, G.; Johansen, T. p62/SQSTM1 binds directly to Atg8/LC3 to facilitate degradation of ubiquitinated protein aggregates by autophagy. *J. Biol. Chem.* **2007**, *282*, 24131–24145. [CrossRef] [PubMed]
41. Ghislat, G.; Lawrence, T. Autophagy in dendritic cells. *Cell. Mol. Immunol.* **2018**, *15*, 944–952. [CrossRef] [PubMed]
42. Carroll, K.C.; Viollet, B.; Suttles, J. AMPKα1 deficiency amplifies proinflammatory myeloid APC activity and CD40 signaling. *J. Leukoc. Biol.* **2013**, *94*, 1113–1121. [CrossRef] [PubMed]

antioxidants

Article

Heme-Oxygenase-1 Attenuates Oxidative Functions of Antigen Presenting Cells and Promotes Regulatory T Cell Differentiation during *Fasciola hepatica* Infection

Monique Costa , Valeria da Costa, Sofía Frigerio, María Florencia Festari, Mercedes Landeira, Santiago A. Rodríguez-Zraquia, Pablo Lores, Paula Carasi and Teresa Freire *

Laboratorio de Inmunomodulación y Desarrollo de Vacunas, Departamento de Inmunobiología, Facultad de Medicina, Universidad de La República, Montevideo 11800, Uruguay; monique.scosta14@gmail.com (M.C.); valedacosta21@gmail.com (V.d.C.); frigeriosofia@gmail.com (S.F.); mffestari@gmail.com (M.F.F.); meche.land@gmail.com (M.L.); santiagorzq@gmail.com (S.A.R.-Z.); pablolores@hotmail.com (P.L.); paulacarasi@gmail.com (P.C.)

* Correspondence: tfreire@fmed.edu.uy; Tel.: +598-2-924-9562

Citation: Costa, M.; da Costa, V.; Frigerio, S.; Festari, M.F.; Landeira, M.; Rodríguez-Zraquia, S.A.; Lores, P.; Carasi, P.; Freire, T. Heme-Oxygenase-1 Attenuates Oxidative Functions of Antigen Presenting Cells and Promotes Regulatory T Cell Differentiation during *Fasciola hepatica* Infection. *Antioxidants* **2021**, *10*, 1938. https://doi.org/10.3390/antiox10121938

Academic Editors: Elias Lianos and Maria G. Detsika

Received: 2 October 2021
Accepted: 10 November 2021
Published: 3 December 2021

Abstract: *Fasciola hepatica* is a fluke that infects livestock and humans causing fasciolosis, a zoonotic disease of increasing importance due to its worldwide distribution and high economic losses. The parasite regulates the host immune system by inducing a strong Th2 and regulatory T (Treg) cell immune response through mechanisms that might involve the expression or activity of heme-oxygenase-1 (HO-1), the rate-limiting enzyme in the catabolism of free heme that also has immunoregulatory and antioxidant properties. In this paper, we show that *F. hepatica*-infected mice upregulate HO-1 on peritoneal antigen-presenting cells (APC), which produce decreased levels of both reactive oxygen and nitrogen species (ROS/RNS). The presence of these cells was associated with increased levels of regulatory T cells (Tregs). Blocking the IL-10 receptor (IL-10R) during parasite infection demonstrated that the presence of splenic Tregs and peritoneal APC expressing HO-1 were both dependent on IL-10 activity. Furthermore, IL-10R neutralization as well as pharmacological treatment with the HO-1 inhibitor SnPP protected mice from parasite infection and allowed peritoneal APC to produce significantly higher ROS/RNS levels than those detected in cells from infected control mice. Finally, parasite infection carried out in gp91^{phox} knockout mice with inactive NADPH oxidase was associated with decreased levels of peritoneal HO-1^{+} cells and splenic Tregs, and partially protected mice from the hepatic damage induced by the parasite, revealing the complexity of the molecular mechanisms involving ROS production that participate in the complex pathology induced by this helminth. Altogether, these results contribute to the elucidation of the immunoregulatory and antioxidant role of HO-1 induced by *F. hepatica* in the host, providing alternative checkpoints that might control fasciolosis.

Keywords: helminth; heme-oxigenase-1; immunoregulation; antigen presenting cell; regulatory T cell; ROS/RNS

1. Introduction

Fasciolosis is a major parasitic disease of livestock caused by the trematode *Fasciola* spp. [1]. Nowadays, the number of infected people around the world is increasing, which makes fasciolosis an emerging zoonosis [1]. The World Health Organization (WHO) estimates that approximately 180 million are at risk of infection and 17 million people are infected, with a high prevalence in humans in Africa and South America [2]. Moreover, the economic losses caused by fasciolosis are estimated at around 3 billion US dollars per year due to livestock infection [1].

During infection, *Fasciola hepatica* modulates the host immune response characterized by the presence of regulatory dendritic cells (DC) [2–5], alternative activated macrophages [6], and an adaptive immune response characterized by Th2 and regulatory T cell (Treg)-associated cytokines [7–10]. Macrophages and DC comprise a highly heterogeneous myeloid cell population specialized in antigen presentation. DC play a crucial role in orchestrating the adaptive immune response by activating naïve T cells and inducing their differentiation into different effector T cells depending on the pathogen [11,12]. Furthermore, both macrophages and DC can promote inflammation by secreting pro-inflammatory cytokines and reactive oxygen or nitrogen species (ROS/RNS), although they can also inhibit inflammation through anti-inflammatory cytokine production [13–15]. Thus, they exhibit functional plasticity that enables them to adapt to various local conditions in order to restore homeostasis after inflammation [11,12,16].

Heme-oxygenase-1 (HO-1) is the inducible rate-limiting enzyme involved in the catabolism of free heme. It participates in several processes, by providing cytoprotection, protecting from heme-induced toxicity, and regulating the host inflammatory response [14,17]. In fact, HO-1 acts as a stress-responsive enzyme induced by the nuclear factor NF-E2-related factor 2 (NRF2), to provide defense against oxidative-induced injury during inflammatory processes [14]. HO-1 also limits the secretion of pro-inflammatory cytokines [18,19] and promotes anti-inflammatory cytokines [20,21], in a process triggered, at least in part, by ROS in macrophages, although its effects depend on the model of study [14]. In the same line, expression of HO-1 inhibits DC-maturation and the production of ROS [22]. In addition, HO-1 expression by DC induces the production of IL-10, an anti-inflammatory cytokine, that inhibits T-cell proliferation [23]. Additionally, IL-10-producing anti-inflammatory M2c macrophages express HO-1 [24], especially on those that express CD163, a hemoglobin scavenger receptor, that in fact mediates IL-10 production and HO-1 synthesis [25]. Therefore, HO-1 acts as a key mediator of anti-inflammatory pathways in both macrophages and DC.

Several reports demonstrate that HO-1 can have both beneficial and detrimental effects for the host immunity against different pathogens [26]. Furthermore, the antioxidant role of HO-1 in infectious diseases is still unclear, especially in helminth parasites. Interestingly, bilirubin, one of the enzymatic byproducts of HO-1, suppresses the killing of bacteria by reducing the neutrophil burst via its antioxidant activity [27]. In addition, HO-1 promotes bacteria survival inside macrophages by decreasing ROS production, as demonstrated for *Salmonella typhimurium* [28] and *Mycobacterium abscessus* [29] via its ROS-diminishing properties. Recently, we showed that the anti-inflammatory effects of HO-1 induction are detrimental in *F. hepatica* infection, which is dominated by a Th2/Treg differentiation profile [17]. However, the antioxidant mechanisms induced by HO-1 during *F. hepatica* infection have not been approached so far.

Thus, in this work, we investigated the role of ROS/RNS production by myeloid HO-1^+ cells during *F. hepatica* experimental infection in mice, and characterized the adaptive cell immune response. Our results indicate that HO-1 expression by myeloid cells during *F. hepatica* infection negatively correlates with the production of ROS or RNS and the increase of antioxidant molecules. Furthermore, the pharmacological inhibition of HO-1 by a well-characterized inhibitor of HO-1 enzymatic activity, tin protoporphyrin IX (SnPP), in *F. hepatica*-infected mice was associated with lower levels of Tregs, in a process that was mediated by IL-10 biological activity. However, the gene expression of the NADPH oxidase subunit $gp91^{phox}$ was decreased when SnPP was administrated to infected mice. Moreover, parasite infection carried out in $gp91^{phox}$ knockout mice with inactive NADPH oxidase showed decreased levels of peritoneal HO-1^+ cells, splenic Tregs, and partially protected mice from the hepatic damage induced by the parasite, revealing that a more complex molecular mechanism involving ROS production participates in the intricate pathology induced by this helminth. Altogether, these results contribute to the elucidation of the immunoregulatory and antioxidant role of HO-1 during *F. hepatica* infection, providing interesting molecular checkpoints that might control fasciolosis.

2. Materials and Methods

2.1. Mice

Female BALB/c mice (six- to eight-week-old) were purchased from DILAVE Laboratories (Uruguay). Gp91^{phox-} knockout (B6.129S-Cybbtm1Din/J) mice were purchased from Jackson Laboratory (USA) and maintained at UATE, Institut Pasteur Montevideo. Six to eight BALB/c, gp91^{phox-}, and C57BL/6 littermates (controls for gp91^{phox-} mice) were used per experiment. Animals were kept in the animal house (URBE, School of Medicine, UdelaR, Uruguay), with water and food supplied *ad libitum*. Mouse handling, care, and experiments were carried out in compliance with institutional guidelines and regulations from the National Committee on Animal Research (Comisión Nacional de Experimentación Animal, CNEA, https://www.cnea.gub.uy/, accessed on 12 November 2021, National Law 18.611, Uruguay). Procedures involving animals were approved by the Universidad de la República's Committee on Animal Research (Comisión Honoraria de Experimentación Animal, CHEA Protocol Number 07153-000817-18).

2.2. Parasite Protein Extract (FhTE)

FhTE was prepared from live adult flukes obtained from infected bovines. Flukes were washed for 1 h at 37 °C with phosphate buffered saline (PBS), pH 7.4, sonicated, and then centrifugated at 40,000× *g* for 60 min [4,30]. Finally, protein lysates were dialyzed against PBS. FhTE protein concentration was measured using the bicinchoninic acid assay (Sigma, St. Louis, MO, USA). In order to remove endotoxin contamination, FhTE was applied to a column containing endotoxin-removing gel (detoxi-gel, Pierce Biotechnology, Waltham, MA, USA), and endotoxin levels were quantified using the Limulus Amebocyte Lysate kit Pyrochrome (Associates of Cape Cod, East Falmouth, MA, USA) and found to be lower than 0.05 EU/mL. Furthermore, at the used concentrations, FhTE did not induce the production of pro-inflammatory cytokines such as IL-12 or IL-6 [4,30]. The concentration of *F. hepatica* extracts used in the in vitro experiments did not modify cell viability, as evaluated by the MTT (2-[4,5-dimethyl-2-thiazolyl]-3,5-diphenyl-2H-tetrazolium bromide) assay [4,17,30].

2.3. *F. hepatica* Infections

2.3.1. Parasite Infections, Animal Treatments, and Sample Obtention

The infection was achieved by orally administrating 10 *F. hepatica* metacercariae (Montevideo, Uruguay) per mouse. Mice were bled and peritoneal exudate cells (PECs), spleens, and livers were removed after 1, 8, 15, and 21 days post-infection (dpi), while non-infected animals were used as controls. Each experimental group contained at least four mice. PECs and hepatic leukocytes were obtained as already described [31]. Red cells were lysed with ammonium chloride potassium buffer.

HO-1 activity was inhibited using SnPP (40 mg/kg), and vehicle (PBS, 200 µL) was used as a control. The SnPP dose was within a range of doses used in previous works [32,33]. Mice received intraperitoneal injections of SnPP 1 day before infection, 1 day after infection, and every 4 days until the end of the experimental protocol (between 19 and 21 dpi). When gp91^{phox-} and non-infected littermates were used (n = 6–8/group), infections were performed in the same conditions as previously described. In order to neutralize IL-10 receptor (IL-10R), BALB/c mice (n = 6–8/group) were intraperitoneally injected with 15 µg of monoclonal rat IgG2a anti-IL-10R (clone 1B1.3A from BioXcell, Lebanon, NH, USA) or an isotype-matched control antibody (clone HRPN from BioXcell, Lebanon, NH, USA), the day before and after infection with *F. hepatica* and every 3 days until animal sacrifice at 20 dpi. Blood samples were obtained, and PECs, spleens, and livers were removed. The infection severity was assessed with a defined clinical score according to the following parameters: presence or absence of peritoneal hemorrhage, presence of macroscopic liver damage and splenomegaly, and the amount of cell content in the peritoneal cavity [17], where the minimum score was 0 and the maximum was 10. The alanine aminotransferase

(ALT) activity in sera was used to quantify liver damage, determined with a commercial kit (Spinreact, Girona, Spain) according to the manufacturer's instructions.

2.3.2. Proliferation Assay and Cell Culture

Splenocytes (0.5 × 10^6/well) from infected mice or uninfected naïve mice (control group) were cultured for 5 days at 37 °C and 5% CO_2, in RPMI-1640 with 400 μg/mL of glutamine (Capricorn, Ebsdorfergrund, Germany) complete medium containing 10% heat-inactivated fetal bovine serum (FBS, Capricorn Scientific, Ebsdorfergrund, Germany), 50 mM of 2-mercaptoethanol, 100 U/mL of penicillin, and 0.1 mg/mL of streptomycin (Merk, Sigma-Aldrich, St. Louis, MO, USA) in the presence or absence of FhTE (75 μg/mL), as previously described [30]. An IFNγ-specific sandwich ELISA assay (Biolegend, San Diego, CA, USA) was used to quantify IFNγ levels in culture supernatants.

RAW264.7 macrophages were cultured at 0.5 × 10^6/mL in complete RPMI medium in the presence or absence of the HO-1 inductor (CoPP) and inhibitor (SnPP) (50 and 100 μM, respectively) or FhTE (75 μg/mL) overnight at 37 °C. Afterwards, the ROS/RNS production was determined as described in the following section.

2.4. Flow Cytometry

Cell suspensions from PECs, livers, and spleens were washed twice with PBS containing 2% FBS and 0.1% sodium azide (FACS buffer), stained with specific antibodies for 30 min at 4 °C as previously published [31]. The following antibodies were used: anti-Sirpα (P-84), -CD11c (N418), -CD86 (GL1), CD8 (53-6.7), -Siglec-F (E50-2440), -F4/80 (BM8), -CD11b (M1/70), -CD40 (HM40-3), -CD80 (16-10A1), and I-A/I-E (M5/114.15.2). Expression of FoxP3, HO-1, and IL-10 was analyzed by intracellular staining. Cells in which IL-10 was analyzed were incubated with Brefeldin-A for 6 h at 37 °C and phorbol myristate acetate (PMA, 200 nM) (Merk, Sigma-Aldrich, USA). After two washes with FACS buffer, cells were incubated with the following antibodies: anti-CD3 (17A2), -CD4 (RM4-5), -CD8 (53-6.7), or -F4/80 (BM8). After permeabilization with Cytofix and Perm wash buffers (Biolegend, USA), cells were incubated with IL-10 (JES5-1E3), FoxP3 (MF14), and HO-1 (clone ab13248 from Abcam, Waltham, MA, USA) specific antibodies. ROS/RNS produced by F4/80^+ cells were determined with 2′,7′-dichlorofluorescein diacetate (DCFDA, Merk, Kenilworth, NJ, USA) probe, a fluorogenic dye that is oxidized into the fluorescent 2′,7′-dichlorofluorescein. Briefly, cells were incubated in PBS for 30 min at 37 °C with DCFDA, washed with FACS buffer, and fluorescence was measured in a flow cytometer. Analyses were performed using a BD Accuri C6 Plus cytometer and software (BD-Biosciences). Antibodies were obtained from Biolegend (USA). Analyses were performed with Accuri C6 Plus software.

2.5. Determination of Oxidative and Antioxidative Genes by qRT-PCR

Nrf2, *catalase*, glutathione peroxidase (*gpx*) 1 and 2, superoxide dismutase (*sod*) 1 and 2, and NADPH-oxidase subunits $p47^{phox}$ and $gp91^{phox}$ mRNA levels were detected using the Eco real-time PCR System (Illumina, San Diego, CA, USA) and Fast SYBR® Green Master Mix (Applied Biosystems, Waltham, MA, USA). ARN purification was performed with Tri-Reagent (Merk, Kenilworth, NJ, USA) of PECs obtained from BALB/c mice at 20 dpi, SnPP-treated or untreated, as previously described [17]. Standard amplification conditions were 10 min at 95 °C, 40 thermal cycles of 15 s at 95 °C, 30 s at 60 °C, and 30 s at 72 °C, with a final extension of 10 min at 72 °C. The following primers were used: *nrf2*-F: 5′-CAGCATGTTACGTGATGAGG-3′, *nrf2*-R: 5′-GCTCAGAAAAGGCTCCATCC-3′, *gpx1*-F: 5′-GGGACTACACCGAGATGAACGA-3′, *gpx1*-R: 5′-ACCATTCACTTCGCACTTCTCA-3′, *gpx2*-F: 5′-GAGGAACAACTACCCGGGACTA-3′, *gpx2-R*: 5′-ACCCCCAGGTCGGACATACT-3′, *sod1*-F: 5′-TGGGTTCCACGTCCATCAGTA-3′, *sod1*-R: 5′-ACCGTCCTT TCCAGCAGTCA-3′, *sod2*-F: 5′-ATTAACGCGCAGATCATGCA-3′, *sod2*-R: 5′-TGTCCCCCACCATTGAACTT-3′, *catalase*-F: 5′-GCGTCCAGTGCGCTGTAGA-3′, *catalase*-R: 5′-TCAGGGTGGACGTCAGTGAA-3′, *p47phox*-F: 5′-GAGGCGGAGGATCCGG-3′, *p47phox*-R: 5′-TCTTCAACAGCAGCGTACGC-3′, *gp91phox*-F: 5′-CCAGTGAAGATGTGTTCAGCT-3′, *gp91phox*-R: 5′-GCACAGCCAGTAGA

AGTAGA-3′, *gapdh*-F: 5′-ATGACATCAAGAAGGTGGTGAAG-3′, *gapdh*-R: 5′-TCCTTGGAGG CCATGTAGG-3′. Results were expressed as the ratio between each gene under study and GAPDH expression. Relative gene expression levels were calculated using the $2^{-\Delta\Delta CT}$ method and normalized to GAPDH. All reactions were performed with at least five biological replicates.

2.6. Statistical Analysis

Results of the experiments were expressed as mean ± SEM. GraphPad Prism version 6.04 for Windows (GraphPad Software, San Diego, CA, USA) was used to perform statistical analyses. Results were analyzed using one-way ANOVA followed by Tukey's test, or the two-tailed Student's *t*-test, depending on the experiment. Significant differences shown by an asterisk were considered when $p < 0.05$.

3. Results

3.1. HO-1 Expression in F4/80$^+$ Peritoneal Cells Inversely Correlate with ROS/RNS Production

In order to confirm the recruitment of PECs expressing HO-1 to the peritoneum of *F. hepatica*-infected mice, we identified HO-1$^+$ cells by flow cytometry at different time points of the infection. As seen in Figure 1A, the clinical score increased upon infection, although ALT in serum significantly increased only after 21 dpi, demonstrating liver dysfunction. In addition, HO-1$^+$ cells significantly increased in the peritoneal cavity during the infection (Figure 1B,C and Supplementary Figure S1). These cells were mainly composed by F4/80$^+$ cells (Figure 1D,E and Supplementary Figure S1), and their increase also correlated with the advanced stages of the infection (after 15 dpi). The expression of HO-1 in F4/80$^+$ cells slightly increased after 1 dpi, while it considerably increased during the infection (Figure 1F). On the other hand, the production of ROS/RNS was significantly increased only at 1 dpi, and decreased during infection (Figure 1G), suggesting that the expression of HO-1 in peritoneal F4/80$^+$ cells inversely correlated with the production of ROS/RNS. In order to provide more evidence in this regard, we incubated RAW 264.7 macrophages with parasite components (FhTE) in the presence of CoPP or SnPP, and analyzed the production of ROS/RNS by these cells. FhTE slightly increased the production of ROS/RNS, while CoPP and SnPP significantly decreased and increased the production of ROS/RNS by FhTE-treated macrophages, respectively (Figure 1H). Of note, FhTE *per se* induced ROS/RNS expression, which could be the result of an active respiratory burst, such as that seen in F4/80$^+$ cells from PECs of infected mice at 1 dpi (Figure 1G). Altogether, these results might indicate that *F. hepatica* induces the expression of HO-1 in F4/80$^+$ cells recruited to the peritoneum, inhibiting ROS/RNS production during the course of the infection.

3.2. The Presence of Peritoneal HO-1$^+$ Cells Associates with Increased Splenic CD4$^+$ CD25$^+$ and CD8$^+$ CD25$^+$ Cells during Infection

Considering that HO-1 can induce regulatory T cells [34–37], we analyzed the presence of both CD4$^+$ and CD8$^+$ cells in spleens of infected mice. Although we could not find any significant differences between the percentage of CD4$^+$ and CD8$^+$ cells during the infection, we did observe that they significantly increased in number after 15 dpi (Figure 2A and Supplementary Figure S2). We also analyzed the presence of splenic CD25$^+$ CD4$^+$ (Figure 2B and Supplementary Figure S2) or CD8$^+$ (Figure 2C and Supplementary Figure S2) T cells. Again, no significant differences were found in the percentage of these cells during the infection, while their number significantly increased after 15 dpi. Finally, we analyzed the presence of CD4$^+$ T cells in livers from infected animals and did not find any difference in their percentage nor their number (Figure 2D and Supplementary Figure S3). However, the number, but not the frequency, of hepatic CD25$^+$ CD4$^+$ T cells was increased in advanced infected mice (Figure 2E and Supplementary Figure S3). Further analyses demonstrated that the number of splenic CD25$^+$ CD4$^+$ and CD25$^+$ CD8$^+$cells positively correlated with the number of peritoneal HO1$^+$ cells (Figure 2F).

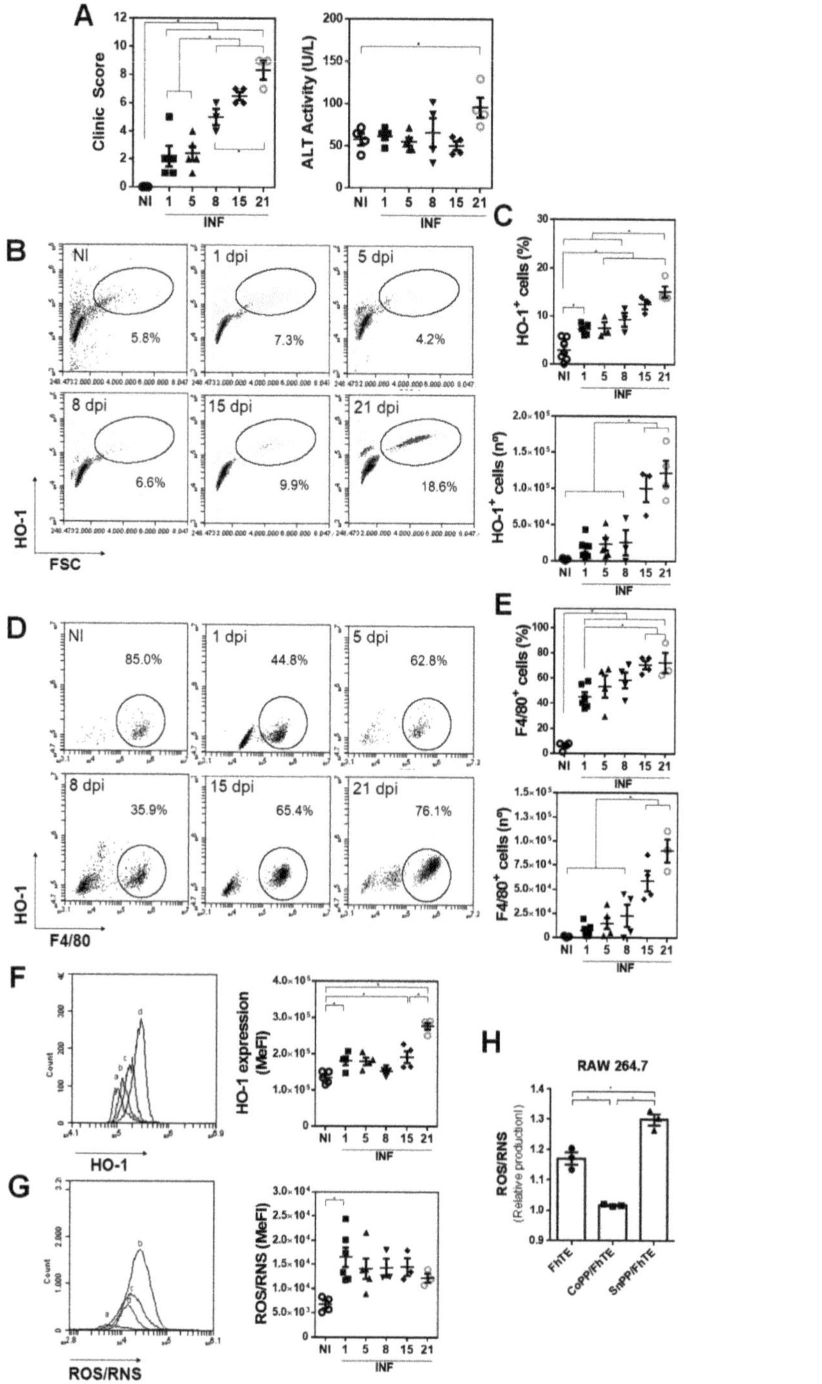

Figure 1. HO-1 expression in F4/80⁺ peritoneal cells inversely correlates with ROS/RNS production. Mice (n = 4–6) were infected with 10 metacercariae and sacrificed at 1, 8, 15, and 21 dpi. Non-infected

mice were used as control (NI). (**A**) Clinical signs including hemorrhage, splenomegaly, and macroscopic liver damage were assessed to evaluate the severity of the disease [17]. ALT activity in sera was used as a marker of liver damage. (**B**) Analysis by flow cytometry of HO-1$^+$ cells in PEC from infected and control (NI) mice. (**C**) Frequency and cell number of HO-1$^+$ cells in the peritoneal cavity of infected and control (NI) animals by flow cytometry. (**D**) Analysis by flow cytometry of F4/80$^+$ HO-1$^+$ cells in PEC from infected and control (NI) mice. (**E**) Frequency and cell number of F4/80$^+$ in the peritoneal cavity of infected and control (NI) animals by flow cytometry. (**F**) HO-1 expression in F4/80$^+$ in the peritoneal cavity of infected and control (NI) mice. Letters on histograms correspond as follows: a: NI, b: 1 dpi, c: 15 dpi, d: 21 dpi. Median fluorescence intensity is shown (MeFI) in the plot. (**G**) ROS/RNS quantification in F4/80$^+$ in the peritoneal cavity of infected and control (NI) mice using the DCFDA probe by flow cytometry. Letters on histograms correspond as follows: a: NI, b: 1 dpi, c: 15 dpi, d: 21 dpi. Median fluorescence intensity is shown (MeFI) in the plot. (**H**) Murine RAW264.7 macrophages were cultured in the presence of 75 µg/mL of FhTE or CoPP (100 µM/mL) and SnPP (50 µM/mL) overnight at 37 °C. Then, cells were collected and incubated for 30 min at 37 °C in PBS with the DCFDA probe and analyzed by flow cytometry. The RNS/ROS levels are shown as the ratio between FhTE/medium (FhTE), CoPP + FhTE/CoPP (CoPP/FhTE), and SnPP + FhTE/SnPP (SnPP/FhTE). Representative experiments are shown. Asterisks indicate significant differences with $p < 0.05$, performed by one-way ANOVA followed by Tukey's test with multiple comparisons.

3.3. HO-1 Activity Decreases the Production of ROS/RNS by F4/80$^+$ Cells and Correlates with an Increase of Splenic Regulatory CD4$^+$ T Cells Induced by F. hepatica Infection

In order to evaluate whether HO-1 interferes with the production of ROS/RNS, we treated *F. hepatica*-infected mice with the HO-1 inhibitor SnPP. SnPP treatment was associated with a decrease in the clinical signs of infected mice (Figure 3A). In addition, SnPP treatment of infected mice abrogated the increase of HO-1$^+$ cells, both in frequency and number, induced by the infection, since no significant difference was found in infected mice with respect to the control group with SnPP treatment (Figure 3B and Supplementary Figure S4). Surprisingly, a significant increase in F4/80$^+$ cell number, but not frequency, was found in both SnPP-treated and non-treated infected mice (Figure 3C and Supplementary Figure S4). Indeed, the F4/80$^+$ cell number was higher in SnPP-treated infected mice. Nevertheless, F4/80$^+$ cells of SnPP-treated infected mice produced higher levels of ROS/RNS than control infected mice (Figure 3D), although they expressed similar levels of HO-1 (Figure 3E). Of note, the expression of ICOSL in peritoneal F4/80$^+$ cells of infected mice was significantly reduced with SnPP treatment (Figure 3F). Lastly, SnPP treatment during *F. hepatica* infection did not induce an increase in the number of splenic CD4$^+$ T cells (Figure 3G) or CD4$^+$/CD25$^+$ T cells (Figure 3H), although these cells expressed higher levels of CTLA4 in the absence of SnPP treatment (Figure 3H). Altogether, these results suggest that HO-1 activity inhibited by SnPP decreases the production of ROS/RNS during fasciolosis and correlates with an increase of splenic regulatory CD4$^+$ T cells in a process that might involve ICOSL in antigen-presenting cells or CTLA4 expression in Tregs.

To complement these results, we evaluated whether the inhibition of HO-1 by SnPP treatment affected the recruitment or the phenotypical characteristics of peritoneal F4/80$^+$ cells at the early stages of *F. hepatica* infection. To this end, we analyzed F4/80$^+$ cells in the peritoneal cavity of SnPP-treated mice at 1 dpi and compared them with both non-treated infected and control mice. We observed the presence of two different cell populations according to F4/80 expression (Figure 4A and Supplementary Figure S1). SnPP treatment increased both the frequency and number of F4/80^{int} cells, while it reduced F4/80^{hi} cells in the peritoneal cavity of infected mice at 1 dpi (Figure 4B). Nevertheless, F4/80^{int} cells expressed similar levels of HO-1 (Figure 4C) and ROS/RNS (Figure 4D) in F4/80^{int} cells regardless of SnPP treatment. Interestingly, peritoneal F4/80^{int} cells expressed higher levels of CCR2 (Figure 4E), while only those from SnPP-treated infected mice expressed significantly increased levels of IL-33R (Figure 4F), which could be related to the initiation of an early immune response against the parasite. Thus, these results suggest that the presence of F4/80^{int} IL-33R$^+$ cells in the peritoneum is induced by SnPP treatment, which in turn protects mice from infection.

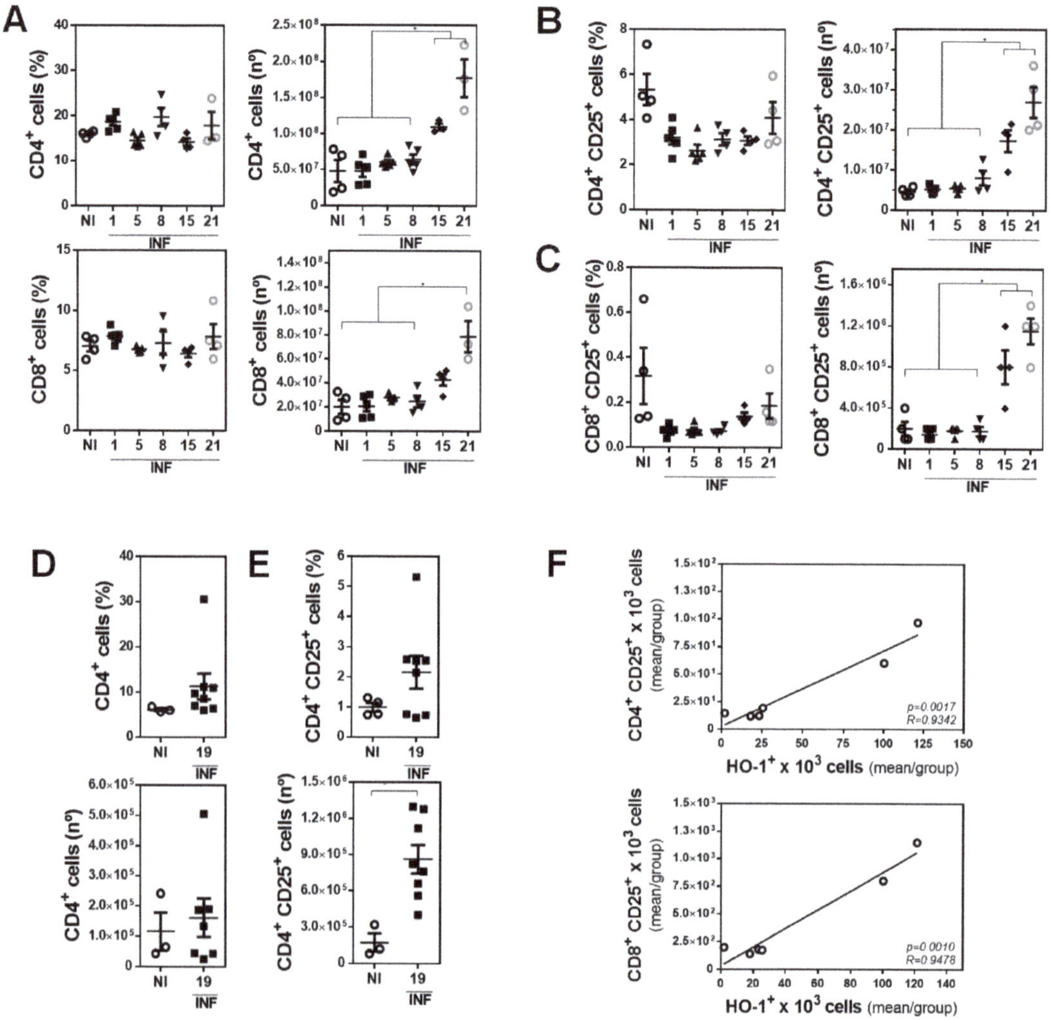

Figure 2. Peritoneal HO-1$^+$ cells correlate with increased splenic CD4$^+$ CD25$^+$ and CD8$^+$ CD25$^+$ cells during *F. hepatica* infection. Mice (n = 4–8/group) were infected with 10 metacercariae and sacrificed at 1, 8, 15, and 21 dpi. Non-infected mice were used as control (NI). (**A**) Frequency and cell number of splenic CD4$^+$ or CD8$^+$ cells from infected and control (NI) mice. (**B**,**C**) Frequency and cell number of splenic CD4$^+$ CD25$^+$ (**B**) or CD8$^+$ CD25$^+$ (**C**) cells from infected and control (NI) mice. (**D**) Frequency and cell number of hepatic CD3$^+$ CD4$^+$ cells from infected and control (NI) mice. (**E**) Frequency and cell number of hepatic CD4$^+$ CD25$^+$ cells from infected and control (NI) mice (right). (**F**) Splenic CD4$^+$ CD25$^+$ or CD8$^+$ CD25$^+$ cell number in function of the number of peritoneal HO-1$^+$ cells in NI and infected mice. The mean of CD4$^+$ CD25$^+$ (in Figure 2B), CD8 CD25$^+$ (in Figure 2C), and HO-1$^+$ cells (in Figure 1C) was plotted. Gate analyses by flow cytometry are shown in Supplementary Figures S2 and S3. The results shown represent one experiment. Asterisks indicate significant differences with $p < 0.05$, performed by one-way ANOVA followed by Tukey's test with multiple comparisons.

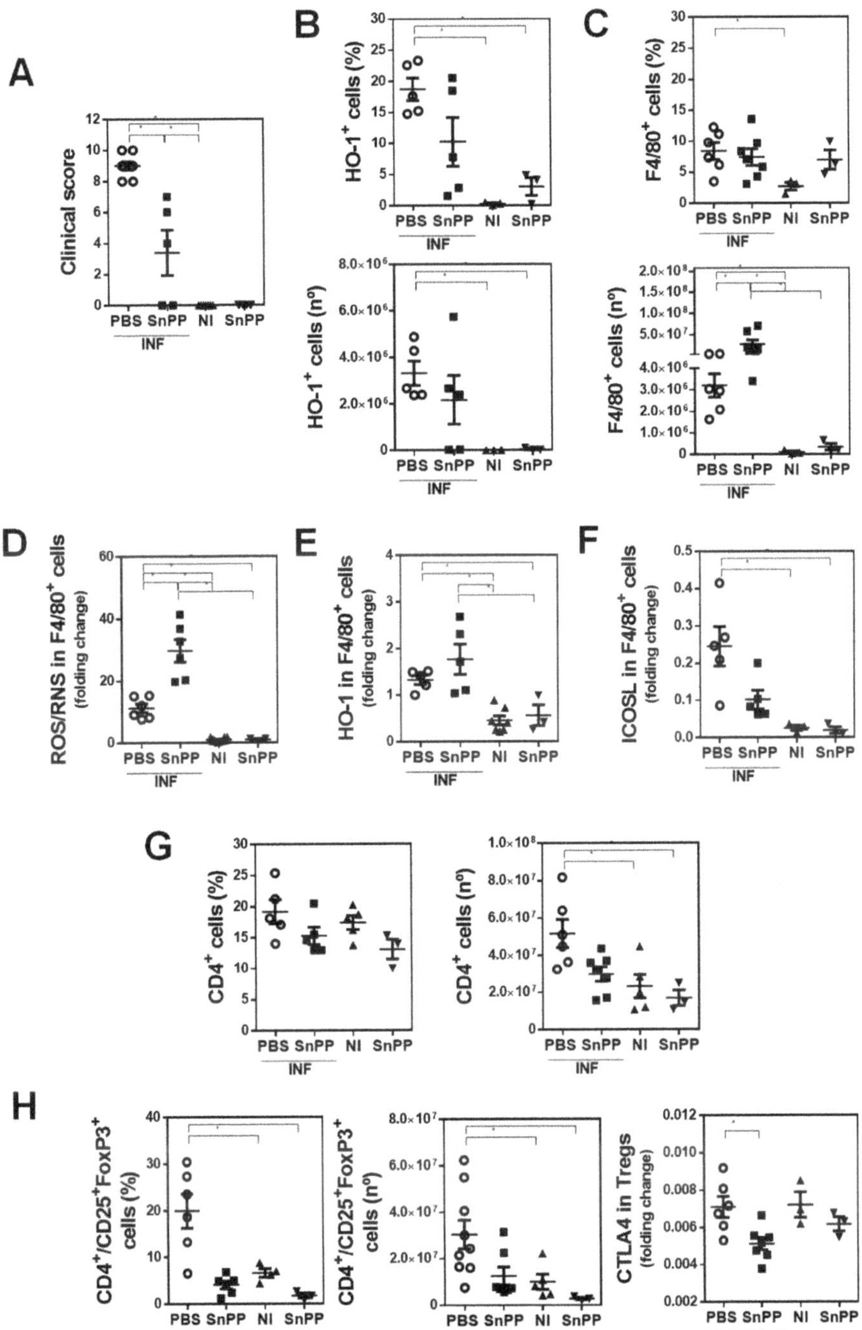

Figure 3. The HO-1 pharmacological inhibitor SnPP decreases the production of ROS/RNS by peritoneal F4/80$^+$ cells and correlates with an increase of splenic FoxP3$^+$CD25$^+$/CD4$^+$ T cells induced by *F. hepatica* infection. The HO-1 pharmacological

inhibitor SnPP decreases the production of ROS/RNS by peritoneal F4/80$^+$ cells and correlates with an increase of splenic FoxP3$^+$CD25$^+$/CD4$^+$ T cells induced by *F. hepatica* infection. Mice were injected with SnPP (40 mg/kg) or vehicle (PBS) one day before infection and every 4 days until the end of the experimental protocol. Mice (n = 5/group) were infected with 10 metacercariae (day 0) and sacrificed at 20 dpi. Non-infected (NI) mice (n = 3/group) both treated and untreated with SnPP were used as control. (**A**) Clinical signs of infected mice were analyzed to assess disease severity [17]. (**B**) Frequency and cell number of HO-1$^+$ cells in the peritoneal cavity of SnPP-treated or untreated infected and control (NI) animals by flow cytometry. (**C**) Frequency and cell number of F4/80$^+$ cells in the peritoneal cavity of SnPP-treated or untreated infected and control (NI) animals by flow cytometry. (**D**) ROS/RNS quantification in F4/80$^+$ cells of the peritoneal cavity of SnPP-treated or untreated infected and control (NI) mice using the DCFDA probe by flow cytometry. HO-1 (**E**) and ICOSL (**F**) expression in F4/80$^+$ cells in the peritoneal cavity of SnPP-treated or untreated infected and control (NI) mice. (**G**) Splenic CD4$^+$ T cell frequency and number from SnPP-treated or untreated infected and control (NI) mice. (**H**) Splenic CD4$^+$/CD25$^+$FoxpP3$^+$ cell frequency and number from SnPP-treated or untreated infected and control (NI) mice (left). CTLA4 expression in splenic CD4$^+$/CD25$^+$FoxpP3$^+$ cells from SnPP-treated or untreated infected and control (NI) mice. Gate analyses by flow cytometry are shown in Supplementary Figures S2 and S4. Results obtained for one representative experiment are shown. Asterisks indicate significant differences with $p < 0.05$, performed by one-way ANOVA followed by Tukey's test with multiple comparisons.

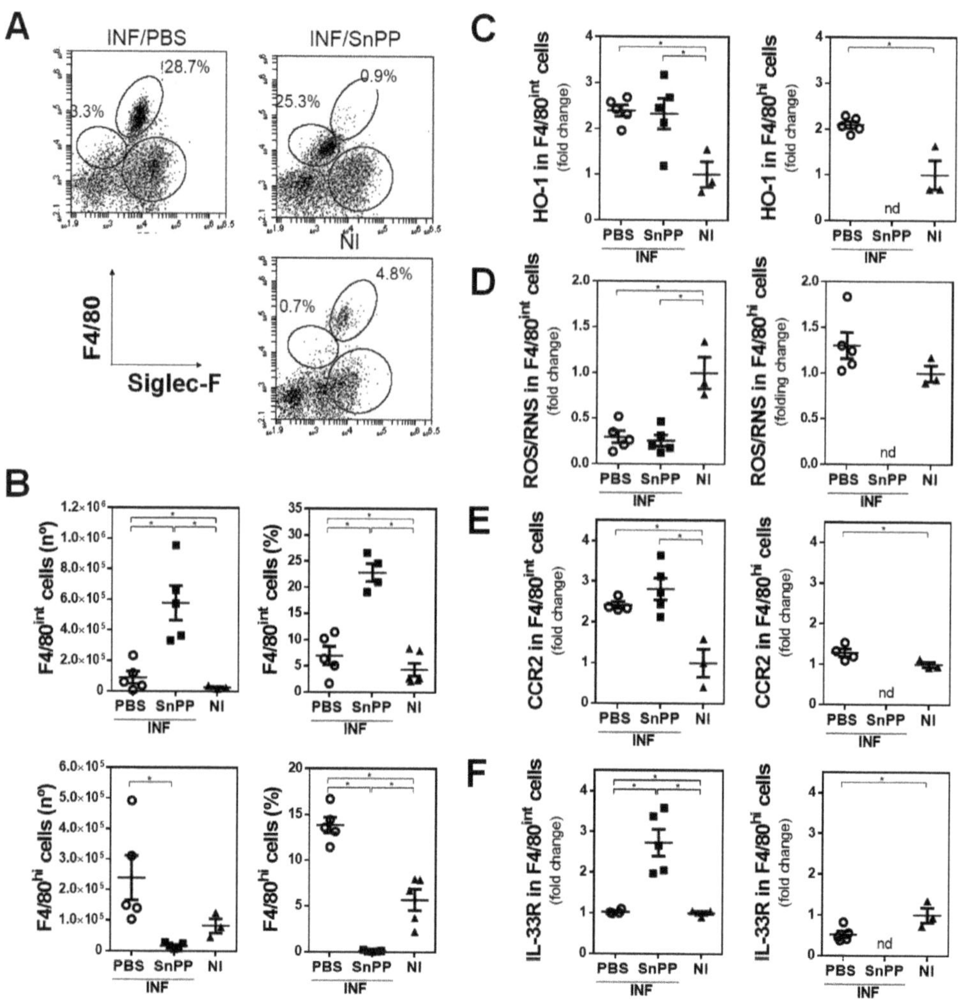

Figure 4. Analyses of peritoneal F4/80$^+$ cells at the early events of *F. hepatica* infection. Mice (n = 5/group) were injected with

SnPP (40 mg/kg) or vehicle (PBS) one day before infection, infected with 10 metacercariae (day 0), and sacrificed at 1 dpi. Non-infected (NI) mice (n = 3/group) both treated and untreated with SnPP were used as a control. (**A**) Analysis of F4/80^{int} and F4/80^{hi} cells in PEC from SnPP-treated or untreated infected and control (NI) mice by flow cytometry. (**B**) Frequency and cell number of F4/80^{int} (upper plots) or F4/80^{hi} (lower plots) cells in the peritoneal cavity of SnPP-treated or untreated infected and control (NI) animals. HO-1 (**C**), ROS/RNS (**D**), CCR2 (**E**), and IL-33R (**F**) expression in peritoneal F4/80^{int} and F4/80^{hi} cells by flow cytometry. "nd" means none detected, since barely any F4/80^{hi} cells in SnPP-treated infected mice were detected. Median fluorescence intensity is shown (MeFI) in the plot. Gate analyses by flow cytometry are shown in (**A**) and Supplementary Figure S1. Representative experiments are shown. Asterisks indicate significant differences with $p < 0.05$, performed by one-way ANOVA followed by Tukey's test with multiple comparisons.

3.4. The Inhibition of HO-1 Activity by SnPP Controls the Gene Expression of Antioxidant Molecules

To deeply analyze the relationship between HO-1 expression in peritoneal cells induced by *F. hepatica* infection with the production of ROS/RNS, the gene expression of different molecules involved in the oxidative and antioxidative responses was evaluated. PECs of SnPP-treated infected mice were characterized by a significant decrease in the mRNA levels of the transcription factor *nrf2* (Figure 5A). Moreover, the SnPP-induced decrease in *nrf2* gene expression levels was associated with decreased mRNA levels in the antioxidant enzymes *catalase*, glutathione peroxidase 2 (*gpx2*), and superoxide dismutase 2 (*sod2*) (Figure 5B). However, no differences were found in the gene expression of *gpx1*, while an increase in *sod1* expression was observed (Figure 5B). Finally, an unexpected decrease in the mRNA levels of the NADPH oxidase subunits *gp91^{phox}* and *p47^{phox}* was observed with SnPP treatment of infected mice (Figure 5C).

3.5. Deficiency of Functional NADPH Oxidase Partially Protects Mice from Liver Damage Induced by F. hepatica and Limits the Production of IL-10

Considering the fact that SnPP treatment protected mice from parasite infection and that reduced levels of *gp91^{phox}* mRNA were found in PECs from infected mice, we analyzed the infection in gp91^{phox} knockout mice. gp91^{phox} deficiency reduced the clinical signs and liver damage induced by *F. hepatica* infection (Figure 6A and Supplementary Figure S5). This partial protection was associated with a lower increase of HO-1^{+} peritoneal cells, both in frequency and number (Figure 6B). Moreover, HO-1^{+} cells from gp91^{phox} knockout infected mice expressed lower levels of MHCII (Figure 6C) and CD40 (Figure 6D), but not CD80 (Figure 6E), than those from wildtype mice, indicating that NADPH oxidase may play a role both in the immune response and the pathogenesis induced by *F. hepatica* infection. Further characterization of the peritoneal cells from these mice indicated that the increase of F4/80^{+} cells was abrogated in the absence of gp91^{phox} (Figure 7A), and as expected, very low levels of ROS/RNS produced by these cells (Figure 7B). Additionally, these cells expressed lower levels of Sirpα (Figure 7C), ICOSL (Figure 7D), and IL-10 (Figure 7E). Finally, lower numbers of CD4^{+} (Figure 7F and Supplementary Figure S6) and CD4^{+}/CD25^{+}FoxP3^{+} Tregs (Figure 7G and Supplementary Figure S6) were found in gp91^{phox} knockout infected mice with respect to wildtype mice. However, no significant differences in the production of IFNγ by splenocytes stimulated with parasite components (FhTE) between gp91^{phox} knockout and wildtype infected mice were detected (Figure 7H).

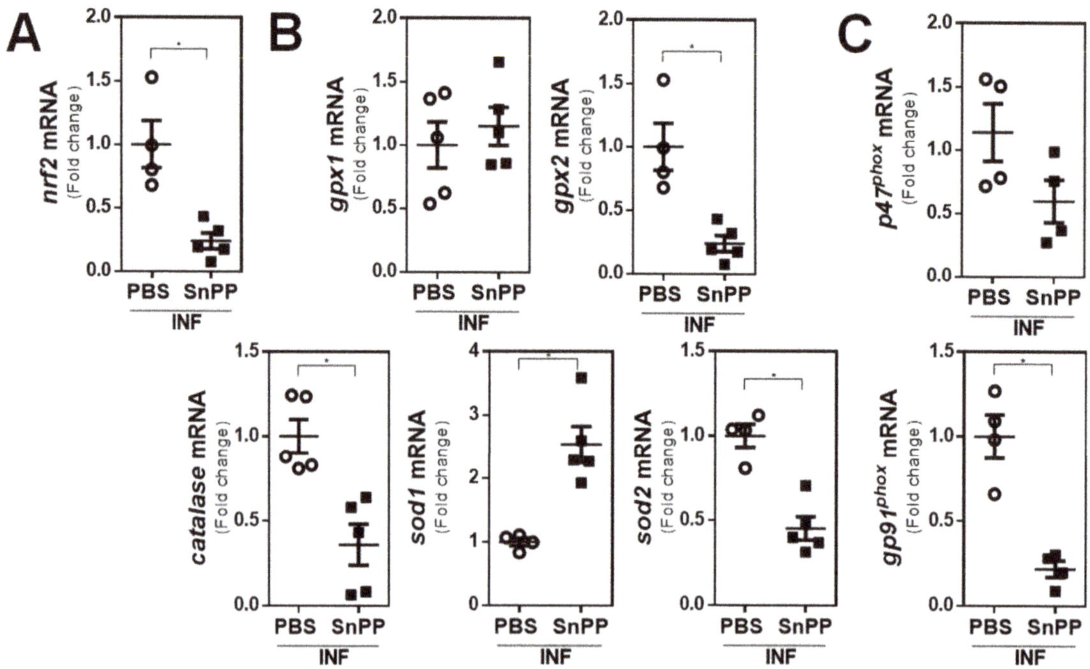

Figure 5. mRNA levels of antioxidative and oxidative genes in the infection by *F. hepatica*. Mice ($n = 5$/group) were injected with SnPP (40 mg/kg) or PBS 1 day before infection, infected with 10 metacercariae (day 0), and sacrificed at 20 dpi. *nrf2* (**A**), *catalase*, *gpx1*, *gpx2*, *sod1*, *sod2* (**B**), and *p47phox* and *gp91phox* (**C**) gene expression in PECs from SnPP-treated and control infected mice evaluated by qRT-PCR. mRNA levels were analyzed by qRT-PCR with respect to *gapdh* expression in PECs from SnPP-treated and control infected mice (PBS). Results were compared to the infected (control) group of mice and represented as the ratio between gene expression in SnPP-treated and control mice. Asterisks indicate significant differences with $p < 0.05$, performed by the Student's *t*-test.

3.6. IL-10 Signaling Is Crucial for HO-1 Expression in F. hepatica-Infected Mice

Considering the fact that IL-10 induces HO-1 expression that can favor the production of IL-10, and that IL-10 is crucial for Treg differentiation [36], we analyzed whether there was a relationship between IL-10 signaling and HO-1 expression during *F. hepatica* infection. To this end, we treated infected mice with a neutralizing antibody of IL-10 receptor (IL-10R). The results demonstrate that IL-10R blocking reduced the clinical signs associated with parasite infection (Figure 8A). Although the recruitment of F4/80$^+$ cells in the peritoneum of infected mice was not affected by IL-10R neutralization (Figure 8B), it abrogated the elevated expression of HO-1 induced by *F. hepatica* infection (Figure 8C). Interestingly, IL-10R neutralization reduced the frequency, but not the number, of CD4$^+$ (Figure 8D and Supplementary Figure S7) and CD4$^+$CD25$^+$ (Figure 8E and Supplementary Figure S7) T cells in the spleens of infected animals. Altogether, these results indicate that IL-10 signaling is essential for HO-1 expression of F4/80$^+$ cells during *F. hepatica* infection, likely affecting the differentiation of regulatory T cells.

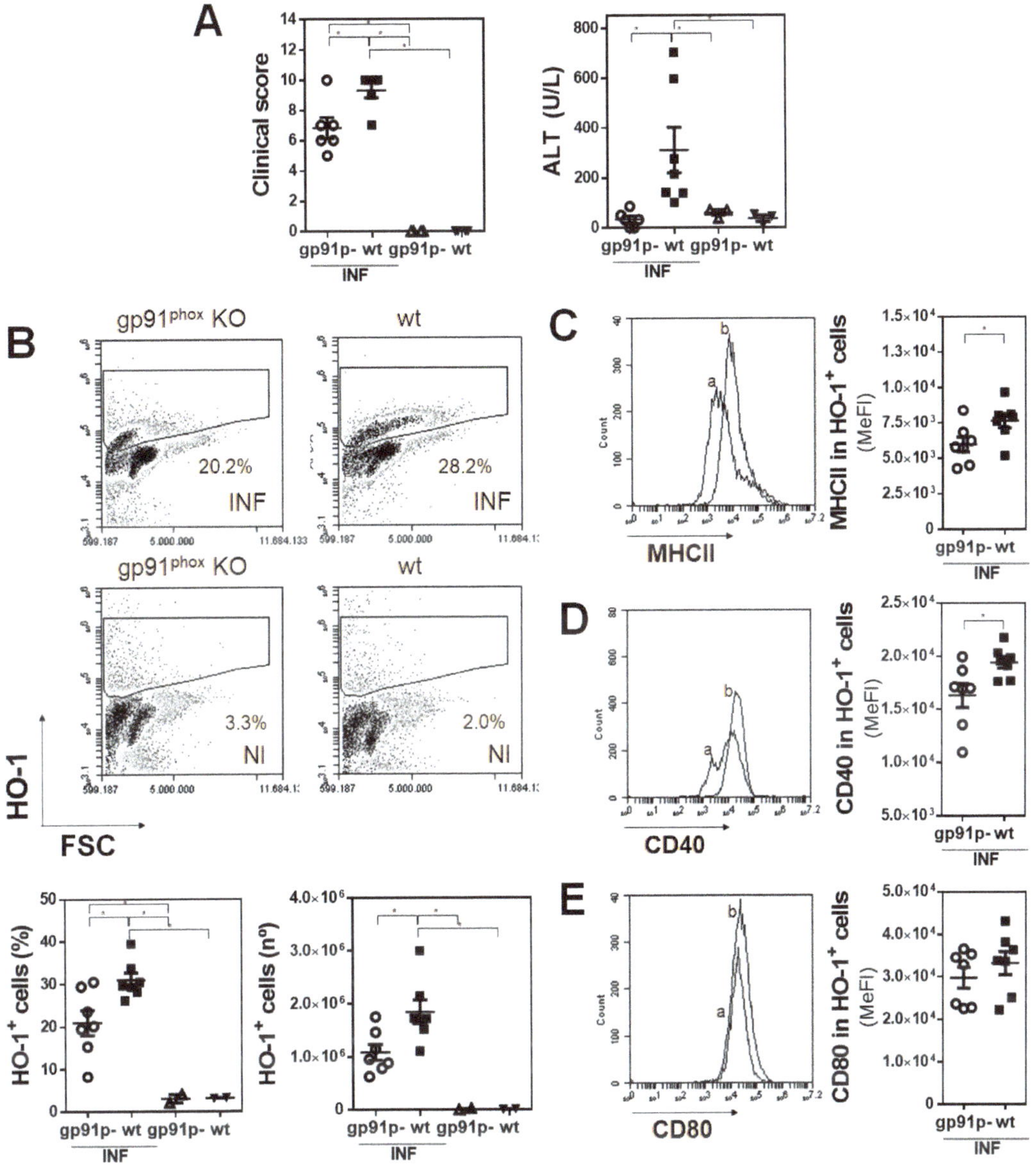

Figure 6. Gp91^{phox} (gp91p-) knockout (n = 6–8/group) and littermate control (n = 2/group) mice were infected with 10 metacercariae and sacrificed at 20 dpi. Non-infected mice were used as a control. (**A**) Clinical signs analyzed to assess disease severity [17]. Quantification of liver damage by ALT activity in sera. (**B**) Flow cytometry analysis of HO-1$^+$ in PECs from gp91^{phox} knockout mice and littermate controls (upper panel). (**C**) Frequency and cell number of HO-1$^+$ cells in the peritoneal cavity of infected and non-infected animals by flow cytometry (lower panel). MHCII (**C**), CD40 (**D**), and CD80 (**E**) expression in HO-1$^+$ cells in the peritoneal cavity of infected and control mice. Letters on histograms correspond as follows: a: infected gp91^{phox}, b: infected wildtype littermates. Median fluorescence intensity is shown (MeFI) in the plot. Asterisks indicate significant differences with $p < 0.05$, performed by one-way ANOVA followed by Tukey's test with multiple comparisons.

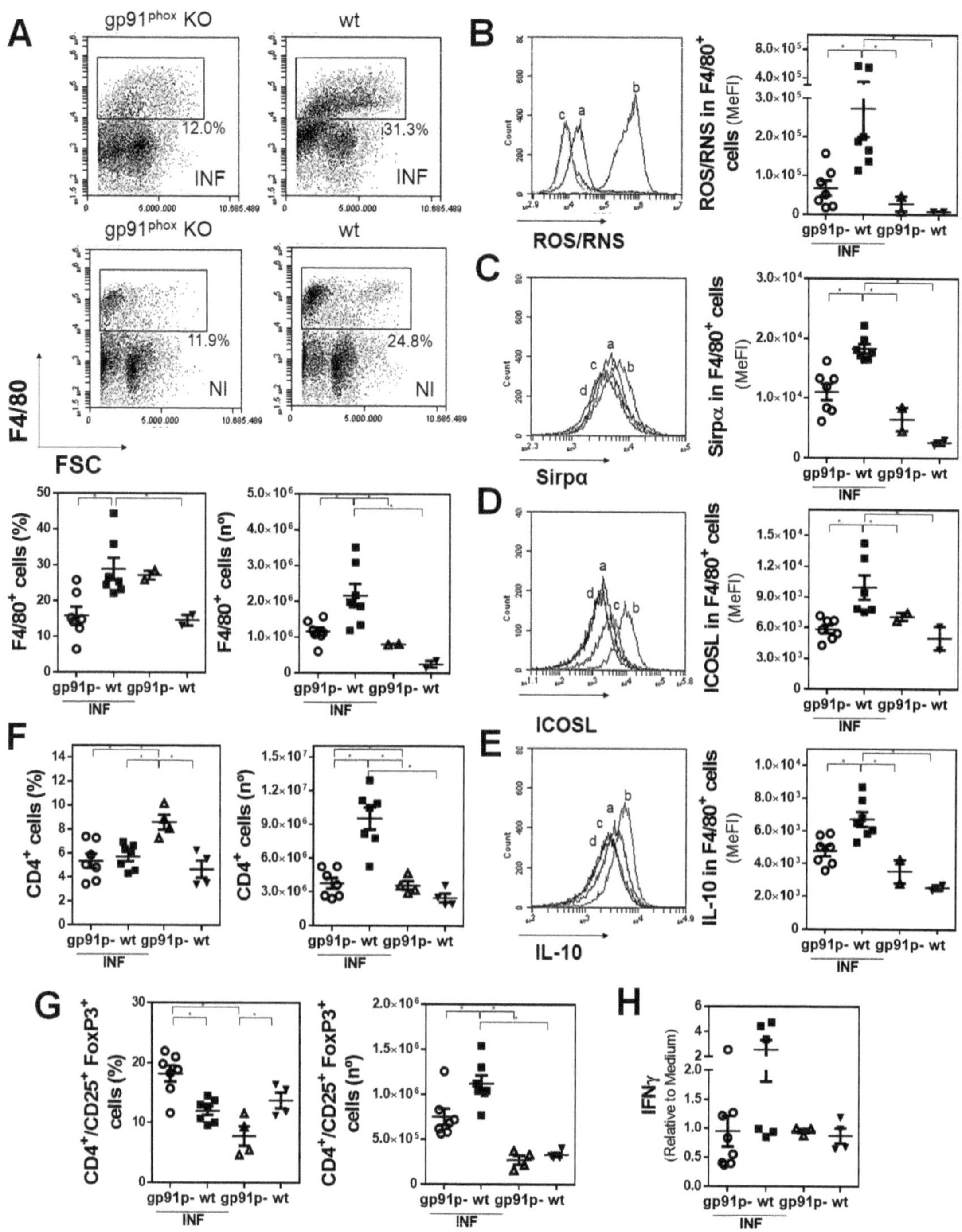

Figure 7. Peritoneal F4/80⁺ cells from wildtype infected mice present a regulatory-like phenotype. Gp91^{phox} (gp91p-) knockout

($n = 7$–8/group) and littermate control ($n = 2$–4/group) mice were infected with 10 metacercariae and sacrificed at 20 dpi. Non-infected mice were used as a control. (**A**) Flow cytometry analysis of F4/80$^+$ cells in PEC from gp91^{phox} knockout mice and littermate controls (upper panel). Frequency and cell number of F4/80$^+$ cells in the peritoneal cavity of infected and non-infected animals by flow cytometry (lower plots). ROS/RNS (**B**), Sirpα (**C**), ICOSL (**D**), and IL-10 (**E**) expression in F4/80$^+$ cells in the peritoneal cavity of infected and uninfected mice. Letters on histograms correspond as follows: a: infected gp91^{phox}, b: infected littermates, c: uninfected gp91^{phox}, and d: uninfected littermates. Median fluorescence intensity is shown (MeFI) in the plot. (**F**) Frequency and cell number of splenic CD4$^+$ cells from infected and non-infected mice. (**G**) Frequency and cell number of splenic CD4$^{+/}$CD25$^+$ Foxp3$^+$ cells from infected and non-infected mice. Gate analyses by flow cytometry are shown in (**A**) and Supplementary Figure S1. (**H**) IFNγ levels in culture supernatants of splenocyte proliferation assay cultured with FhTE for 5 days at 37 °C. The results shown represent one experiment. Asterisks indicate significant differences with $p < 0.05$, performed by one-way ANOVA followed by Tukey's test with multiple comparisons.

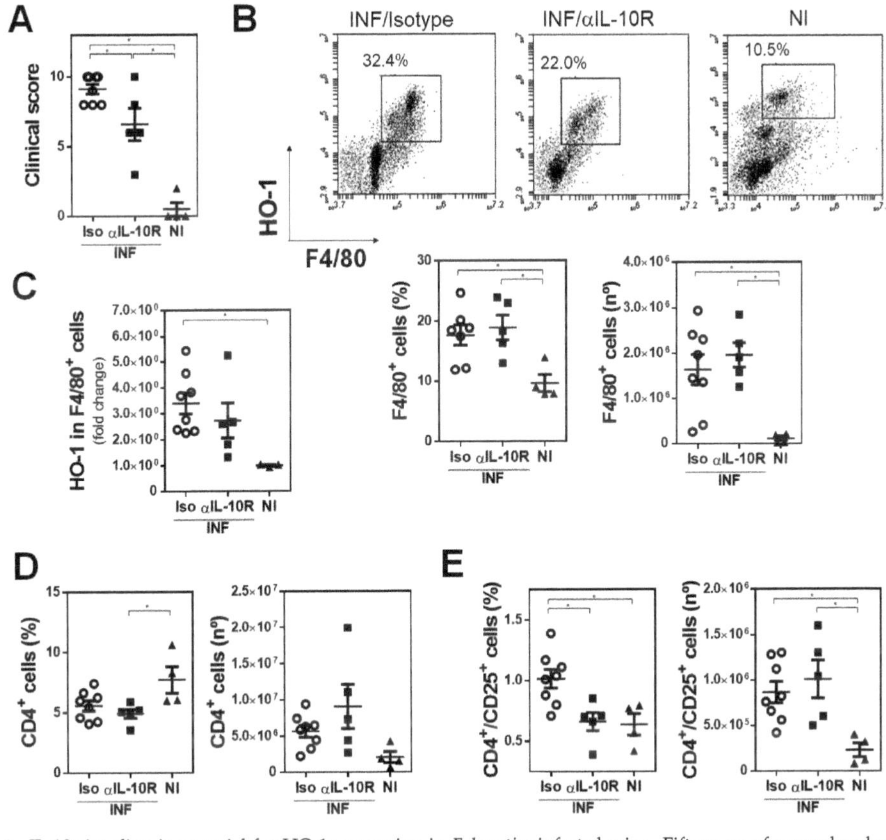

Figure 8. IL-10 signaling is essential for HO-1 expression in *F. hepatica*-infected mice. Fifteen μg of monoclonal rat IgG$_{2a}$ anti-IL-10R antibody was administrated by intraperitoneal injection the day before and after infection with *F. hepatica* and every 3 days until sacrifice ($n = 5$–8/group). The control group ($n = 4$/group) received an isotype control antibody. At day 20 post-infection, animals were sacrificed and splenocytes were analyzed by flow cytometry. (**A**) Clinical signs were analyzed to assess disease severity [17]. (**B**) Analysis by flow cytometry of F4/80$^+$ cells in PEC from infected and non-infected (NI) mice showing frequency and number of F4/80$^+$ cells in PECs. (**C**) HO-1 expression in F4/80$^+$ cells in the peritoneal cavity of infected and uninfected mice. (**D**) Frequency and cell number of CD4$^+$ T cells in spleens from infected and non-infected (NI) mice. (**E**) Frequency and cell number of CD4$^+$ CD25$^+$ T cells in spleens from infected and non-infected (NI) mice. Gate analyses by flow cytometry are shown in (**B**) and Supplementary Figure S7. Representative results of one representative are shown. Asterisks indicate significant differences with $p < 0.05$, performed by one-way ANOVA followed by Tukey's test with multiple comparisons.

4. Discussion

In this work, we have examined the cellular and molecular mechanisms that govern the expansion or differentiation of Tregs induced by HO-1$^+$ cells in *F. hepatica* infection. We presented evidence showing that HO-1 activity results in decreased ROS/RNS production by F4/80$^+$ antigen-presenting cells, thereby enhancing the pathological effects caused by *F. hepatica* and promoting parasite infection. Furthermore, apart from its antioxidant capacity, HO-1 has other functions, such as its immunoregulatory properties and controlling gene expression as a transcription factor [14,21,26,32,38]. Indeed, HO-1 inhibition promotes IFNγ- and NOS2-mediated control of *M. tuberculosis* infection in mice [39]. Furthermore, it has been previously reported that HO-1 has a role in suppressing pro-inflammatory Th1 immune responses in experimental colitis, and sickle cell alloimmunization has been reported, and it protects from atherosclerosis [40,41]. Finally, HO-1 can impair the immunity against other pathogens, such as *Plasmodium yoelii* [42].

Indeed, we demonstrated that during *F. hepatica* experimental infection in mice, there is an increase in the expression of HO-1 in F4/80$^+$ cells in the peritoneal cavity and it inversely correlates with ROS/RNS production. Furthermore, we demonstrated an association between the expression of HO-1 and the presence of putative Tregs in the spleens of infected animals (Figure 9A). These results were also confirmed when using the HO-1 inhibitor SnPP, which inhibits its enzymatic activity. At first sight, the inhibition of HO-1 activity by SnPP would suggest that its effects are caused by the heme-catabolizing activity rather than by its expression and function as a transcription factor. Indeed, F4/80$^+$ peritoneal cells from SnPP-treated mice did not show a decrease in HO-1 expression, although a significant increase in ROS/RNS production was detected. SnPP is a metalloporphyrin that acts as a competitive inhibitor of HO-1 both in vitro and in vivo. Its efficiency can be explained by its higher binding affinity to HO-1/2 than to heme [43,44]. However, enzymatically inactive HO-1 can still mediate protection against hydrogen peroxide-induced toxicity, probably by promoting the gene expression of antioxidant proteins [14,45], although the mechanisms underlying these effects are still unclear. Thus, the possibility that HO-1 would act as transcription factor cannot be discarded, since the nuclear localization of HO-1 in F4/80$^+$ cells derived from *F. hepatica*-infected mice with or without SnPP treatment was not investigated. Furthermore, it is unlikely that the protective outcome of SnPP treatment represents a direct effect on *F. hepatica*, since the degree of infection and pathological effects induced by the parasite were also related to an increase in Tregs, evidencing that HO-1 activity influences the host adaptive immunity in vivo. Indeed, our results indicate that the increase of the mRNA levels of *nrf2*, a transcription factor responsible for the regulation of cellular redox balance and protecting antioxidant responses [46,47], is accompanied by an increase in some antioxidant enzyme genes, demonstrating that the infection, HO-1, Tregs, and the Nrf2 master regulator comprise a complex axis of antioxidant and immunoregulatory properties in *F. hepatica* infection. However, the function of these enzymes should be determined in order to confirm their antioxidant role during *F. hepatica* infection. On the other hand, heme-activated murine macrophages have functional anti-inflammatory features that are dependent on the enzymatic activity of HO-1 [38]. Thus, the immunoregulatory and immunosuppressive properties of HO-1 together with its antioxidant properties demonstrate that its function during *F. hepatica* infection goes far beyond heme degradation itself.

The role of ROS/RNS in helminth parasite killing is still controversial. Some reports showed that the infection by *Strongyloides papillosus* induced an oxidative/nitrosative stress in sheep [48], although its effect on the parasite itself has not been demonstrated. On the other hand, *Schistosome* infection relates to an immense oxidative stress by the host that is not sufficient to control infection [49]. Further data demonstrated that excretory/secretory factors from *Mesocestoides corti* inhibit ROS-induced neutrophil extracellular traps, showing that the parasite could use this mechanism to attenuate the effects induced by ROS [50]. It should be highlighted, however, that although oxidative mechanisms are induced by helminth parasite infections, their detrimental role in the parasite itself as well as in the host surroundings is not well-established yet [51–53]. A recent report has demonstrated a high oxidative status in serum and liver in rabbits infected with *F. gigantica*, together with a decline in the SOD and catalase gene expression and enzyme activity in sera from infected animals [54], which is not in agreement with data from our work in *F. hepatica* experimentally infected mice. However, the authors came to the conclusion that the disruption of antioxidant and detoxification cascades by *F. gigantica* likely leads to the pathogenic response from the host [54].

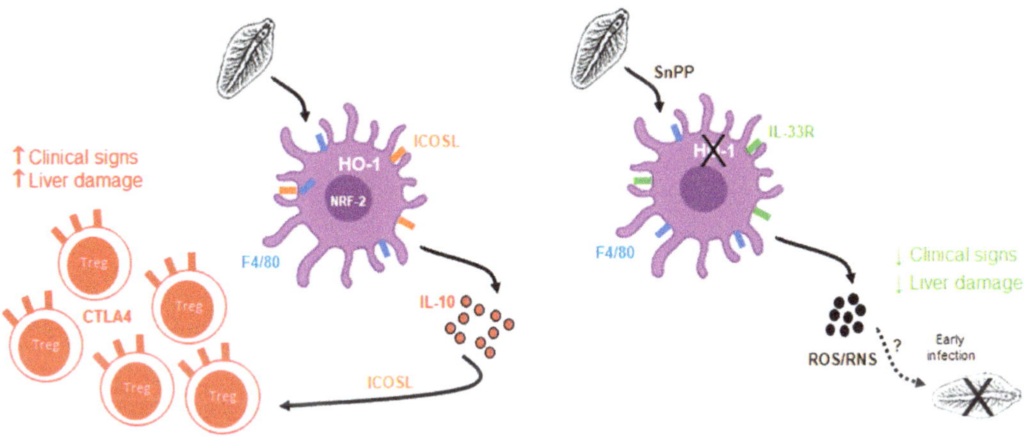

(**A**) (**B**)

Figure 9. General hypothesis. (**A**) ICOSL$^+$ F4/80$^+$ APC express HO-1 induced by parasite infection that promotes IL-10 production and differentiation or expansion of CTLA4$^+$ Tregs. (**B**) SnPP treatment inhibiting HO-1 activity in IL-33R$^+$ F4/80$^+$ APC allows ROS/RNS production, that induce parasite damage in early stages of the infection.

It is worth noting that in our work, we used a DCFDA fluorescent probe that does not distinguish between ROS and RNS. Therefore, these studies should be complemented with others using ROS-specific probes such as DHE or specific inhibitors of nitric oxide production (such as L-Name). In order to analyze the ROS produced by NADPH-oxidase, we used, instead, gp91phox knockout mice. Interestingly, the fact that mice that are deficient in NADPH oxidase function, with a considerable decrease in ROS production, were partially protected against *F. hepatica* infection, suggests that the moment when ROS is produced by NADPH oxidase might be crucial to limit *F. hepatica*-induced damage (Figure 9B). Indeed, an exacerbated ROS production induced by a pro-inflammatory immune response can be detrimental to leukocyte cell function or viability and induced damage to the immune system [54]. Thus, a prolonged and not regulated production of ROS by F4/80$^+$ cells could benefit the parasite, and not the host. Of note, these cells expressed higher levels of ICOSL and IL-10 than those from gp91^{phox} knockout mice, which could be associated with the differentiation or expansion of a higher number of splenic Tregs, which in turn express higher levels of CTLA4. Indeed, both ICOSL [55,56] and CTLA4 [57] are key mediators of Treg differentiation. In the same line, macrophages can suppress T cell responses and favor the expansion of Tregs [58]. Furthermore, ROS levels on T cell activation seem to be important, since small quantities of ROS result in antigen hypo-responsiveness, while high doses lead to oxidative stress-induced apoptosis [59]. Further analysis of the role of IL-10 produced by antigen-presenting cells in the differentiation or expansion of Tregs showed that IL-10 signaling is essential to increase HO-1 expression in peritoneal F4/80$^+$ cells and likely the production of Tregs. Interestingly, it would seem that the parasite exploits the host defense mechanisms, on the one hand by recruiting HO-1$^+$ cells with less antioxidative functions that produce IL-10, and on the other hand by in turn inducing the differentiation to Tregs. Nevertheless, the production of IL-10 by the host would also protect host cells in the acute pro-inflammatory immune response, caused either by damage induced by the parasite in the early state of the infection or by liver damage, at least in this experimental model. However, more experiments are needed in order to confirm these results, and to determine the role of ROS in the induction of Tregs and its relationship with IL-10.

One hypothesis that can explain these results might be the fact that ROS/RNS production is (partially) effective only during early stages of *F. hepatica* infection (Figure 9B). After ingestion of metacercariae by the mammalian host, juvenile flukes penetrate the host intestine wall and reach the liver through the peritoneal cavity between 4 and 6 days in livestock, although it is thought that it takes around 24 h in mice [3]. To further understand the early events that take place during *F. hepatica* infection in mice, we analyzed HO-1 expression and F4/80^{+} cell recruitment at 1 dpi, finding that two different populations expressing different levels of F4/80 are present in the peritoneum, and those elicited in SnPP-treated mice expressed higher levels of IL33R (Figure 9B). IL33 is an alarmin that participates in the type 2 innate immune response, promoting innate lymphoid cells type 2. However, during *Schisotosoma* infection, IL-33 seems to contribute to the development of pathology via the induction of type 2 innate lymphoid cells and alternative activation of macrophages, thus favoring the infection [60–62]. Therefore, the functions of IL-33 during *F. hepatica* infection in mice, and in particular the overexpression of its receptor in antigen-presenting cells at the early events of the infection, remain to be elucidated.

In conclusion, our work showed that HO-1 is a key molecule that favors *F. hepatica* infection, by which HO-1 could control ROS/RNS production and Treg differentiation and how the parasite elicits/triggers these mechanisms. Altogether, these results contribute to the elucidation of the immunoregulatory and antioxidant roles of HO-1 induced by *F. hepatica* in the host, providing interesting checkpoints that might control fasciolosis.

Supplementary Materials: The following are available online at https://www.mdpi.com/article/10.3390/antiox10121938/s1, Figure S1: Gates used for flow cytometry analyses of PECs from infected and control mice corresponding to Figures 1B,D and 4A; Figure S2: Gates used for flow cytometry analyses of splenocytes from infected and control mice corresponding to Figures 2A–C and 3G,H; Figure S3: Gates used for flow cytometry analyses of hepatic leukocytes from infected and control mice corresponding to Figure 2D,E; Figure S4: Gates used for flow cytometry analyses of splenocytes from infected and control mice corresponding to Figure 3B,C; Figure S5: Representative images of livers from gp91^{phox} KO and C57BL/6 liitermates infected and control mice; Figure S6: Gates used for flow cytometry analyses of splenocytes from infected and control mice corresponding to Figure 7F,G; Figure S7: Gates used for flow cytometry analyses of splenocytes from IL 10R or isotype treated infected and control mice corresponding to Figure 8D,E.

Author Contributions: M.C. performed all the experiments, with the exception of the anti-IL-10R experiment, and wrote the original draft of the manuscript; V.d.C. and S.F. performed biological sample collection and flow cytometry analyses, and reviewing the manuscript; M.F.F., M.L., S.A.R.-Z. and P.L. helped with cell cultures, and performed flow cytometry analyses and interpretation; M.F.F. critically reviewed the manuscript; P.C. performed the IL-10R experiment, conceptualization, and reviewed the manuscript; T.F. performed conceptualization, experiment design and supervision, analyses of data, and writing, review, and editing of the manuscript. All authors have read and agreed to the published version of the manuscript.

Funding: This research was funded by Programa de Desarrollo de Ciencias Básicas (PEDECIBA), Agencia Nacional de Investigación e Innovación (SNI-ANII, FCE_1_2017_1_136094 and FCE_1_2019_1_156295) to Teresa Freire.

Institutional Review Board Statement: Animal experimentation was carried out according to the International Guiding Principles for Biomedical Research Involving Animals, as issued by the Council for the International Organizations of Medical Sciences. Procedures involving animals were approved by the Universidad de la República's Committee on Animal Research (Comisión Honoraria de Experimentación Animal, CHEA Protocol Number 07153-000817-18).

Informed Consent Statement: Not applicable.

Data Availability Statement: Data is contained within the article and supplementary materials.

Acknowledgments: We are grateful to Unidad de Reactivos Biológicos de experimentación at Facultad de Medicina, UdelaR, for animal housing and care and to Ignacio Anegón for his help.

Conflicts of Interest: The authors declare no conflict of interest.

References

1. Mas-Coma, S.; Valero, M.A.; Bargues, M.D. Fascioliasis. *Adv. Exp. Med. Biol.* **2019**, *1154*, 71–103.
2. Cwiklinski, K.; O'Neill, S.M.; Donnelly, S.; Dalton, J.P. A prospective view of animal and human Fasciolosis. *Parasite Immunol.* **2016**, *38*, 558–568. [CrossRef]
3. Moazeni, M.; Ahmadi, A. Controversial aspects of the life cycle of Fasciola hepatica. *Exp. Parasitol.* **2016**, *169*, 81–89. [CrossRef] [PubMed]
4. Rodríguez, E.; Carasi, P.; Frigerio, S.; Da Costa, V.; van Vliet, S.; Noya, V.; Brossard, N.; Van Kooyk, Y.; García-Vallejo, J.J.; Freire, T. Fasciola hepatica Immune Regulates CD11c+ Cells by Interacting with the Macrophage Gal/GalNAc Lectin. *Front. Immunol.* **2017**, *8*, 264. [CrossRef] [PubMed]
5. Dowling, D.J.; Hamilton, C.M.; Donnelly, S.; La Course, J.; Brophy, P.M.; Dalton, J.; O'Neill, S.M. Major Secretory Antigens of the Helminth Fasciola hepatica Activate a Suppressive Dendritic Cell Phenotype That Attenuates Th17 Cells but Fails to Activate Th2 Immune Responses. *Infect. Immun.* **2010**, *78*, 793–801. [CrossRef]
6. Adams, P.N.; Aldridge, A.; Vukman, K.V.; Donnelly, S.; O'Neill, S.M. Fasciola hepatica tegumental antigens indirectly induce an M2 macrophage-like phenotype in vivo. *Parasite Immunol.* **2014**, *36*, 531–539. [CrossRef]
7. Walsh, K.P.; Brady, M.T.; Finlay, C.M.; Boon, L.; Mills, K.H. Infection with a helminth parasite attenuates autoimmunity through TGF-beta-mediated suppression of Th17 and Th1 responses. *J. Immunol.* **2009**, *183*, 1577–1586. [CrossRef] [PubMed]
8. Donnelly, S.; Stack, C.M.; O'Neill, S.M.; Sayed, A.A.; Williams, D.L.; Dalton, J.P. Helminth 2-Cys peroxiredoxin drives Th2 responses through a mechanism involving alternatively activated macrophages. *FASEB J.* **2008**, *22*, 4022–4032. [CrossRef]
9. Flynn, R.J.; Mannion, C.; Golden, O.; Hacariz, O.; Mulcahy, G. Experimental Fasciola hepatica Infection Alters Responses to Tests Used for Diagnosis of Bovine Tuberculosis. *Infect. Immun.* **2007**, *75*, 1373–1381. [CrossRef]
10. O'Neill, S.M.; Brady, M.T.; Callanan, J.; Mulcahy, G.; Joyce, P.; Mills, K.; Dalton, J.P. Fasciola hepatica infection downregulates Th1 responses in mice. *Parasite Immunol.* **2000**, *22*, 147–155. [CrossRef]
11. Joardar, N.; Mondal, C.; Babu, S.P.S. A review on the interactions between dendritic cells, filarial parasite and parasite-derived molecules in regulating the host immune responses. *Scand J. Immunol.* **2021**, *93*, e13001. [CrossRef] [PubMed]
12. Zanna, M.; Yasmin, A.; Omar, A.; Arshad, S.; Mariatulqabtiah, A.; Nur-Fazila, S.; Mahiza, I.N. Review of Dendritic Cells, Their Role in Clinical Immunology, and Distribution in Various Animal Species. *Int. J. Mol. Sci.* **2021**, *22*, 8044. [CrossRef]
13. Gordon, S.; Plüddemann, A. Tissue macrophages: Heterogeneity and functions. *BMC Biol.* **2017**, *15*, 1–18. [CrossRef] [PubMed]
14. Vijayan, V.; Wagener, F.A.; Immenschuh, S. The macrophage heme-heme oxygenase-1 system and its role in inflammation. *Biochem. Pharmacol.* **2018**, *153*, 159–167. [CrossRef] [PubMed]
15. Davies, L.C.; Jenkins, S.J.; Allen, J.E.; Taylor, P.R. Tissue-resident macrophages. *Nat. Immunol.* **2013**, *14*, 986–995. [CrossRef]
16. Jenkins, S.J.; Allen, J.E. The expanding world of tissue-resident macrophages. *Eur. J. Immunol.* **2021**, *51*, 1882–1896. [CrossRef]
17. Carasi, P.; Rodríguez, E.; Da Costa, V.; Frigerio, S.; Brossard, N.; Noya, V.; Robello, C.; Anegón, I.; Freire, T. Heme-Oxygenase-1 Expression Contributes to the Immunoregulation Induced by Fasciola hepatica and Promotes Infection. *Front. Immunol.* **2017**, *8*, 883. [CrossRef]
18. Greil, J.; Verga-Falzacappa, M.V.; Echner, N.E.; Behnisch, W.; Bandapalli, O.R.; Pechanska, P.; Immenschuh, S.; Vijayan, V.; Balla, J.; Tsukahara, H.; et al. Mutating heme oxygenase-1 into a peroxidase causes a defect in bilirubin synthesis associated with microcytic anemia and severe hyperinflammation. *Haematologica* **2016**, *101*, e436–e439. [CrossRef]
19. Tzima, S.; Victoratos, P.; Kranidioti, K.; Alexiou, M.; Kollias, G. Myeloid heme oxygenase–1 regulates innate immunity and autoimmunity by modulating IFN-β production. *J. Exp. Med.* **2009**, *206*, 1167–1179. [CrossRef]
20. Hull, T.; Agarwal, A.; George, J.F. The Mononuclear Phagocyte System in Homeostasis and Disease: A Role for Heme Oxygenase-1. *Antioxid. Redox Signal* **2014**, *20*, 1770–1788. [CrossRef]
21. Lee, T.-S.; Chau, L.-Y. Heme oxygenase-1 mediates the anti-inflammatory effect of interleukin-10 in mice. *Nat. Med.* **2002**, *8*, 240–246. [CrossRef]
22. Kutty, R.K.; Maines, M.D. Purification and characterization of biliverdin reductase from rat liver. *J. Biol. Chem.* **1981**, *256*, 3956–3962. [CrossRef]
23. Chauveau, C.; Rémy, S.; Royer, P.J.; Hill, M.; Tanguy-Royer, S.; Hubert, F.-X.; Tesson, L.; Brion, R.; Beriou, G.; Grégoire, M.; et al. Heme oxygenase-1 expression inhibits dendritic cell maturation and proinflammatory function but conserves IL-10 expression. *Blood* **2005**, *106*, 1694–1702. [CrossRef]
24. Sierra-Filardi, E.; Vega, M.A.; Sánchez-Mateos, P.; Corbí, A.L.; Puig-Kröger, A. Heme Oxygenase-1 expression in M-CSF-polarized M2 macrophages contributes to LPS-induced IL-10 release. *Immunobiology* **2010**, *215*, 788–795. [CrossRef]

25. Philippidis, P.; Mason, J.C.; Evans, B.J.; Nadra, I.; Taylor, K.M.; Haskard, D.O.; Landis, R.C. Hemoglobin scavenger receptor CD163 mediates interleukin-10 release and heme oxygenase-1 synthesis: Antiinflammatory monocyte-macrophage responses in vitro, in resolving skin blisters in vivo, and after cardiopulmonary bypass surgery. *Circ. Res.* **2004**, *94*, 119–126. [CrossRef]
26. Chung, S.W.; Hall, S.R.; Perrella, M.A. Role of haem oxygenase-1 in microbial host defence. *Cell. Microbiol.* **2009**, *11*, 199–207. [CrossRef]
27. Arai, T.; Yoshikai, Y.; Kamiya, J.; Nagino, M.; Uesaka, K.; Yuasa, N.; Oda, K.; Sano, T.; Nimura, Y. Bilirubin Impairs Bactericidal Activity of Neutrophils through an Antioxidant Mechanism in Vitro. *J. Surg. Res.* **2001**, *96*, 107–113. [CrossRef] [PubMed]
28. Mitterstiller, A.-M.; Haschka, D.; Dichtl, S.; Nairz, M.; Demetz, E.; Talasz, H.; Soares, M.; Einwallner, E.; Esterbauer, H.; Fang, F.C.; et al. Heme oxygenase 1 controls early innate immune response of macrophages toSalmonellaTyphimurium infection. *Cell. Microbiol.* **2016**, *18*, 1374–1389. [CrossRef] [PubMed]
29. Abdalla, M.Y.; Ahmad, I.; Switzer, B.; Britigan, B.E. Induction of heme oxygenase-1 contributes to survival of Mycobacterium abscessus in human macrophages-like THP-1 cells. *Redox Biol.* **2015**, *4*, 328–339. [CrossRef] [PubMed]
30. Rodriguez, E.; Noya, V.; Cervi, L.; Chiribao, M.L.; Brossard, N.; Chiale, C.; Carmona, C.; Giacomini, C.; Freire, T. Glycans from Fasciola hepatica Modulate the Host Immune Response and TLR-Induced Maturation of Dendritic Cells. *PLOS Negl. Trop. Dis.* **2015**, *9*, e0004234. [CrossRef]
31. Frigerio, S.; Da Costa, V.; Costa, M.; Festari, M.F.; Landeira, M.; Rodríguez-Zraquia, S.A.; Härtel, S.; Toledo, J.; Freire, T. Eosinophils Control Liver Damage by Modulating Immune Responses Against Fasciola hepatica. *Front. Immunol.* **2020**, *11*, 2276. [CrossRef]
32. Paul, G.; Bataille, F.; Obermeier, F.; Bock, J.; Klebl, F.; Strauch, U.; Lochbaum, D.; Rummele, P.; Farkas, S.; Scholmerich, J.; et al. Analysis of intestinal haem-oxygenase-1 (HO-1) in clinical and experimental colitis. *Clin. Exp. Immunol.* **2005**, *140*, 547–555. [CrossRef]
33. Sardana, M.K.; Kappas, A. Dual control mechanism for heme oxygenase: Tin(IV)-protoporphyrin potently inhibits enzyme activity while markedly increasing content of enzyme protein in liver. *Proc. Natl. Acad. Sci. USA* **1987**, *84*, 2464–2468. [CrossRef]
34. Van Nguyen, T.; Piao, C.H.; Fan, Y.J.; Shin, D.U.; Kim, S.Y.; Song, H.J.; Song, C.H.; Shin, H.S.; Chai, O.H. Anti-allergic rhinitis activity of alpha-lipoic acid via balancing Th17/Treg expression and enhancing Nrf2/HO-1 pathway signaling. *Sci. Rep.* **2020**, *10*, 12528. [CrossRef]
35. Yan, S.C.; Wang, Y.J.; Li, Y.J.; Cai, W.Y.; Weng, X.G.; Li, Q.; Chen, Y.; Yang, Q.; Zhu, X.X. Dihydroartemisinin Regulates the Th/Treg Balance by Inducing Activated CD4+ T cell Apoptosis via Heme Oxygenase-1 Induction in Mouse Models of Inflammatory Bowel Disease. *Molecules* **2019**, *24*, 2475. [CrossRef] [PubMed]
36. Zhang, Q.; Cui, T.; Chang, Y.; Zhang, W.; Li, S.; He, Y.; Li, B.; Liu, L.; Wang, G.; Gao, T.; et al. HO-1 regulates the function of Treg: Association with the immune intolerance in vitiligo. *J. Cell. Mol. Med.* **2018**, *22*, 4335–4343. [CrossRef] [PubMed]
37. Yoon, S.-J.; Kim, S.-J.; Lee, S.-M. Overexpression of HO-1 Contributes to Sepsis-Induced Immunosuppression by Modulating the Th1/Th2 Balance and Regulatory T-Cell Function. *J. Infect. Dis.* **2017**, *215*, 1608–1618. [CrossRef]
38. Hualin, C.; Wenli, X.; Dapeng, L.; Xijing, L.; Xiuhua, P.; Qingfeng, P. The Anti-inflammatory Mechanism of Heme Oxygenase-1 Induced by Hemin in Primary Rat Alveolar Macrophages. *Inflammation* **2011**, *35*, 1087–1093. [CrossRef] [PubMed]
39. Costa, D.L.; Amaral, E.P.; Namasivayam, S.; Mittereder, L.R.; Fisher, L.; Bonfim, C.C.; Sardinha-Silva, A.; Thompson, R.W.; Hieny, S.E.; Andrade, B.B.; et al. Heme oxygenase-1 inhibition promotes IFNγ-and NOS2-mediated control of Mycobacterium tuberculosis infection. *Mucosal Immunol.* **2021**, *14*, 253–266. [CrossRef]
40. Takagi, T.; Naito, Y.; Mizushima, K.; Hirai, Y.; Harusato, A.; Okayama, T.; Katada, K.; Kamada, K.; Uchiyama, K.; Handa, O.; et al. Heme oxygenase-1 prevents murine intestinal inflammation. *J. Clin. Biochem. Nutr.* **2018**, *63*, 169–174. [CrossRef]
41. Zhong, H.; Bao, W.; Friedman, D.; Yazdanbakhsh, K. Hemin Controls T Cell Polarization in Sickle Cell Alloimmunization. *J. Immunol.* **2014**, *193*, 102–110. [CrossRef]
42. Harding, C.L.; Villarino, N.F.; Valente, E.; Schwarzer, E.; Schmidt, N.W. Plasmodium Impairs Antibacterial Innate Immunity to Systemic Infections in Part Through Hemozoin-Bound Bioactive Molecules. *Front. Cell. Infect. Microbiol.* **2020**, *10*, 328. [CrossRef] [PubMed]
43. Schulz, S.; Wong, R.J.; Vreman, H.J.; Stevenson, D.K. Metalloporphyrins—An update. *Front Pharmacol.* **2012**, *3*, 68. [CrossRef] [PubMed]
44. Fernández-Fierro, A.; Funes, S.C.; Rios, M.; Covián, C.; González, J.; Kalergis, A.M. Immune Modulation by Inhibitors of the HO System. *Int. J. Mol. Sci.* **2020**, *22*, 294. [CrossRef]
45. Hori, R.; Kashiba, M.; Toma, T.; Yachie, A.; Goda, N.; Makino, N.; Soejima, A.; Nagasawa, T.; Nakabayashi, K.; Suematsu, M. Gene Transfection of H25A Mutant Heme Oxygenase-1 Protects Cells against Hydroperoxide-induced Cytotoxicity. *J. Biol. Chem.* **2002**, *277*, 10712–10718. [CrossRef] [PubMed]
46. Loboda, A.; Damulewicz, M.; Pyza, E.; Jozkowicz, A.; Dulak, J. Role of Nrf2/HO-1 system in development, oxidative stress response and diseases: An evolutionarily conserved mechanism. *Cell. Mol. Life Sci.* **2016**, *73*, 3221–3247. [CrossRef] [PubMed]
47. Pibiri, M.; Leoni, V.P.; Atzori, L. Heme oxygenase-1 inhibitor tin-protoporphyrin improves liver regeneration after partial hepatectomy. *Life Sci.* **2018**, *204*, 9–14. [CrossRef]
48. Dimitrijević, B.; Borozan, S.; Katić-Radivojević, S.; Stojanovic, S. Effects of infection intensity with Strongyloides papillosus and albendazole treatment on development of oxidative/nitrosative stress in sheep. *Veter. Parasitol.* **2012**, *186*, 364–375. [CrossRef]
49. Masamba, P.; Kappo, A. Immunological and Biochemical Interplay between Cytokines, Oxidative Stress and Schistosomiasis. *Int. J. Mol. Sci.* **2021**, *22*, 7216. [CrossRef]

50. Chauhan, A.; Sharma, A.; Tripathi, J.K.; Sun, Y.; Sukumran, P.; Singh, B.B.; Mishra, B.B.; Sharma, J. Helminth derived factors inhibit neutrophil extracellular trap formation and inflammation in bacterial peritonitis. *Sci. Rep.* **2021**, *11*, 1–9. [CrossRef]
51. de Oliveira, R.B.; Senger, M.R.; Vasques, L.M.; Gasparotto, J.; dos Santos, J.P.A.; Pasquali, M.A.; Moreira, J.C.; Silva, F.P., Jr.; Gelain, D.P. Schistosoma mansoni infection causes oxidative stress and alters receptor for advanced glycation endproduct (RAGE) and tau levels in multiple organs in mice. *Int. J. Parasitol.* **2013**, *43*, 371–379. [CrossRef]
52. Derda, M.; Wandurska-Nowak, E.; Hadaś, E. Changes in the level of antioxidants in the blood from mice infected with Trichinella spiralis. *Parasitol. Res.* **2004**, *93*, 207–210. [CrossRef]
53. Sánchez-Campos, S.; Tuñón, M.J.; González, P.; González-Gallego, J. Oxidative stress and changes in liver antioxidant enzymes induced by experimental dicroceliosis in hamsters. *Parasitol. Res.* **1999**, *85*, 468–474. [CrossRef] [PubMed]
54. Rehman, A.; Rehman, L.; Ullah, R.; Beg, M.A.; Khan, M.H.; Abidi, S. Oxidative status and changes in the adenosine deaminase activity in experimental host infected with tropical liver fluke, Fasciola gigantica. *Acta Trop.* **2020**, *213*, 105753. [CrossRef]
55. Li, D.-Y.; Xiong, X.-Z. ICOS+ Tregs: A Functional Subset of Tregs in Immune Diseases. *Front. Immunol.* **2020**, *11*, 2104. [CrossRef] [PubMed]
56. Li, D.-Y.; Xiong, X.-Z. Corrigendum: ICOS+ Tregs: A Functional Subset of Tregs in Immune Diseases. *Front. Immunol.* **2021**, *12*, 701515. [CrossRef]
57. Sobhani, N.; Tardiel-Cyril, D.; Davtyan, A.; Generali, D.; Roudi, R.; Li, Y. CTLA-4 in Regulatory T Cells for Cancer Immunotherapy. *Cancers* **2021**, *13*, 1440. [CrossRef] [PubMed]
58. Kraaij, M.D.; van der Kooij, S.W.; Reinders, M.; Koekkoek, K.; Rabelink, T.; van Kooten, C.; Gelderman, K.A. Dexamethasone increases ROS production and T cell suppressive capacity by anti-inflammatory macrophages. *Mol. Immunol.* **2011**, *49*, 549–557. [CrossRef] [PubMed]
59. Devadas, S.; Zaritskaya, L.; Rhee, S.G.; Oberley, L.; Williams, M.S. Discrete generation of superoxide and hydrogen peroxide by T cell receptor stimulation: Selective regulation of mitogen-activated protein kinase activation and fas ligand expression. *J. Exp. Med.* **2002**, *195*, 59–70. [CrossRef]
60. Yu, Y.; Deng, W.; Lei, J. Interleukin-33 promotes Th2 immune responses in infected mice with Schistosoma japonicum. *Parasitol. Res.* **2015**, *114*, 2911–2918. [CrossRef]
61. Peng, H.; Zhang, Q.; Li, X.; Liu, Z.; Shen, J.; Sun, R.; Wei, J.; Zhao, J.; Wu, X.; Feng, F.; et al. IL-33 Contributes to Schistosoma japonicum-induced Hepatic Pathology through Induction of M2 Macrophages. *Sci. Rep.* **2016**, *6*, 29844. [CrossRef] [PubMed]
62. Mchedlidze, T.; Waldner, M.; Zopf, S.; Walker, J.; Rankin, A.L.; Schuchmann, M.; Voehringer, D.; McKenzie, A.N.; Neurath, M.F.; Pflanz, S.; et al. Interleukin-33-Dependent Innate Lymphoid Cells Mediate Hepatic Fibrosis. *Immunity* **2013**, *39*, 357–371. [CrossRef] [PubMed]

antioxidants

Article

Heme Oxygenase-1 and Blood Bilirubin Are Gradually Activated by Oral D-Glyceric Acid

O. Petteri Hirvonen *, Maarit Lehti, Heikki Kyröläinen and Heikki Kainulainen

Faculty of Sport and Health Sciences, Neuromuscular Research Center, University of Jyväskylä, 40014 Jyväskylä, Finland
* Correspondence: petteri.hirvonen@replicon.fi

Abstract: It has been shown that small doses of oral D-glyceric acid (DGA) activate mitochondrial metabolism and reduce inflammation among 50–60-year-old healthy volunteers. The present results with the same small doses reveal that after a 4-day DGA regimen, a dose of DGA activated the HO-1 pathway acutely, while enhanced inflammatory status after the 4-day DGA regimen seemed to be able to downregulate the HO-1 pathway in non-acute measurement. Blood bilirubin was strongly upregulated towards the end of the altogether 21-day study period with positive associations towards improved inflammation and reduced blood triglycerides. After the 4-day DGA regimen, hepatic inflow of blood bilirubin with albumin as the carrier was clearly upregulated in the lower-aerobic-capacity persons. At the same time also, blood triglycerides were down, pointing possibly to the activation of liver fatty acid oxidation. The combination of activated aerobic energy metabolism with transient HO-1 pathway activation and the upregulation of blood bilirubin may reduce the risks of chronic diseases, especially in aging. Furthermore, there exist certain diseases with unsatisfactorily-met medical needs, such as fatty and cholestatic liver diseases, and Parkinson's disease, that can be possibly ameliorated with the whole-body mechanism of the action of the DGA regimen.

Keywords: HO-1; bilirubin; HIF-1α; subclinical inflammation; ROS

Citation: Hirvonen, O.P.; Lehti, M.; Kyröläinen, H.; Kainulainen, H. Heme Oxygenase-1 and Blood Bilirubin Are Gradually Activated by Oral D-Glyceric Acid. *Antioxidants* **2022**, *11*, 2319. https://doi.org/10.3390/antiox11122319

Academic Editor: Maria Detsika and Elias Lianos

Received: 27 October 2022
Accepted: 15 November 2022
Published: 23 November 2022

1. Introduction

Bilirubin is a product of heme catabolism. The cytosolic heme oxygenase (HO) reaction produces biliverdin, iron (Fe), and carbon monoxide (CO) [1,2]. Biliverdin is further reduced into (unconjugated) bilirubin via the biliverdin reductase (BVR) enzyme [2]. In the BVR reaction, one cytosolic NADH is oxidized into NAD^+. The majority of blood bilirubin is made from the breakdown of hemoglobin from senescent red blood cells, but a significant part of bilirubin also originates from the turnover of various heme-containing proteins found in other tissues, primarily in the liver and muscles. Heme oxygenase is the rate-limiting factor in bilirubin production [3].

Water insoluble unconjugated bilirubin is excreted from cells and carried by albumin in plasma for hepatic conjugation and further excretion via bile ducts into the intestines [4]. An increase in bilirubin concentration, both in plasma and in tissues, has been shown to possess significant antioxidant [5] and anti-inflammatory effects, as well as therapeutic effects in neurodegenerative diseases, such as Parkinson's disease [6,7]. Additionally in recent studies, bilirubin has been shown to activate fatty acid metabolism via the peroxisome proliferator-activated receptor alpha (PPARα) [8].

Inducible heme oxygenase (HO-1) activity has also been found relevant for antioxidant and anti-inflammatory protection independently of bilirubin. Furthermore, it is beneficial in therapies of many chronic diseases, such as NAFLD [9]. HO-1 is activated by stress factors e.g., by increased reactive oxygen species (ROS) generation via the NF-E2–related factor2 (Nrf2 transcription factor). The Nrf2 transcription factor upregulates the mRNA, protein, and enzymatic activity of HO-1 [10]. Nrf2/HO-1 is induced, e.g., by oxidative stress from mitochondrial oxidative phosphorylation (OXPHOS). Nrf2 related HO-1 activation is additionally linked to the upregulation of mitochondrial biogenesis [10]. For simplicity, we later refer to combined HO-1 and BVR reactions that generate bilirubin and Fe as the HO-1 pathway.

Too much Nrf2 or HO-1 activation may also cause deleterious effects [1]. In fact, all the products of the HO-1 pathway are toxic at higher concentrations, but even at physiological levels CO may possess important protective effects in certain pathologic situations [11,12]. Endogenous CO is eventually excreted from the body via the lungs [13]. The liver stores and regulates excess endogenous blood Fe [14].

HO-1 is upstream of bilirubin while the hypoxia inducible factor 1 alpha (HIF-1α) expression is induced by bilirubin in the physiological oxygen content [15], i.e., HIF-1α may be downstream of bilirubin when bilirubin concentration is increased in normoxia. Furthermore, HO-1 has been recognized as a downstream gene of HIF-1α pathway in many tissues during hypoxia [16,17]. Finally, HO-1 expression has been shown to be transcriptionally regulated by PPARα [18]. Due to multiple direct and indirect connections and feedback mechanisms, it is difficult or sometimes even impossible to separate independent therapeutic effects of HO-1 activation from the effects of bilirubin, and vice versa [19]. This is especially valid in studies related to whole physiological systems.

Physical exercise activates ATP generation by cytosolic glycolysis and mitochondrial OXPHOS, and also ROS generation [20]. Induced ROS generation activates a counterbalancing antioxidant HO-1 pathway, and as a follow-up the generation of bilirubin and Fe are upregulated. It has been found that conducting strenuous exercises elevates the plasma bilirubin level [21]. Furthermore, some studies indicate that plasma bilirubin is elevated in highly exercising subjects due to the reduced rate of hepatic bilirubin conjugation and subsequent reduced excretion of bilirubin from the body [22,23].

Effective counterbalancing of elevated oxidative stress and inflammation makes HO-1 pathway studies somewhat challenging because HO-1 activation aims at suppressing itself via neutralizing ROS that originally caused the activation. Additionally, all details of bilirubin in the entire body metabolism are not fully understood [7]. Nevertheless, it is wildly accepted that a gradual basal increase in bilirubin concentration within the physiological range and mild or transient HO-1 activation forms therapeutic effects throughout the body [24,25].

D-glyceric acid (DGA) is a trace metabolite present in vertebrates [26]. An oral DGA regimen has been shown to possess similar signaling effects to physical exercise and may promote health benefits related to mitochondrial activation, at least in older persons [27].

In this study we aim to find out whether the HO-1 pathway is activated and the blood bilirubin level is elevated after an acute oral DGA dose. The same research questions are studied in the non-acute 4-day and 4 + 14-day DGA administration of doses in healthy 50–60-year-old humans. Additionally, shorter 4-day whole-body effects of the DGA regimen on the HO-1 pathway are analyzed separately in high-capacity and lower-capacity persons. Because possible ROS scavenging (antioxidant) effects can be analyzed only in cell cultures, we report the impact of DGA administration on ROS generation from a human primary hepatocyte study and from a rat primary astrocyte study in the Supplementary Materials.

2. Materials and Methods

2.1. Participants

Altogether 27 healthy 50–60-year-old Caucasian participants were carefully selected for the study group. This age group was chosen because systemic inflammation markers are on average somewhat elevated even in healthy persons at that age [28]. All the participants were informed of the experimental design, and the benefits and possible risks that could be associated with the study prior to signing an informed consent to voluntarily participate in the study. Detailed selection criteria and other characteristics beyond the information presented in Table 1B can be found in Hirvonen et al., (2021) [27]. All studies were conducted in line with the statement of the Ethical Committee of the Central Finland Health Care District (Dnro 1U/2019, KSSHP). Trial registration number (14 January 2021), ClinicalTrials.gov Identifier: NCT04713319.

Table 1. (**A**) Study phases and timings, (**B**) characteristics of the study group and analysed subgroups modified from our earlier publication [27].

(**A**)

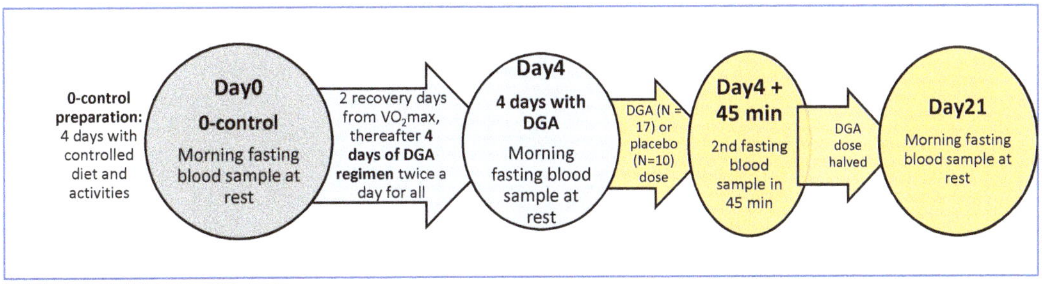

(**B**)

	0-Control	4-Day Health, HC-LC		DGA vs. Placebo	
Study Group	**Whole Group (*N* = 27)**	**LC Subgroup (*N* = 17)**	**HC Subgroup (*N* = 10)**	**DGA Group (*N* = 17)**	**Placebo Group (*N* = 10)**
Mean age (years)	56 years (range 50.3–60.9)	55	56	56	55
Mean VO_2max (ml/kg/min)	35.5 (range 21.8–48.8)	31.3	43.8	35.1	36.1
Mean BMI	25.3 (range 20.1–31.7)	26.2	23.5	25.0	25.8
Female/Male	16 females/11 males	10/7	6/4	10/7	6/4

2.2. Study Setup and Monitoring

There were altogether three measurement days (Day0, Day4 and Day21) in the study (Table 1A). First fasting and resting blood samples at Day0, Day4 and Day21 were taken 12 h after the last DGA or placebo dose (non-acute). Blood samples for each participant were always taken at the same time in the morning (±2 min). On Day4, an acute blood sample was additionally drawn at the study site 45 min after the morning sample. In between, a dose of DGA or placebo was taken immediately after the morning blood sample (Table 1A). Furthermore, on Day0 after the morning blood sample, an indirect VO_2max test with a bicycle ergometer was performed for all participants to assess aerobic capacity.

The test setup was double-blinded. Measurements were always performed on the same weekday (Friday or Saturday) for each participant. To avoid any bias from the strenuous VO_2max test at Day0, we added two full recovery days after it before initiating the oral DGA regimen. For simplicity, we call the second measurement day "Day4" because it was taken after the 4-days of the DGA regimen. The different phases of the study are illustrated in Table 1A.

Normal lifestyle and stable behavior were encouraged through the use of personal diaries and reminder emails. Morning interviews were carried out individually when participants arrived at the study site at a minimum of 30 min before the first blood sample. Moreover, participant's health status and the timing of the last dose were always checked. All the participants, who came to the Day0 measurement, completed the whole study.

2.3. Characteristics of the Study Group and Analysed Subgroups

The study group consisted of healthy 50–60-year-old males and females. Its characteristics are presented on the first column of Table 1B (later also called the "whole group"). In the Day4 analyses, the whole group was further divided into the high aerobic capacity (HC) and lower aerobic capacity (LC) subgroups based on seven scale classification [29]. The HC subgroup consisted of the two best aerobic capacity classes with "Excellent" and "Very Good" aerobic capacity and the LC subgroup consisted of participants from five lower classification groups. From our study group, ten participants (6 females and 4 males) ended in the HC subgroup and 17 participants (10 females and 7 males) in the LC subgroup (Table 1B).

The placebo group was chosen randomly among females and males separately beforehand. In practice, the number of placebo-treated participants was zero until Day4 morning measurement, which enabled HC–LC subgroup division with a sufficient number of observations for statistical comparison. The priming of the placebo group with the 4-day DGA regimen may have had an impact on the 45-min results on Day4, and this was considered when analyzing them.

2.4. Test Substances and Doses, Preparations, and Administration

D-glyceric acid calcium salt and the placebo (E509/calcium chloride) were dissolved into 1.5 L bottles of water beforehand for each participant. The calculated dose of DGA or placebo was to be drunk in the morning and in the evening. In the placebo group there were equal molar amounts of calcium with water.

Selected doses, regimens, and measurement timings in the present study were based on earlier in vitro and in vivo pre-tests related to the regulatory acceptance processes of DGA. The dose of DGA used was 2 times 3.33 mg per body mass (kg) per day. The acute dose of DGA received on the morning of Day4 was the same as the previous 8 doses, and the placebo group received an equimolar amount of calcium within calcium chloride, also dissolved into water. More detailed information can be found in the study of Hirvonen et al. [27].

2.5. Blood Samples and Biomarker Measurements

Blood samples were drawn from the antecubital vein of each participant always at the same time in the morning. The samples were immediately cooled and centrifuged in heparin plasma tubes and stored into 2 mL portions at −80 °C. Plasma total bilirubin and Fe were measured with clinically accredited standard methods (Synlab, Helsinki, Finland). Total bilirubin represents plasma unconjugated bilirubin in our study model, and in the following, the term plasma bilirubin represents plasma total bilirubin. Plasma glycoprotein acetyls (GlycA) [30], triglycerides (TGs), free fatty acids, and albumin were measured (Nightingale Health Oy, Kuopio, Finland) using nuclear magnetic resonance (NMR) technology with regulatory approval for diagnostics [31]. GlycA, TGs, free fatty acids, and albumin were measured only from non-acute samples. Plasma interleukin 6 (IL-6) was measured using the IMMULITE® 2000 immunoassay system and alkaline phosphatase (ALP) with the IFCC method (kits 981832–981833, Thermo Scientific). The mRNA expressions of inducible heme oxygenase (HO-1) and hypoxia inducible factor 1 alpha (HIF-1α) were measured from white blood cells (WBCs). More detailed information can be found in Hirvonen et al. [27].

2.6. In Vitro Side Studies with Human Primary Hepatocytes and Rat Primary Cortical Astrocytes

In our studies with human primary hepatocytes and rat primary astrocytes, net ROS generation with DGA was compared to 0-controls, and additionally to the positive antioxidant controls [32]. On top of negative and positive controls, dose response studies were conducted to confirm consistent results.

We used two different reagents to detect oxidation caused by ROS to avoid possible pitfalls from selection of only a single detection method. In the hepatocyte studies we used a reactive oxygen species (ROS) assay kit ab113851 (Abcam). It uses the cell permeant reagent 2′,7′—dichlorofluorescein diacetate (DCFDA), a fluorogenic dye that measures hydroxyl, peroxyl and other ROS activity within the cell. In the astrocyte study we used CellROX Green staining. CellROX® Oxidative Stress Reagents are fluorogenic probes designed to reliably measure ROS in live cells.

In both hepatocytes and astrocytes, mild metabolic stress was induced by new nutrition i.e., by measuring cellular ROS generation 1–2 h after the change of culture media with new nutrition and tested concentrations. The main results, including a brief methods description, are now for the first time scientifically reported in Supplementary Materials.

2.7. Statistical Methods

Each person was used as her/his own control. In that way, individual "noise factors" could be eliminated and we were able to pairwise test whether the intra-individual responses were similar among the persons of the study. Furthermore, to achieve statistically unambiguous results all blinded participants were in the same comparison group during the first week. Pre-assessment on the sufficient group size was based on the results from our pilot tests and related volatilities. Among our relatively homogenous study group with fully comparable intra-individual measurement points, selected group size turned out to be clearly sufficient.

Statistical tests were conducted using IBM SPSS statistics software (ver. 26) and Microsoft Excel. Statistical tests for group averages of intra-individual changes were based on paired Student's *t*-tests. Unpaired t-tests were used when testing the HC and LC subgroups or the DGA and placebo groups against each other. When testing intraday 45 min changes in the DGA group, possible circadian variation was corrected by deducting the mean change in the placebo group from the DGA group observations. Squared linear correlations (R^2) from scatter diagrams were calculated between selected studied variables. Technically, R^2 can be interpreted as the ratio of explained variation to the total variation of the explanatory variable. The significance of R^2 was statistically tested by the F-test. When $N < 10$, we checked the normality of the underlying data visually or using the Kolmogorov–Smirnov test. A *p*-value lower than or equal to 0.05 in a one-sided t-test was considered statistically significant (marking = * or #) and a *p*-value lower than or equal to 0.01 as statistically very significant (marking = ** or ##). A non-significant *p*-value is marked by "n.s." or > 0.05, and $p >> 0.05$ in quite clear cases. All presented tests were predetermined or derived from predetermined test settings.

3. Results

3.1. Changes in Bilirubin and Iron Concentrations during the Entire Study Period

Blood bilirubin concentration increased by 18.6% from Day0 to Day21 in the DGA administration group ($p = 0.009$) while in the placebo group, no statistically significant changes were noticed (Figure 1A). Furthermore, the Day21 changes from Day0 in the DGA group differed statistically significantly ($p = 0.023$) from the respective changes in the placebo group (Figure 1A). There was an interesting decline by 8.8% in blood Fe from Day0 to Day4 ($p = 0.042$). Blood bilirubin and Fe concentration shared a similar pattern during the study period (Figure 1A,B).

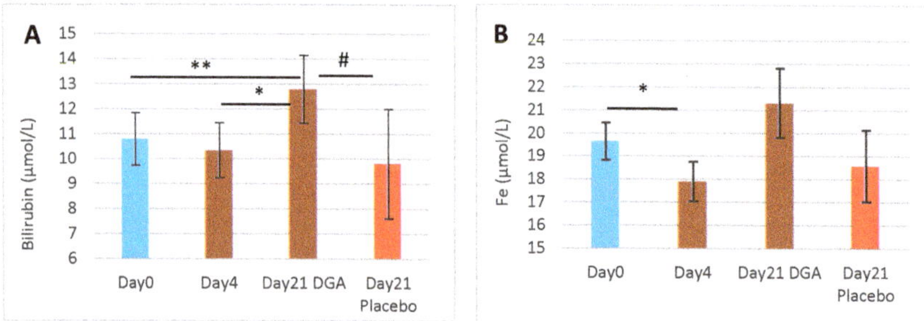

Figure 1. End products of the HO-1 pathway in blood 12 h after the last DGA or placebo dose: (**A**) Mean (±SEM) blood bilirubin concentration, (**B**) Mean (±SEM) blood Fe concentration. **Notes:** (1) The Day21 DGA and Placebo averages are indexed to Day0 so that Day21 fully reflects the change from Day0. (2) * and ** indicate paired *t*-test, # indicates non-paired *t*-test between the%-changes from Day0 to Day21.

3.2. Biomarkers Downstream of Bilirubin and the HO-1 Pathway during the Entire Study Period

Figure 2A shows that the relative changes between GlycA and bilirubin from Day0 to Day21 correlated significantly (R^2 = 0.353, p = 0.015). Respectively, the changes between triglycerides and bilirubin correlated statistically significantly (R^2 = 0.305, p = 0.032) (Figure 2B) and a not-significant correlation was noticed between the changes Day4 to Day21 in IL-6 and bilirubin (Figure 2C).

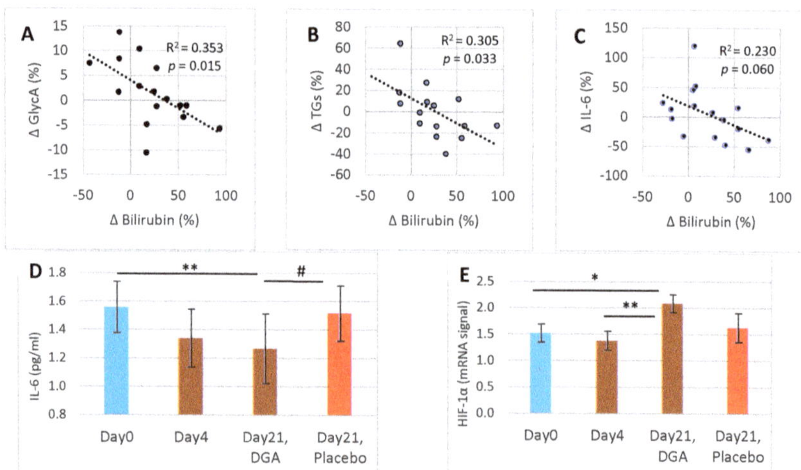

Figure 2. Bilirubin associated Day21 anti-inflammatory, fat metabolic, and mRNA expression effects: (**A**) scatter diagram and R^2 of the percent changes from Day0 to Day21 between GlycA and bilirubin, (**B**) scatter diagram and R^2 of the percent changes from Day0 to Day21 between blood triglycerides and bilirubin, (**C**) scatter diagram and R^2 of the percent changes from Day4 to Day21 between IL-6 and bilirubin, (**D**) mean (±SEM) blood IL-6 concentration, (**E**) mean (±SEM) mRNA expression of HIF-1α from collected WBCs. **Notes:** (1) In (**D**,**E**), the Day21 DGA and placebo averages are indexed to Day0 so that Day21 fully reflects the change from Day0. (2) * and ** indicate paired *t*-test, # indicates non-paired *t*-test between the %-changes from Day0 to Day21. In (**A**–**C**), the *p*-values are based on an F-test. (3) One Day21 observation was deducted from (**B**) as an outlier. Additionally, one Day0 IL-6 observation was excluded from (**D**) when comparing the changes from Day0 to Day21 in the placebo group against the DGA group.

Strong upregulation of blood bilirubin from Day4 to Day21 was typically accompanied by a reduction of subclinical inflammation (IL-6) at the individual level (Figure 2C). IL-6 declined by 19.0% from Day0 to Day21 in the DGA administration group ($p = 0.002$), while in the placebo group it declined only by 2.6% (Figure 2D). Furthermore, the individual changes of IL-6 in the DGA compared to the placebo group differed statistically significantly (Figure 2D) [27].

HIF-1α mRNA expression in WBCs (Figure 2E) shared a similar pattern in all observation points to blood bilirubin in the DGA group (Figure 1A). HIF-1α mRNA expression increased by 37.0% from Day0 to Day21 in the DGA administration group ($p = 0.014$) while in the placebo group there was no change (Figure 2D). Furthermore, the 49% increase ($p = 0.004$) in HIF-1α mRNA expression resembles the 24% increase in blood bilirubin from Day4 to Day21.

HO-1 mRNA expression in WBCs was very weak on Day0. Eight of the successfully analyzed twenty-five WBC samples showed a lower signal than the 0.29 detection limit and were set to zero (graph not shown). On Day4, the number of zeros increased to 11, which may suggest that the HO-1 mRNA expression was downregulated during the 4-day DGA regimen. Further in line with blood bilirubin (Figure 1A), average HO-1 mRNA expression increased in the DGA group by 34.0% on Day21 from Day0. However, poor mRNA signals destroyed the possibility of drawing statistical conclusions on the changes of HO-1 mRNA expression in WBCs.

3.3. Acute 45 Min Activation of the HO-1 Pathway on Day4

Blood bilirubin concentration rose by 2.3% 45 min after the acute dose in the DGA group ($p = 0.066$) but not in the placebo group (Figure 3A). Blood Fe rose by 9.1% ($p < 0.001$) in the DGA group but not in the placebo group (Figure 3B). Furthermore, there existed a statistically very significant positive correlation between the 45-min %-changes of Fe and bilirubin at the intra-individual level in the DGA group (Figure 3C).

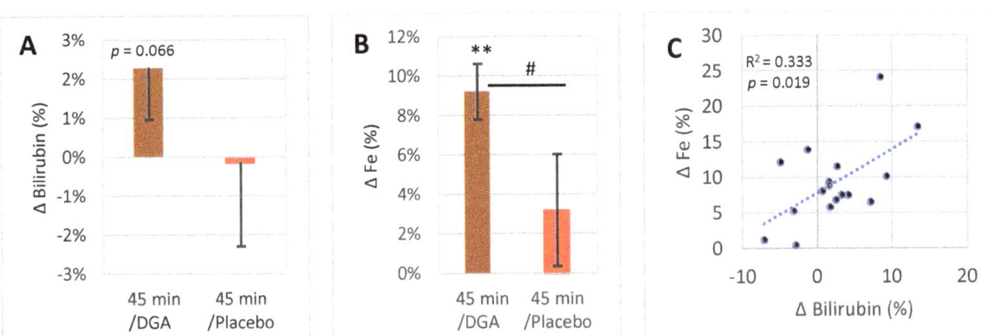

Figure 3. HO-1 pathway end products excluding CO in blood 45 min after acute DGA dose: (**A**) 45 min average intraindividual percent change in bilirubin (±SEM), (**B**) similar 45 min change in Fe (±SEM), (**C**) scatter diagram between 45 min %-changes of bilirubin and Fe after acute DGA dose. **Notes:** ** indicate paired *t*-test, # indicates non-paired *t*-test between the %-changes in 45 min. In (**C**), the *p*-value is based on an F-test.

The average relative increase in blood bilirubin (Figure 3A) was much lower than in Fe (Figure 3B) as some of the individual relative changes in blood bilirubin concentrations were negative, unlike in Fe (Figure 3C).

3.4. 4-Day Changes of Blood Bilirubin Are Tightly Linked to the Changes of Its Carrier in Blood

On Day0, bilirubin concentration in the HC subgroup was on average 12.8 μmol/L, while in the LC subgroup it was 9.6 μM ($p = 0.085$). Higher bilirubin in the HC compared to the LC subgroup remained throughout the 4-day DGA regimen until Day4 (Figure 4A). Conversely, Day0 blood albumin concentration was very similar in the HC and LC subgroups (Figure 4B). Interestingly, both blood albumin and bilirubin declined in the LC subgroup during the 4-day DGA regimen. Moreover, there existed a very tight association ($p < 0.001$) at an individual level between molar *changes* of albumin and bilirubin during the 4-day DGA regimen (Figure 4C). In the HC subgroup, there was no correlation between the changes of albumin and bilirubin during the 4-day DGA regimen (graph not shown). Furthermore, there was no association between the changes of Fe and albumin ($R^2 = 0.017$, $p >> 0.05$) in the LC subgroup. We additionally measured the association between the 4-day changes of blood free fatty acids (= total fatty acids minus TGs) and blood albumin, and found only a mild positive correlation ($R^2 = 0.075$, $p > 0.05$, graph not shown).

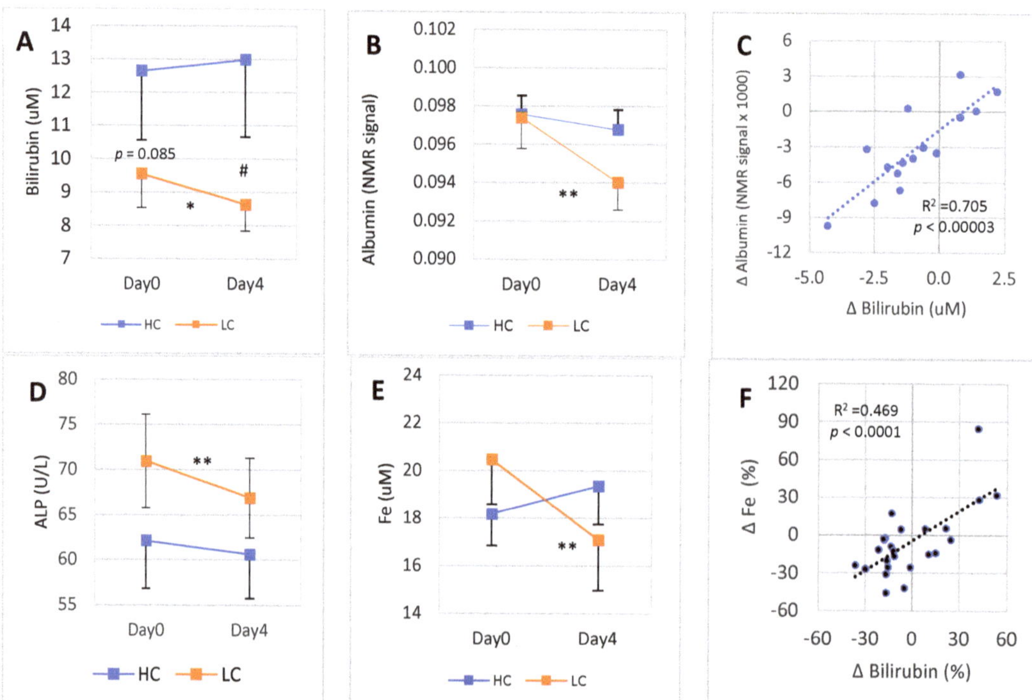

Figure 4. 4-day DGA regimen activates hepatic bilirubin metabolism and Fe storage in the LC subgroup: (**A**) mean (±SEM) blood bilirubin in the HC and LC subgroups separately, (**B**) mean (±SEM) blood albumin in HC and LC subgroups separately, (**C**) cross correlation between intraindividual 4-day molar changes of albumin (vertical scale) and bilirubin (horizontal scale) in the LC subgroup. (**D**) mean (±SEM) blood ALP in the HC and LC subgroups separately, (**E**) mean (±SEM) blood Fe in the HC and LC subgroups separately, and (**D**) cross correlation between intra-individual four-day percent changes of Fe (vertical scale) and bilirubin (horizontal scale) in the whole group. **Notes:** * and ** indicate paired *t*-test from Day0 to Day4, # indicates non-paired *t*-test between the bilirubin concentration on Day4. In (**C**,**F**), the *p*-values are based on an F-test.

Remarkably, plasma alkaline phosphatase (ALP) was reduced in each LC participant during the 4-day DGA regimen ($p < 0.0001$, Figure 4D). Also, blood Fe concentration was reduced statistically very significantly in the LC subgroup (Figure 4E). The reduction of average blood Fe was 16.3% in the LC subgroup. Finally, there was a very strong correlation between the 4-day relative changes of blood Fe and bilirubin (Figure 4F, $p < 0.0001$).

3.5. 4-Day Responses of Blood Bilirubin and Fe Concentrations Are "Inversely" Associated with the Changes of Blood TGs and Subclinical Inflammation (GlycA)

Blood TGs concentration deviated strongly between the HC and LC subgroups at Day0 (Figure 5A). During the 4-day DGA regimen, the difference in TGs between the HC and LC subgroups reduced. Especially in the LC subgroup, there was a significant reduction in blood TGs. The Day0 concentrations in the HC and LC subgroups were opposite compared to each other in blood bilirubin (Figure 4A) and TGs. Nevertheless, the responses in bilirubin and TGs to the 4-day DGA regimen were in the same directions (Figures 4A and 5A). Moreover, there existed a statistically significant positive correlation between the 4-day individual percent changes of blood TGs and blood bilirubin (Figure 5B), and with the 4-day percent changes of blood Fe (Figure 5C).

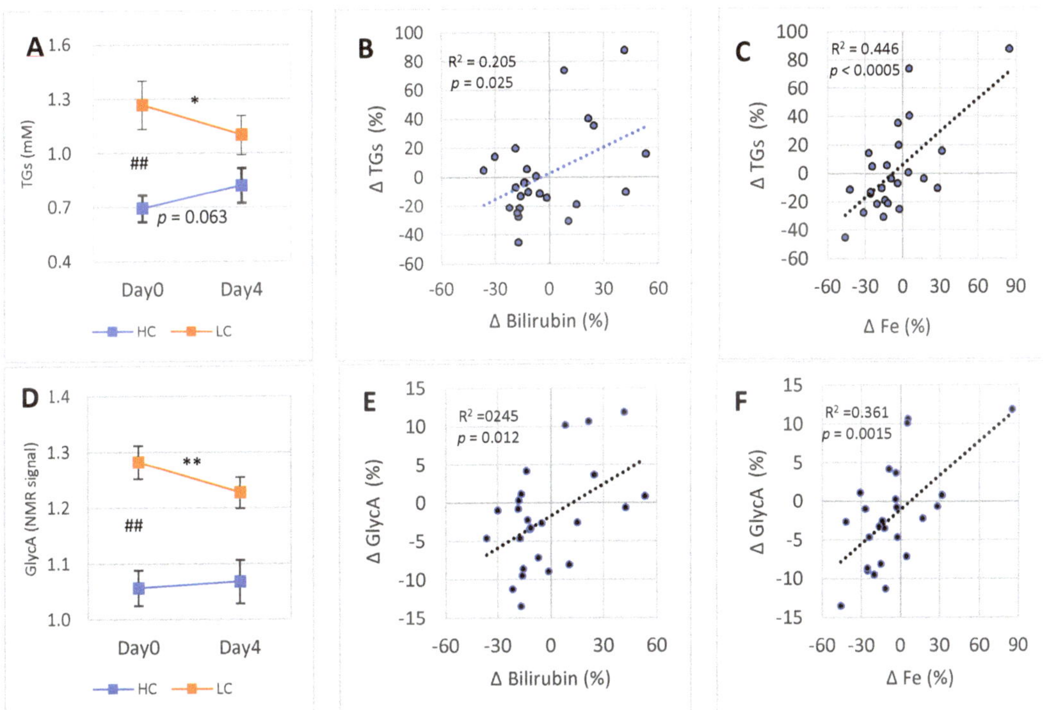

Figure 5. 4-day DGA regimen activates the LC subgroup: (**A**) mean (±SEM) blood TGs in the HC and LC subgroups separately, (**B**,**C**) cross correlation between intra-individual four-day percent changes of TGs (vertical scale) and (**B**) blood bilirubin (horizontal scale) and (**C**) blood Fe (horizontal scale) in the whole group. (**D**) Mean (±SEM) blood GlycA in the HC and LC subgroups separately. (**E**,**F**) Cross correlation between intraindividual four-day percent changes of GlycA (vertical scale) and € blood bilirubin (horizontal scale) and (**F**) blood Fe (horizontal scale) in the whole group. **Notes:** * and ** indicate paired *t*-test from Day0 to Day4, # indicates non-paired *t*-test between the TGs and GlycA concentrations on Day0. In (**B**,**C**) and (**E**,**F**), the *p*-values are based on an F-test.

Blood subclinical inflammation (GlycA concentration) deviated strongly between the HC and LC subgroups at Day0 (Figure 5D). During the 4-day DGA regimen, the difference reduced and, especially in the LC subgroup, there occurred a remarkable reduction in blood GlycA (Figure 5D). There existed a statistically significant positive correlation at an individual level between the 4-day relative changes of GlycA and bilirubin (Figure 5E). Moreover, the 4-day relative changes of GlycA correlated strongly with the relative changes of Fe (Figure 5F). Finally, responses in GlycA during the 4-day DGA regimen resemble the respective changes in TGs (Figure 5A,D). There existed a very strong association between the 4-day relative changes of blood TGs and GlycA ($R^2 = 0.686$, $p < 0.001$, graph not shown).

3.6. *In Vitro Reduction of ROS*

The DGA regimen effectively hinders excessive ROS generation and possesses a similar or superior equimolar antioxidant effect to glutathione, and vitamins C and E in hepatocytes (Supplementary Materials). Similar ROS scavenging results were obtained from rat primary optic astrocytes, both in naturally induced and hydrogen peroxide induced ROS models (Supplementary Materials).

4. Discussion

Blood bilirubin concentration increased significantly 12 h after the last DGA on Day21 in the DGA group but not in the placebo group. However, the increase in blood Fe and the mRNA expression of the HO-1 gene in WBCs were not statistically significant. As expected, the increase in blood bilirubin during the 21-day DGA regimen was associated with reduced blood TGs and systemic inflammation. In the acute 45 min test, HO-1 enzyme activity seemed to be activated, based on responses in blood bilirubin and Fe levels. In the 4-day comparisons intra-individual responses of blood bilirubin and albumin correlated very strongly in the LC subgroup, indicating an activation of hepatic bilirubin intake. Furthermore, hepatic inflow of Fe was possibly activated in the LC subgroup during the 4-day DGA regimen. Finally, the changes in blood TGs and GlycA between the days 4 and 21 correlated inversely with bilirubin (and Fe).

4.1. *Acute 45 Min Activation of the HO-1 Pathway*

Based on the increasing acute 45-min responses in blood bilirubin and Fe, it can be interpreted that the acute DGA dose seems to upregulate whole-body enzymatic activity of the HO-1 pathway. The suggestion is based on the fact that the HO-1 pathway is the only physiologically relevant source of bilirubin and Fe in blood [3,5,33]. Furthermore, statistically significant positive correlation between the changes of blood bilirubin and Fe was observed at the intra-individual level, which underlines that the main source of the acute responses is the same.

Interestingly, the acute increase in blood bilirubin concentration (+2.3%, Figure 3A) was much smaller than the 9.2% increase in blood Fe. At the same time, multistep HO-1 pathway reactions produce one iron and one bilirubin per one heme substrate, and the molar concentrations of bilirubin (10.3 μM) and Fe (17.9 μM) in blood on Day4 morning were rather close to each other. These somewhat contradictory results may be explained by a temporary increase in intracellular biliverdin/bilirubin concentration during the 45 min after the acute dose of DGA.

The increase in tissue biliverdin/bilirubin concentrations may materialize in the peripheral tissues that generate biliverdin from the HO-1 reaction [1] and/or in the liver that handles the excretion of bilirubin from the body [3,4]. In the latter case, liver sinusoids import bilirubin from the blood rapidly, which may be possible because after the 4-day DGA regimen the hepatic import of bilirubin was increased. In both cases, a possible cause of the increased concentrations is the intracellular antioxidant properties of the BVR enzyme activity [5]. In a reverse reaction, BVR is able to convert formed bilirubin back to biliverdin and induce a reducing antioxidant reaction in each cycle [5]. The acute DGA dose increases DGA concentration in cells which induces similar pathways as in exercise [27], while exercise signaling may induce transient ROS generation [20]. Scavenging of formed extra ROS may need antioxidant BVR activity and thus postpone the release of bilirubin into blood circulation [23] after the acute DGA dose.

4.2. Day21 Non-Acute Activation of the HO-1 Pathway

An important take away from the acute 45 min experiment is the transient activation of the HO-1 pathway. It happens after each DGA dose, even after the priming with earlier DGA doses. This pattern may eventually lead to the permanent (non-acute) upregulation in blood bilirubin and at least temporary whole-body enzymatic activation of the HO-1 pathway after each DGA dose. The 34% increase, although statistically non-significant, in the mRNA expression of HO-1 in WBCs was in line with transient whole-body activation of the HO-1 pathway.

We cannot exclude the possibility that the strong increase in blood bilirubin level from Day0 to Day21 in the DGA group was partially based on a downregulation of the hepatic conjugation activity of bilirubin. It has been shown that exercising persons possess elevated bilirubin levels [21], and further that rats with higher exercise capacity possess lower hepatic conjugation activity and thus higher plasma bilirubin concentration [22]. The DGA regimen activates similar pathways as an exercise regimen, e.g., the concentration of blood beta-hydroxybutyrate increases by 20% [27]. Thus, the observed almost 20% increase in bilirubin concentration, and also the higher bilirubin in the HC subgroup (Figure 4A), were consistent with earlier findings [23]. Interestingly, PPARα is one of the key transcription factors taking part in hepatic beta-hydroxybutyrate biosynthesis [34] and as already mentioned, bilirubin has been shown to activate PPARα nuclear translocation [8].

4.3. Day21 Non-Acute Activation of Downstream Metabolites and Pathways

Transient HO-1 pathway activation and bilirubin independently possess antioxidant and anti-inflammatory effects [5,10]. There existed a statistically significant association ($p = 0.015$) between the relative changes of bilirubin and GlycA from Day0 to Day21; the higher the individual rise in blood bilirubin, the greater is the reduction in individual subclinical systemic inflammation. Furthermore, relative changes in IL-6 also correlated negatively with the changes of bilirubin from Day4 to Day21. These results indicate that the positive anti-inflammatory effects of the long term DGA regimen may be partially mediated by the HO-1 pathway. Based on the present findings, it seems that the blood bilirubin level plays a direct role in the reduction of systemic inflammation. Persistent elevated systemic inflammation is often the cause of many chronic diseases [1] and thus its reduction may be of utmost importance.

According to a recent study, bilirubin activates cellular fatty acid use via the PPARα pathway [8]. The increase of blood bilirubin tends to reduce excretion of bilirubin from tissues and thus intracellular concentrations rise. The increase in bilirubin concentration should increase cellular use of TGs for mitochondrial fatty acid oxidation. Indeed, there was a statistically significant correlation between the relative changes in bilirubin and TG from Day0 to Day21. The higher the individual rise in blood bilirubin, the greater is the reduction in individual blood TGs concentration. Increased use of fatty acids in energy metabolism can be important in NAFLD [35].

Finally, HIF-1α expression is induced by bilirubin in the physiological oxygen content, i.e., normoxia [15]. Also, HO-1 activation and related CO generation may stabilize HIF-1α and result in cytoprotection [11]. Current human studies were performed at rest showing that HIF-1α mRNA expression was strongly upregulated in WBCs on Day21. This was well in line with bilirubin upregulation. Furthermore, the increase in HIF-1α mRNA upregulation by 49% in WBCs from Day4 to Day21 was statistically very significant and again in line with the statistically very significant upregulation of bilirubin during the same period. However, we could not find a statistically significant correlation between the changes of blood bilirubin level and the changes of HIF-1α mRNA expression in WBCs in the DGA group. Furthermore, HIF-1α intracellular proteasomal degradation is strongly increased in normoxia situation [36]. Thus, the strongly increased mRNA expressions of HIF-1α could also compensate for the increased HIF-1α enzyme degradation rate due to the improved inflammatory state in the DGA group (Figure 2D). Nevertheless, HIF-1α mRNA upregulation on Day21 was consistent with energy metabolic activation by the

DGA regimen [27], which also likely caused the upregulation of the HO-1 pathway [10]. It may even be that the energy metabolic activation by the DGA regimen induced a marginal additional need for oxygen that gradually induced HIF-1α upregulation [17,37].

4.4. Four-Day Reduction in Bilirubin in the LC Subgroup Is Directly Associated with the Change in Plasma Albumin

There existed a difference in blood bilirubin concentrations between the HC and the LC subgroups on Day0 and the difference remained throughout the 4-day DGA regimen. Average bilirubin concentration was clearly higher in the HC subgroup. As already said, higher bilirubin possesses antioxidant effects [5,21], thus the HC subgroup is in a better position in this respect. Why the blood bilirubin concentration was reduced by the 4-day DGA regimen in the LC subgroup is still without answers.

Albumin serves as a carrier of blood unconjugated bilirubin. It also facilitates hepatic intake of unconjugated bilirubin. During the process of hepatic intake, the albumin-bilirubin complex is bound to the sinusoidal surfaces of hepatocytes and the process is relatively time consuming. Nevertheless, the albumin-bilirubin complex is not in blood circulation. Thus, when there is a change in the host's hepatic demand for bilirubin from blood, it is expected to be reflected as a simultaneous change in blood albumin. Furthermore, the liver functions as the final destiny of blood bilirubin to be subsequently conjugated and excreted from the body with bile acids [4].

Blood bilirubin level declined by 9.6% in the LC subgroup during the 4-day DGA regimen, and albumin declined only by 3.0%; however, while this change in albumin is not physiologically relevant, it was nevertheless statistically very significant. Furthermore, there existed a very strong association (R^2 = 0.705) between the 4-day molar changes in bilirubin and albumin in the LC subgroup. Because one albumin protein transports several bilirubin molecules [3,38], it seems clear that the very small relative reduction in the abundant pool of blood albumin is directly related to the hepatic inflow of bilirubin. All in all, after only eight doses of DGA in four days, it seems that hepatic demand for blood unconjugated bilirubin increased significantly in the LC subgroup. This result would be in line with the findings of our earlier study on the main energy metabolites, in which we demonstrated that hepatic demand for glycerol (and alanine) was strongly activated during the 4-day DGA regimen [27]. Interestingly, glycerol metabolism is positively regulated by PPARα in the liver [39].

4.5. Also, Other Bilirubin Related Blood Biomarkers Indicate Strong Hepatic Activation in Four Days

Blood Fe declined strongly in the LC subgroup during the 4-day DGA regimen. Furthermore, there was a very strong association ($p < 0.0001$) between the relative changes in bilirubin and Fe in blood. These two events indicate that hepatic iron demand was also increased [38]. However, it should be noticed that also HO-1 pathway downregulation could explain, at least partly, the strong correlation between the changes in bilirubin and Fe [1]. The strong 16.3% reduction in Fe in the LC subgroup may be additionally explained by the mitochondrial activation during the 4-day DGA regimen [27]. Mitochondrial activation increases the use of Fe in OXPHOS heme proteins [33]. This is because the liver especially seems to be activated significantly during the 4-day DGA regimen [27], and the demand of blood Fe for mitochondrial heme protein synthesis may have increased the most in the liver [40]. The blood Fe pool is only 3–4 mg and must turn over several times daily to meet the demand, mostly from erythropoiesis [33]. All in all, hepatic activation could explain most of the strong decline in blood Fe during the 4-day DGA regimen.

The four-day changes in blood TGs were also associated with the changes in Fe (R^2 = 0.446, Figure 5C). This surprisingly strong relationship may indicate that the hepatic activation [41] in the LC subgroup may also be behind the decline in blood TGs in the LC subgroup. Change in blood TGs concentration is directly related to the net release of TGs from the liver [42], and a more active liver consumes a bigger part of formed TGs itself [43].

Finally, unconjugated bilirubin presumably flows into the liver with albumin and at the same time also the HO pathway was likely activated in the liver. To avoid excessive concentration of potentially toxic bilirubin, it is likely that hepatic conjugation and excretion of the conjugated bilirubin increased simultaneously. Conjugated bilirubin is excreted from hepatocytes as bile pigment via the bile ducts. ALP may be increased even if only a few small bile ducts are obstructed, and serum bilirubin is normal [44,45]. Remarkably, plasma ALP was reduced in each LC participant during the 4-day DGA activation ($p < 0.0001$). The extremely consistent reduction in ALP and the reduction in blood bilirubin in the LC subgroup may point to some unspecified hepatobiliary enhancement, such as the bile acid secretion to the small intestine [3], during the 4-day DGA regimen in the LC subgroup.

4.6. In Four Days, Systemic Inflammation (GlycA) and TGs Correlated Inversely with Blood Bilirubin

The 21-day changes in blood bilirubin and GlycA correlated negatively. During the 4-day DGA regimen, correlation between the changes in bilirubin and GlycA was positive, i.e., totally opposite to that on Day21. A technical explanation for this inverse 4-day bilirubin–GlycA correlation arises from the strong reduction of systemic inflammation in the LC subgroup and simultaneous modest increase in the HC subgroup, and from similar subgroup dynamics in blood bilirubin.

What happens during the 4-day DGA regimen that temporarily distorts the canonical relationship that has been observed also in the literature [1,23,24]? We do not know the exact reason, but the hepatic activation that is accompanied with the reduction in systemic low-grade inflammation may be an important explanatory factor during the 4-day DGA regimen. Additionally, it is possible that in the LC subgroup, the HO-1 pathway activity was reduced due to reduction in inflammation, and as a follow-up, the release of bilirubin and Fe from tissues to blood was reduced causing the inverse correlation. All in all, the positive 4-day correlations between GlycA and bilirubin, and GlycA and Fe can be likely explained by the activation of the liver and the downregulation of whole-body HO-1 pathway activity due to reduction of systemic inflammation.

Also, the correlation between the changes in bilirubin and TGs during the 4-day DGA regimen was positive, i.e., fully the opposite than during the period from Day0 to Day21. Partial explanation for the positive correlation likely arises from the presumed inflow of bilirubin into the liver in the LC subgroup. It possibly increased hepatic fatty acid oxidation via PPARα [8] and tended to reduce blood TGs more in participants with higher hepatic bilirubin inflow from blood, and vice versa. Further explanation for the positive correlation between bilirubin and TGs during the 4-day DGA regimen arises from the strong correlation between the changes in TGs and GlycA during the same 4-day period ($R^2 = 0.686$), i.e., the correlations between the changes in blood TGs, bilirubin, and Fe were partially due to very strong association between the changes in GlycA and TGs.

4.7. DGA Regimen Hinders Excessive ROS Generation under Metabolic Stress in Hepatocytes and Astrocytes

The DGA regimen effectively hindered excessive ROS generation (Supplementary Materials) [46]. The exact reason why the two-day DGA priming and acute DGA dose with fresh cell culture media 2 h earlier is able reduce net ROS generation remains elusive, but it is likely that the activation on the HO-1 pathway plays an important role in both hepatocytes [9] and astrocytes [7].

5. Conclusions

The inducible heme oxygenase pathway was activated transiently after each DGA dose during the 21-day study period. Simultaneously, blood bilirubin level was strongly upregulated towards Day21 under the DGA regimen. The positive associations of blood bilirubin and systemic inflammation (GlycA and IL-6) and blood lipids (TGs) were observed. Also, HIF-1α mRNA expression was shown to be strongly activated in WBCs by the 3-week

DGA regimen. All these activations likely relate to the mitochondrial energy metabolic activation by the DGA [27].

During the first four days under the DGA regimen, the reduction in blood TGs in the LC subgroup may be partially linked to the hepatic inflow of bilirubin with plasma albumin. Simultaneously, subclinical inflammation was markedly improved in the LC subgroup, i.e., in the participants with somewhat elevated systemic inflammation before the study. Fast improvement of the anti-inflammatory status likely reduced HO-1 pathway activity in four days in the LC subgroup.

It seems that the DGA regimen is able to both up- and downregulate the HO-1 pathway in humans. This possibly materializes via temporary activation of ROS generation by OXPHOS after each DGA dose, and a more permanent reduction of oxidative stress inter alia via HO-1 pathway activation. In vitro studies with hepatocytes and astrocytes showed that the DGA regimen efficiently reduces ROS generation in metabolic stress.

All in all, the DGA regimen seems to be able to activate whole-body aerobic energy metabolism without excessive ROS generation. Benefits materialize in the whole body, but the liver is activated the most due to its central role in maintaining glucose homeostasis and its other vital homeostatic tasks. Increased plasma bilirubin and beta-hydroxybutyrate, and the activation of the HIF-1α pathway are beneficial also for neurodegenerative diseases. The therapeutic combination activated by the DGA regimen poses benefits for chronic diseases related to elevated systemic inflammation.

Supplementary Materials: The following supporting information can be downloaded at: https://www.mdpi.com/article/10.3390/antiox11122319/s1, Supplementary S1. Human primary hepatocyte ROS study model and results; Supplementary S2. Astrocyte ROS study, study outline, main results and materials and methods [46].

Author Contributions: O.P.H.: conceptualization, methodology, formal analyses, writing the original draft. H.K. (Heikki Kyröläinen): supervision, review and editing, validation. H.K. (Heikki Kainulainen): supervision, review and editing, validation. M.L.: review and editing, validation. All authors have read and agreed to the published version of the manuscript.

Funding: University of Jyväskylä (JyU) provided the study site and equipment without cost. Replicon Health Oy paid the salary for the corresponding author.

Institutional Review Board Statement: The study was conducted in accordance with the Declaration of Helsinki and approved by the Ethical committee of the Central Finland Health Care District (Dnro 1U/2019, Approval date 4 June 2019).

Informed Consent Statement: Informed consent was obtained from all subjects involved in the study.

Data Availability Statement: The original data contributions presented in the study are included in the graphs of the article. Anonymous data on presented biomarkers is available from the corresponding author upon reasonable request. Additional anonymous data can be found in Hirvonen et al., (2021) [27] Supplement A.

Conflicts of Interest: Replicon Health Oy develops DGA as a health beneficial agent. Corresponding Author is an owner of this company and H.K. (Heikki Kainulainen) owns shares in it. The remaining authors declare that the research was conducted in the absence of any commercial or financial relationships that could be construed as a potential conflict of interest.

References

1. Araujo, J.A.; Zhang, M.; Yin, F. Heme Oxygenase-1, Oxidation, Inflammation, and Atherosclerosis. *Front. Pharmacol.* **2012**, *3*, 119. [CrossRef] [PubMed]
2. Hinds, T.D., Jr.; Stec, D.E. Bilirubin, a Cardiometabolic Signaling Molecule. *Hypertension* **2018**, *72*, 788–795. [CrossRef] [PubMed]
3. Kalakonda, A.; Bianca, A.J.; John, S. Physiology, Bilirubin. 2021. Available online: https://www.ncbi.nlm.nih.gov/books/NBK470290/ (accessed on 1 October 2022).
4. Memon, N.; Weinberger, B.i.; Hegyi, T.; Aleksunes, L.M. Inherited disorders of bilirubin clearance. *Pediatr. Res.* **2015**, *79*, 378–386. [CrossRef]

5. Sedlak, T.W.; Saleh, M.; Higginson, D.S.; Paul, B.D.; Juluri, K.R.; Snyder, S.H. Bilirubin and glutathione have complementary antioxidant and cytoprotective roles. *Proc. Natl. Acad. Sci. USA* **2009**, *106*, 5171–5176. [CrossRef]
6. Jayanti, S.; Moretti, R.; Tiribelli, C.; Gazzin, S. Bilirubin: A Promising Therapy for Parkinson's Disease. *Int. J. Mol. Sci.* **2021**, *22*, 6223. [CrossRef] [PubMed]
7. Vasavda, C.; Kothari, R.; Malla, A.P.; Tokhunts, R.; Lin, A.; Ji, M.; Ricco, C.; Xu, R.; Saavedra, H.G.; Sbodio, J.I.; et al. Bilirubin Links Heme Metabolism to Neuroprotection by Scavenging Superoxide. *Cell Chem. Biol.* **2019**, *26*, 1450–1460.e7. [CrossRef]
8. Gordon, D.; Hong, S.; Kipp, Z.; Hinds, T. Identification of Binding Regions of Bilirubin in the Ligand-Binding Pocket of the Peroxisome Proliferator-Activated Receptor-A (PPARalpha). *Molecules* **2021**, *26*, 2975. [CrossRef] [PubMed]
9. Stec, D.E.; Hinds, J.T.D. Natural Product Heme Oxygenase Inducers as Treatment for Nonalcoholic Fatty Liver Disease. *Int. J. Mol. Sci.* **2020**, *21*, 9493. [CrossRef]
10. Piantadosi, C.A.; Carraway, M.S.; Babiker, A.; Suliman, H.B. Heme Oxygenase-1 Regulates Cardiac Mitochondrial Biogenesis via Nrf2-Mediated Transcriptional Control of Nuclear Respiratory Factor-1. *Circ. Res.* **2008**, *103*, 1232–1240. [CrossRef]
11. Chin, B.Y.; Jiang, G.; Wegiel, B.; Wang, H.J.; MacDonald, T.; Zhang, X.C.; Gallo, D.; Cszimadia, E.; Bach, F.H.; Lee, P.J.; et al. Hypoxia-inducible factor 1α stabilization by carbon monoxide results in cytoprotective preconditioning. *Proc. Natl. Acad. Sci. USA* **2007**, *104*, 5109–5114. [CrossRef]
12. Bauer, I.; Pannen, B.H. Bench-to-bedside review: Carbon monoxide—From mitochondrial poisoning to therapeutic use. *Crit. Care* **2009**, *13*, 220. [CrossRef] [PubMed]
13. Pan, K.-T.; Leonardi, G.S.; Croxford, B. Factors Contributing to CO Uptake and Elimination in the Body: A Critical Review. *Int. J. Environ. Res. Public Health* **2020**, *17*, 528. [CrossRef] [PubMed]
14. Wang, C.-Y.; Babitt, J.L. Review Series Iron Metabolism and Its Disorders Liver Iron Sensing and Body Iron Homeostasis. 2019. Available online: http://ashpublications.org/blood/article-pdf/133/1/18/1551567/blood815894.pdf (accessed on 17 October 2022).
15. Kim, S.G.; Ahn, S.-Y.; Lee, E.S.; Kim, S.; Na, K.Y.; Chae, D.-W.; Chin, H.J. Bilirubin Activates Transcription of HIF-1α in Human Proximal Tubular Cells Cultured in the Physiologic Oxygen Content. *J. Korean Med. Sci.* **2014**, *29*, S146–S154. [CrossRef]
16. Zhang, Z.; Yan, J.; Chang, Y.; Yan, S.S.; Shi, H. Hypoxia Inducible Factor-1 as a Target for Neurodegenerative Diseases. *Curr. Med. Chem.* **2011**, *18*, 4335–4343. [CrossRef]
17. Lee, J.W.; Ko, J.; Ju, C.; Eltzschig, H.K. Hypoxia signaling in human diseases and therapeutic targets. *Exp. Mol. Med.* **2019**, *51*, 1–13. [CrossRef]
18. Krönke, G.; Kadl, A.; Ikonomu, E.; Blüml, S.; Fürnkranz, A.; Sarembock, I.J.; Bochkov, V.N.; Exner, M.; Binder, B.R.; Leitinger, N. Expression of Heme Oxygenase-1 in Human Vascular Cells Is Regulated by Peroxisome Proliferator-Activated Receptors. *Arter. Thromb. Vasc. Biol.* **2007**, *27*, 1276–1282. [CrossRef]
19. Jeon, C.; Lee, J.-Y.; Lee, S.J.; Jung, K.J.; Kimm, H.; Jee, S.H. Bilirubin and risk of ischemic heart disease in Korea: A two-sample Mendelian randomization study. *Epidemiol. Health* **2019**, *41*, e2019034. [CrossRef] [PubMed]
20. Powers, S.K.; Deminice, R.; Ozdemir, M.; Yoshihara, T.; Bomkamp, M.P.; Hyatt, H. Exercise-induced oxidative stress: Friend or foe? *J. Sport Health Sci.* **2020**, *9*, 415–425. [CrossRef] [PubMed]
21. Swift, D.L.; Johannsen, N.M.; Earnest, C.; Blair, S.N.; Church, T.S. Effect of Different Doses of Aerobic Exercise Training on Total Bilirubin Levels. *Med. Sci. Sports Exerc.* **2012**, *44*, 569–574. [CrossRef] [PubMed]
22. Hinds, T.; Creeden, J.; Gordon, D.; Spegele, A.; Britton, S.; Koch, L.; Stec, D. Rats Genetically Selected for High Aerobic Exercise Capacity Have Elevated Plasma Bilirubin by Upregulation of Hepatic Biliverdin Reductase-A (BVRA) and Suppression of UGT1A1. *Antioxidants* **2020**, *9*, 889. [CrossRef]
23. Thomas, D.T.; DelCimmuto, N.R.; Flack, K.D.; Stec, D.E.; Hinds, T.D., Jr. Reactive Oxygen Species (ROS) and Antioxidants as Immunomodulators in Exercise: Implications for Heme Oxygenase and Bilirubin. *Antioxidants* **2022**, *11*, 179. [CrossRef] [PubMed]
24. Zhao, X.; Li, L.; Li, X.; Li, J.; Wang, D.; Zhang, H. The Relationship between Serum Bilirubin and Inflammatory Bowel Disease. *Mediat. Inflamm.* **2019**, *2019*, 5256460. [CrossRef] [PubMed]
25. Tsuchihashi, S.-I.; Livhits, M.; Zhai, Y.; Busuttil, R.W.; Araujo, J.A.; Kupiec-Weglinski, J.W. Basal rather than induced heme oxygenase-1 levels are crucial in the antioxidant cytoprotection. *J. Immunol.* **2006**, *177*, 4749–4757. [CrossRef] [PubMed]
26. Hoffmann, G.; Seppel, C.K.; Holmes, B.; Mitchell, L.; Christen, H.; Hanefeld, F.; Rating, D.; Nyhan, W. Quantitative organic acid analysis in cerebrospinal fluid and plasma: Reference values in a pediatric population. *J. Chromatogr. B Biomed. Sci. Appl.* **1993**, *617*, 1–10. [CrossRef]
27. Hirvonen, O.P.; Kyröläinen, H.; Lehti, M.; Kainulainen, H. Randomized Trial: D-Glyceric Acid Activates Mitochondrial Metabolism in 50–60-Year-Old Healthy Humans. *Front. Aging* **2021**, *2*, 752636. [CrossRef]
28. Wyczalkowska-Tomasik, A.; Czarkowska-Paczek, B.; Zielenkiewicz, M.; Paczek, L. Inflammatory Markers Change with Age, but do not Fall Beyond Reported Normal Ranges. *Arch. Immunol. Ther. Exp.* **2015**, *64*, 249–254. [CrossRef]
29. Shvartz, E.; Reibold, R.C. Aerobic fitness norms for males and females aged 6 to 75 years: A review. *Aviat. Space Environ. Med.* **1990**, *61*, 3–11.
30. Connelly, M.A.; Otvos, J.D.; Shalaurova, I.; Playford, M.P.; Mehta, N.N. GlycA, a novel biomarker of systemic inflammation and cardiovascular disease risk. *J. Transl. Med.* **2017**, *15*, 219. [CrossRef]
31. Nightingale Health UK Biobank Initiative; Julkunen, H.; Cichońska, A.; Slagboom, P.E.; Würtz, P. Blood biomarker score identifies individuals at high risk for severe COVID-19 a decade prior to diagnosis metabolic profiling of 105000 adults in the UK Biobank. *Front. Immunol.* **2021**, *12*, 809937. [CrossRef]

32. Hirvonen, P.; Eriksson, P.C.J.; Kaksonen, R. Method for Enhancing Energy Production and Metabolism in Cells. WO2015036656A3, 2014.
33. Camaschella, C.; Nai, A.; Silvestri, L. Iron metabolism and iron disorders revisited in the hepcidin era. *Haematologica* **2020**, *105*, 260–272. [CrossRef]
34. Grabacka, M.; Pierzchalska, M.; Dean, M.; Reiss, K. Regulation of ketone body metabolism and the role of PPARα. *Int. J. Mol. Sci.* **2016**, *17*, 2093. [CrossRef] [PubMed]
35. Pawlak, M.; Lefebvre, P.; Staels, B. Molecular mechanism of PPARα action and its impact on lipid metabolism, inflammation and fibrosis in non-alcoholic fatty liver disease. *J. Hepatol.* **2015**, *62*, 720–733. [CrossRef] [PubMed]
36. Weidemann, A.; Johnson, R.S. Biology of HIF-1α. *Cell Death Differ.* **2008**, *15*, 621–627. [CrossRef] [PubMed]
37. O'Hagan, K.A.; Cocchiglia, S.; Zhdanov, A.V.; Tambuwala, M.M.; Cummins, E.P.; Monfared, M.; Agbor, T.A.; Garvey, J.F.; Papkovsky, D.B.; Taylor, C.T.; et al. PGC-1α is coupled to HIF-1α-dependent gene expression by increasing mitochondrial oxygen consumption in skeletal muscle cells. *Proc. Natl. Acad. Sci. USA* **2009**, *106*, 2188–2193. [CrossRef]
38. Merlot, A.M.; Kalinowski, D.S.; Richardson, D.R. Unraveling the mysteries of serum albumin—More than just a serum protein. *Front. Physiol.* **2014**, *5*, 299. [CrossRef]
39. Patsouris, D.; Mandard, S.; Voshol, P.J.; Escher, P.; Tan, N.S.; Havekes, L.M.; Koenig, W.; März, W.; Tafuri, S.; Wahli, W.; et al. PPARα governs glycerol metabolism. *J. Clin. Investig.* **2004**, *114*, 94–103. [CrossRef]
40. Anderson, E.R.; Shah, Y.M. Iron Homeostasis in the Liver. *Compr. Physiol.* **2013**, *3*, 315–330. [CrossRef]
41. La, P.; Oved, J.H.; Ghiaccio, V.; Rivella, S. Mitochondria Biogenesis Modulates Iron–Sulfur Cluster Synthesis to Increase Cellular Iron Uptake. *DNA Cell Biol.* **2020**, *39*, 756–765. [CrossRef]
42. Melzer, K. Carbohydrate and fat utilization during rest and physical activity. *e-SPEN Eur. e-J. Clin. Nutr. Metab.* **2011**, *6*, e45–e52. [CrossRef]
43. Alves-Bezerra, M.; Cohen, D.E. Triglyceride Metabolism in the Liver. *Compr. Physiol.* **2017**, *8*, 1–22. [CrossRef] [PubMed]
44. Agganis, B.; Lee, D.; Sepe, T. Liver enzymes: No trivial elevations, even if asymptomatic. *Clevel. Clin. J. Med.* **2018**, *85*, 612–617. [CrossRef] [PubMed]
45. Ghazanfar, S.; Qureshi, S. PTH-054 Ercp in Patients with Mildly Raised Alkaline Phosphatase and Normal Biliary Imaging. *Gut* **2013**, *62*, A233. [CrossRef]
46. Scialò, F.; Fernández-Ayala, D.J.; Sanz, A. Role of mitochondrial reverse electron transport in ROS signaling: Potential roles in health and disease. *Front. Physiol.* **2017**, *8*, 428. [CrossRef]

antioxidants

Article

A Novel Class of Dual-Acting DCH-CORMs Counteracts Oxidative Stress-Induced Inflammation in Human Primary Tenocytes

Federico Appetecchia [1,†], Sara Consalvi [1,†], Emanuela Berrino [1], Marialucia Gallorini [2], Arianna Granese [1], Cristina Campestre [2], Simone Carradori [2,*], Mariangela Biava [1,*] and Giovanna Poce [1,*]

1 Department of Chemistry and Technologies of Drug, Sapienza University of Rome, piazzale A. Moro 5, 00185 Rome, Italy; federico.appetecchia@uniroma1.it (F.A.); Sara.consalvi@uniroma1.it (S.C.); emanuela.berrino@uniroma1.it (E.B.); arianna.granese@uniroma1.it (A.G.)

2 Department of Pharmacy, "G. d'Annunzio" University of Chieti-Pescara, Via dei Vestini 31, 66100 Chieti, Italy; marialucia.gallorini@unich.it (M.G.); cristina.campestre@unich.it (C.C.)

* Correspondence: simone.carradori@unich.it (S.C.); mariangela.biava@uniroma1.it (M.B.); giovanna.poce@uniroma1.it (G.P.)

† F. Appetecchia and S. Consalvi contributed equally to the work.

Abstract: Carbon monoxide (CO) can prevent cell and tissue damage by restoring redox homeostasis and counteracting inflammation. CO-releasing molecules (CORMs) can release a controlled amount of CO to cells and are emerging as a safer therapeutic alternative to delivery of CO in vivo. Sustained oxidative stress and inflammation can cause chronic pain and disability in tendon-related diseases, whose therapeutic management is still a challenge. In this light, we developed three small subsets of 1,5-diarylpyrrole and pyrazole dicobalt(0)hexacarbonyl (DCH)-CORMs to assess their potential use in musculoskeletal diseases. A myoglobin-based spectrophotometric assay showed that these CORMs act as slow and efficient CO-releasers. Five selected compounds were then tested on human primary-derived tenocytes before and after hydrogen peroxide stimulation to assess their efficacy in restoring cell redox homeostasis and counteracting inflammation in terms of PGE_2 secretion. The obtained results showed an improvement in tendon homeostasis and a cytoprotective effect, reflecting their activity as CO-releasers, and a reduction of PGE_2 secretion. As these compounds contain structural fragments of COX-2 selective inhibitors, we hypothesized that such a composite mechanism of action results from the combination of CO-release and COX-2 inhibition and that these compounds might have a potential role as dual-acting therapeutic agents in tendon-derived diseases.

Keywords: CO-releasing molecules; tenocytes; PGE_2; 1,5-diarylpyrrole; 1,5-diarylpyrazole; carbon monoxide

Citation: Appetecchia, F.; Consalvi, S.; Berrino, E.; Gallorini, M.; Granese, A.; Campestre, C.; Carradori, S.; Biava, M.; Poce, G. A Novel Class of Dual-Acting DCH-CORMs Counteracts Oxidative Stress-Induced Inflammation in Human Primary Tenocytes. *Antioxidants* **2021**, *10*, 1828. https://doi.org/10.3390/antiox10111828

Academic Editors: Elias Lianos and Maria G. Detsika

Received: 14 October 2021
Accepted: 16 November 2021
Published: 18 November 2021

1. Introduction

CO-releasing molecules (CORMs) can release carbon monoxide (CO) either spontaneously, enzymatically, or triggered by an external stimulus [1]. Their therapeutic potential relies on the release of a limited amount of CO. Along with NO and H_2S, CO is the third small signaling molecule and is produced endogenously by enzymes of the Heme Oxygenase (HO) class through heme oxidative degradation. The expression of the HO inducible isoform (HO-1) is triggered by cell responses toward oxidative stress and inflammation and results in cyto- and tissue protection. HO-1 metabolites, including CO, are important in restoring redox homeostasis and resolution of inflammation, and it has been widely demonstrated that the HO-1/CO axis can help to prevent cellular and tissue damage. Therefore, the manipulation of the HO-1/CO system is an attractive strategy to treat conditions linked to oxidative-stress-induced inflammation, such as lung hyper-inflammation in cystic fibrosis, sepsis and modulation of chronic pain [2–7]. The chemistry of CO is unique: unlike NO and H_2S that react indiscriminately with intracellular targets, CO offers the advantage of binding only to transition metals in a low oxidation state. Such preferential reactivity, along with its greater stability, makes it a more versatile candidate for the development of gaseous-based pharmaceuticals [8]. Indeed, gaseous CO has great potential as a therapeutic tool and has been found beneficial in the treatment of several inflammatory, cardiovascular, and neurological diseases [9–12]. For low-dose CO inhalation, the feasibility of the first clinical trials has been recently assessed [13]. However, the accurate delivery of gaseous CO to its molecular targets through inhalation is challenging, and inhalation therapy is hampered by CO low bioavailability and high affinity to hemoglobin, with consequent toxicity [14]. In this scenario, CORMs have emerged as a safer and attractive therapeutic strategy to deliver a controlled amount of CO to cells. To date, most of the developed CORMs are metal carbonyl complexes (MCCs) [15,16]. Indeed, considering the preferential reactivity of CO for transition metals in a low oxidation state, organometallic complexes have emerged as suitable models to safely deliver CO in vivo and generate innovative therapeutic agents with reasonable pharmacological properties. These molecules have an octahedral shape with six ligands around a central metal and can release CO spontaneously, mainly through hydrolysis in biological buffers. Romão and co-workers [17] introduced a conceptual model to rationalize and improve the design of MCCs with appropriate pharmaceutical properties. This model comprises three portions: (i) a metal core, which accounts for toxicity and the main properties of the MCC; (ii) a coordination-sphere, which influences the electronic density around the metal, tuning the stability and the chemical behavior of the whole complex and triggering CO release under specific conditions; and (iii) a drug-sphere, obtained through modulation of the distal sites of the metal complexes and accounting for pharmacological properties and drug-likeness.

The choice of the transition metal is crucial to design metal-based CORMs. CORMs containing an atom of cobalt (dicobalt(0)hexacarbonyl complexes, DCH) are innovative CO-releasing agents with interesting biological features and good CO-release kinetics [18–25]. The DCH metal core is a hexacarbonyl dicobalt moiety (Co_2CO_6) coordinated through an alkyne bond, which is in turn linked to the drug sphere. One of the main advantages of DCH-CORMs is their synthetic accessibility. Indeed, this highly versatile chemical scaffold is easy to synthesize, facilitating the chemical manipulation of the drug sphere. A series of dual acting DCH-CORMs-carbonic anhydrase inhibitors (CAI-CORMs) have very recently shown promising anti-inflammatory properties under oxidative-stress conditions in different oxidative-based disease models [22,23]. Interestingly, Gallorini et al. [26] demonstrated that some of these compounds were able to differentially modulate inflammation and counteract the H_2O_2-induced stress in rotator-cuff-derived human tenocytes, which activate the nuclear factor erythroid 2 [NF-E2]-related factor 2 (Nrf2)/HO-1/CO pathway to mitigate oxidative stress. It has also been reported that sustained oxidative stress causes aberrant cytokine secretion in a model of rotator cuff disease (RCD) in vitro [27]. Oxidative stress endurance and, consequently, inflammation occurrence, are considered the major factors causing the failure of tendon healing in clinical practice and can lead to chronic pain and disability. Moreover, the benefits of non-steroidal anti-inflammatory drug (NSAID)-based therapy in the acute phase are broadly accepted, but their use in chronic tendon-related diseases is still controversial [28]. Therefore, an innovative therapeutic approach for the treatment of tendon-derived diseases is urgently needed, as their therapeutic management remains a challenge.

In this light, we synthesized a small set of 1,5-diarylpyrrole and 1,5-diarylpyrazole-based DCH-CORMs linked through a propargylic chain (compounds **1–9**, Figure 1). According to the Romão model, the first aim of this study was to analyze the influence of different electronic and steric properties of the drug sphere on the CO release rate. Five selected compounds (**1–5**) were then tested on human primary tendon-derived cells stimulated with a low concentration of hydrogen peroxide (H_2O_2), using the NSAID Meloxicam as a reference compound. The present work aims to assess their efficacy in restoring cell redox homeostasis and counteracting inflammation in terms of PGE_2 secretion and at investigating their potential use in vitro to manage musculoskeletal diseases.

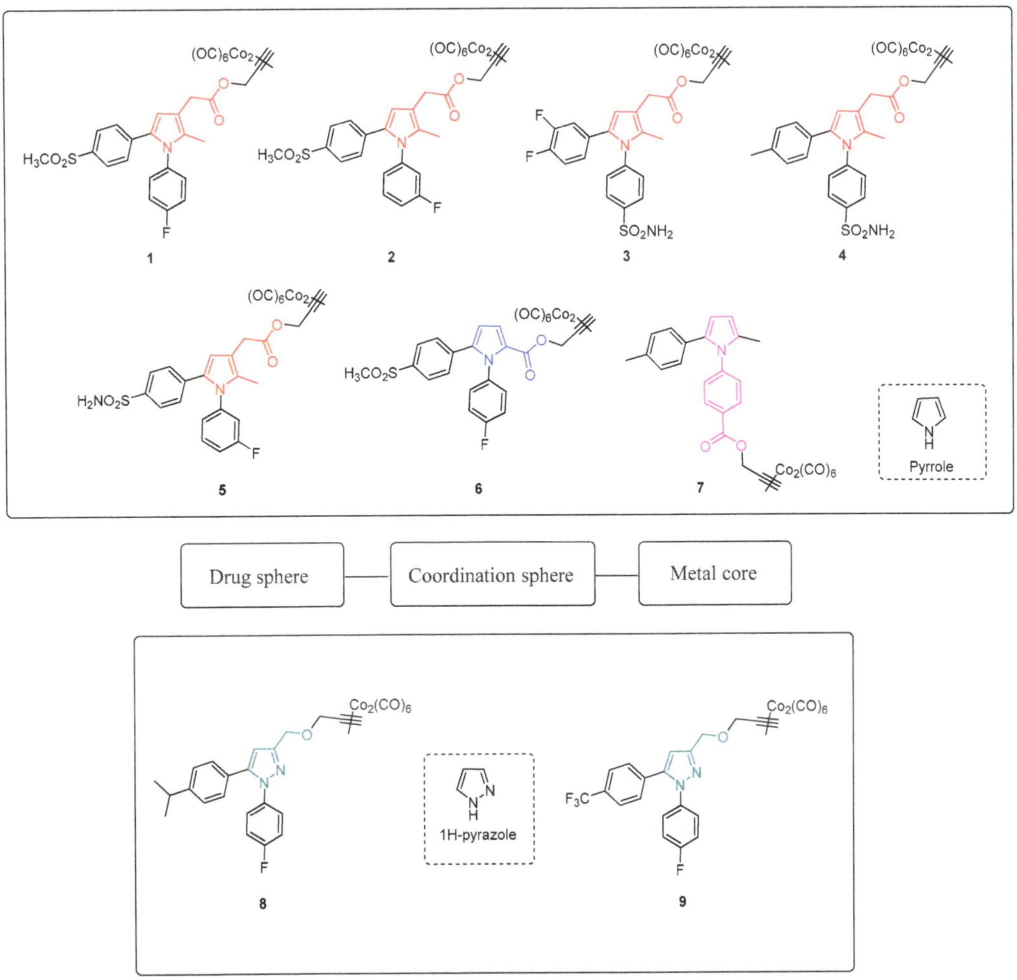

Figure 1. Chemical structures and conceptual model of compounds **1–9**.

2. Materials and Methods

2.1. Chemistry

All chemicals used were obtained from commercial sources (Merck, Acros, Syngene) and were used as supplied without further purification. Merck silica gel 60 (230–400 mesh) and Merck aluminum oxide (activity II–III, according to Brockmann) were used for chromatographic columns with the indicated solvents. Merck TLC plates (silica gel 60 F254 and aluminum oxide F254) were used to monitor all operations, and then compounds were visualized under UV light (254 and 365 nm) and/or stained with the relevant reagents. The yields were not optimized and refer to the purified products. ^{13}C NMR and ^{1}H NMR spectra were recorded on a Bruker Avance III NMR 400 spectrometer with reference to tetramethylsilane (TMS) in the indicated solvent. Chemical shift values are expressed in parts per million (ppm). Coupling constants (*J*) are reported in hertz with signal multiplicities indicated as singlet (s), doublet (d), triplet (t), quadruplet (q) and multiplet (m). When specified, ChemDraw Professional 16.0 was used to generate systematic compound names following IUPAC conventions. Detailed synthetic procedures and spectroscopic data are reported in the Supplementary Materials.

2.2. CO-Release Assay

All reagents were of analytical grade and purchased from Merck. Gaseous CO was obtained from Rivoira (Milan, Italy). A Shimadzu UV1900 UV-Vis Spectrophotometer from 275 to 700 nm at the scanning rate of 200 nm/min was used to record UV-Vis absorption spectra in a disposable plastic cuvette (path length 0.44 cm). An Origin Lab software generated second derivative spectra, and the Savitzky–Golay method was applied using 25 data points for the differentiation process. Neither an increase nor a decrease in the number of points caused changes in the wavelength or in the bandwidth. Lyophilized horse heart Mb was dissolved in phosphate buffered saline flushed with N_2 (PBS, 0.01 M, pH 7.4 to a 20–22 μM final concentration). Two milliliters of this freshly prepared stock solution were placed in a cuvette to record the UV-Vis absorption spectrum of met-Mb. Next, the solutions were divided into two: 10 μL of sodium dithionite (30 mg/mL) were added to the first half (reference) and the UV-Vis spectrum of deoxy-Mb was registered. After that, the solution was flushed with CO gas, and the Mb-CO spectrum was acquired. Sodium dithionite was added to the second half (sample), and a spectrum was recorded. Afterwards, a CORM DMSO solution was added to a final CORM concentration of 3.33 μM and gently mixed. The solution was covered with 300 μL of light mineral oil to avoid CO escaping and oxygenation of Mb, and the absorption spectrum was recorded at t = 0. Spectra were acquired every 30 min for 210 min, keeping the sample at 37 °C. When necessary, a freshly prepared sodium dithionite solution was added. After 210 min, the total Mb concentration at the end of the assay was determined by flushing the sample with CO gas. Mb-CO concentration at each time point was determined as previously reported [23]. Each experiment was replicated three times, and the data were expressed as mean ± SEM.

2.3. Cell Culture

Human tenocytes (#TEN-F; ZenBio Inc.; Durham, NC, USA) were maintained in complete alpha-MEM (EuroClone, Milan, Italy) supplemented with 10% of heat-inactivated FBS (Gibco, ThermoFisher Scientific, Waltham, MA, USA) and 1% penicillin/streptomycin (EuroClone, Milan, Italy) at 37 °C and 5% CO_2 and used from passage 3 up to passage 6.

2.4. Cell Treatment

Cells were seeded in 96-well plates (0.5×10^4/well) (ThermoFisher Scientific, Waltham, MA, USA) and left to adhere overnight at 37 °C and 5% CO_2. In a first set of experiments, tendon-derived cells were treated with increasing concentrations of compounds **1–5** in the range 0–100 μM for 24 and 72 h. Compounds were dissolved in DMSO to obtain a 200 mM stock solution, and they were afterwards diluted in complete alpha-MEM (DMSO final concentration = 0.1%) for further analyses. In a second set of experiments, tenocytes were pre-incubated with 100 μM H_2O_2 for 3 h. After that, the pre-incubation medium was discarded and replaced with a fresh one containing the proper CORM at increasing concentrations for 24 and 72 h. At the established time points, samples were processed for further analyses.

2.5. Cell Metabolic Activity

At the established time points (24 and 72 h), the incubation medium was harvested for further analyses, and complete alpha-MEM containing 0.5 mg/mL MTT (3-[4,5-dimethylthiazol-2-yl]-2,5-diphenyl tetrazolium bromide) (Sigma-Aldrich, St. Louis, MO, USA) was added to each well. Afterwards, cells were incubated for 5 h at 37 °C and cell metabolic activity was measured as already reported [29].

2.6. PGE_2 Secretion

Cell supernatants were collected from the 96-well plates used for the metabolic activity assay (MTT) after 24 and 72 h, and PGE_2 secretion was analyzed. A commercial ELISA kit (Enzo Life Sciences, Farmingdale, NY, USA) was used to measure the amount (pg/mL) of

PGE_2 in the culture media, according to the manufacturer's instructions. The PGE_2 concentration in each sample was determined following a previously reported procedure [30].

3. Results and Discussion

3.1. Chemistry

Compounds **1-9** were easily synthesized in good yields by reacting the terminal alkyne of propargylic intermediates and dicobalt(0)carbonyl octacarbonyl in tetrahydrofuran (THF). Detailed synthetic procedures are reported in the Supporting Information.

3.2. CO Release Assay

The CO releasing behaviors of compounds **1–9** were evaluated through a myoglobin (Mb)-based spectrophotometric assay, considered the gold standard to analyze the CO releasing kinetics and a key criterion to select CORM structures [17,31]. This method analyzes the release of CO from CORMs by following the conversion of deoxy-myoglobin (deoxy-Mb(II)) into CO-myoglobin (CO-Mb(II)) over time by UV-Vis spectroscopy. A 3.3 μM solution of compounds **1–9** was incubated with a 20 μM solution of deoxy-Mb (CORM/Mb 1:6 ratio), and a reducing agent (sodium dithionite) was added to prevent oxidation of deoxy-Mb(II) to Met-Mb(III). Changes in the absorption band in the Soret region of deoxy-Mb and Mb-CO were recorded every 30 min for 210 min, and a second derivative approach was applied to clearly discriminate between the three forms of Mb (deoxy-Mb, Mb-CO and Met-Mb). The relative amount of CO produced over time was calculated following a previously reported equation [23,32], and a correction factor was applied to account for Mb degradation induced by sodium dithionite. Concentration values of Mb-CO formed over time by compounds **1–9** are reported in Table 1. Their CO-release profiles are shown in Figure 2A–C, along with their $T_{1/10}$ values, defined as the time necessary for a CO-RM to produce a concentration of CO-Mb of 1/10 of the initial (Figure 2D). The number of CO units released by CORMs **1-9** after 210 min is reported in Table 2. For selected compounds, the assay was performed using a 1:1 CORM/Mb ratio (Figure 3A–D). The aim was to explore their ability to release CO in a less favored condition, considering that CO release from CORMs is stimulated by an excess of Mb.

Table 1. MbCO formed at each time point when compounds **1–9** were analyzed at 1:6 CORM:Mb ratio.

Time (min)	MbCO Formed (μM)								
	1	**2**	**3**	**4**	**5**	**6**	**7**	**8**	**9**
0	0	0	0	0	0	0	0	0	0
30	0.56	0.58	0.50	0.52	0.63	0.82	0.57	0.56	0.08
60	0.88	1.03	1.01	1.15	1.09	1.46	0.91	0.77	0.45
90	1.49	1.59	1.50	1.42	1.35	2.09	1.27	0.95	0.59
120	1.90	1.76	1.90	1.86	1.57	2.45	1.52	1.24	0.74
150	2.32	2.06	2.07	2.37	1.72	3.12	1.68	1.37	0.99
180	2.28	2.20	2.31	2.35	1.85	3.08	1.88	1.57	1.09
210	2.25	2.22	2.48	2.51	2.12	3.05	2.00	1.75	1.22

Table 2. CO units released by compounds **1–9** after 210 min of incubation working at 1:6 CORM-Mb ratio and by compounds **1**, **5** and **6** at 1:1 CORM-Mb ratio.

CO Units Released after 210 min	1	2	3	4	5	6	7	8	9
3.33 μM (1:6)	0.68	0.67	0.74	0.75	0.64	0.92	0.60	0.53	0.37
20 μM (1:1)	0.11	-	-	-	0.11	0.12	-	-	-

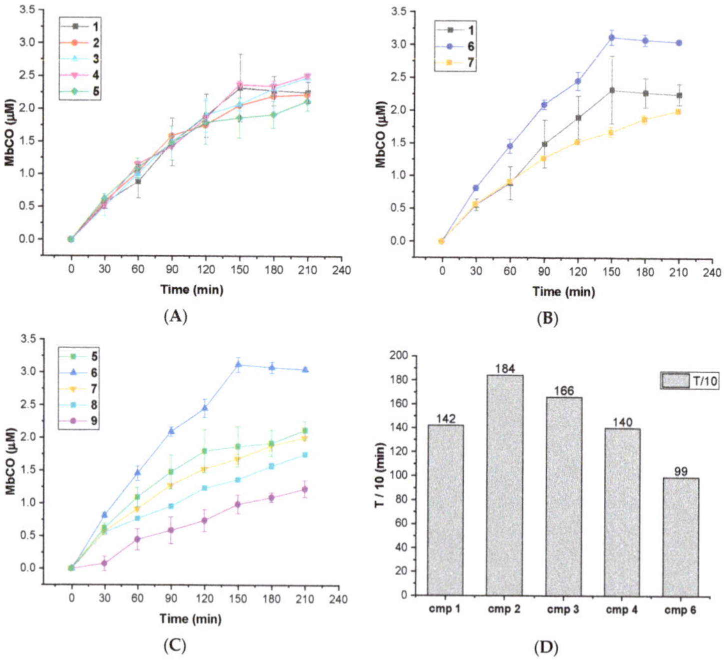

Figure 2. (**A–C**): CO-release profiles of compounds **1–9** as Mb-CO formed over time (1:6 CORM:Mb ratio); (**D**): $T_{1/10}$ values for compounds **1–4** and **6** (defined as the time necessary for a CORM solution to produce a Mb-CO concentration of 1/10 of the initial).

All the tested compounds were effective CO releasers, with CO release kinetics comparable to previously reported DCH-CORMs (slow CO release up to 210 min) [22,23,33]. The obtained data showed considerable differences among the analyzed compounds, depending on the heterocyclic nucleus and the different alkyne system linked to the DCH group (Table 1).

As shown in Figure 1, compounds **1–5** bear a propynyl-pyrrol-3-yl-acetate moiety, compound **6** a propynyl-pyrrole-2-carboxylate motif and compound **7** a propynyl benzoate group at position 1 of the pyrrole ring. Compounds **8** and **9** are characterized by a 3-((prop-2-yn-1-yloxy)methyl)-1*H*-pyrazole scaffold. Derivatives **1–5** showed comparable kinetics, with a sustained release of CO over time (Figure 2A). The kinetics of release were almost superimposable over the first 120 min, with negligible differences over the last 90 min. Despite slight differences in their $T_{1/10}$ (Figure 2D), the comparison of the units of CO released over time (Table 2) displayed a very similar behavior within the series, suggesting that the substitution pattern on the aryl rings only slightly impacts their CO-releasing abilities. In particular, compounds **1**, **2** and **5**, bearing a sulfonyl group on the aryl ring at C5, released almost the same amount of CO at the end of the assay (0.64–0.68 CO units, Table 2). Compounds **3** and **4**, both decorated with a sulfamoylphenyl moiety at position 1 of the pyrrole, displayed very similar behavior and released a comparable amount of CO after 210 min (0.74–0.75 CO units, Table 2). Pyrrole **6** showed a completely different CO releasing profile (Figure 2B,C) and was the fastest and most efficient CO-releaser of the series, with a $T_{1/10}$ value of 99.01 min (Figure 2D) and 0.92 CO units released after 210 min (Table 2). This compound showed a fast release of CO over the first 120 min, which reaches a maximum at 150 min and then slows down. When compared to its analogue **1**, compound **6** produced a 1.35-fold higher amount of Mb-CO at each time point until the end of the assay. These data suggest that the group bearing the DCH moiety and the chemical space around it strongly influences the CO release kinetic. The

CO releasing behavior of compound **7** supported this hypothesis: indeed, it releases CO slower than compounds **1-5**, although reaching almost the same amount of Mb-CO after 210 min of incubation (Figure 2B,C). Differently from compounds **1-6**, the DCH portion of this compound is linked to a propynyl benzoate group at position 1 of the pyrrole ring, suggesting that the electronic structure can induce different CO releasing properties. Pyrazoles **8** and **9** showed different CO releasing profiles (Figure 2C). At the end of the assay, these compounds produced much smaller values of Mb-CO (1.75 μM and 1.22 μM for **8** and **9**, respectively, Table 1) than compounds **1–7** at the same time point. Therefore, it is interesting to note that the 3-((prop-2-yn-1-yloxy)-methyl)-1-pyrazole moiety is probably detrimental in terms of CO releasing efficiency when compared to acetate, carboxylate or benzoate moieties decorating compounds **1–7**. Previous studies reported the impact of the group bearing the DCH moiety in determining the CO releasing properties [22,23]. The different CO release kinetics observed for compounds **1–9** confirm the influence of the drug sphere on CO releasing properties and suggest that both the electronic density around the pyrrole/pyrazole ring and the group bearing the DCH moiety strongly impact the rate of CO release, yet further studies are needed to better characterize this phenomenon. To further explore the releasing properties of these derivatives, we selected compounds **1**, **5** and **6** to be studied at different CO-RM:Mb ratios (Figure 3).

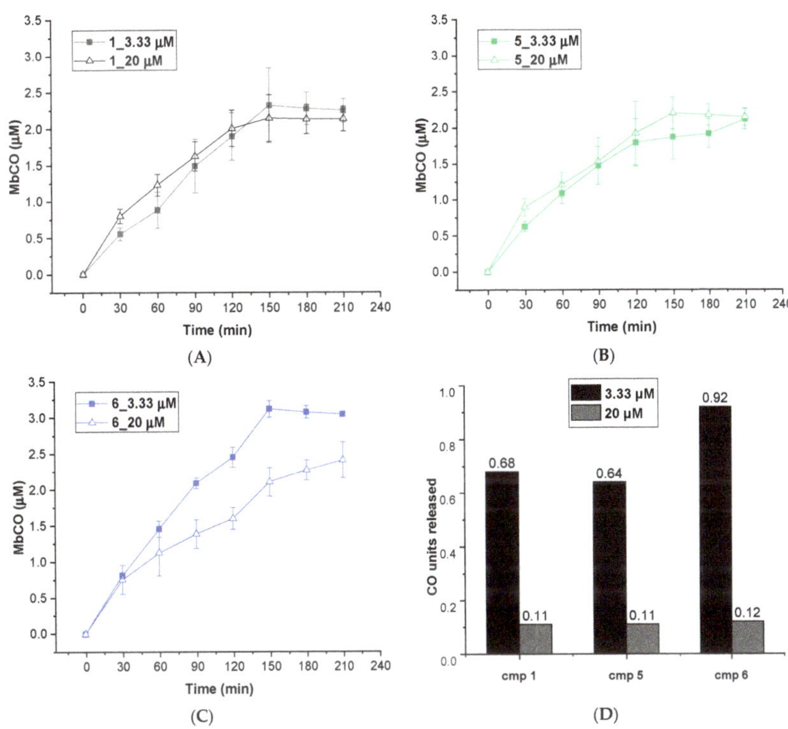

Figure 3. CO release profiles of compounds **1** (**A**), **5** (**B**) and **6** (**C**) analyzed at 1:6 (filled square) and 1:1 (empty triangles) CORM-Mb ratios; (**D**): CO units released by compounds **1**, **5** and **6** after 210 min of incubation working at 1:6 (blue columns) and 1:1 (orange columns) CORM-Mb ratios.

As mentioned above, an excess of the CO acceptor (Mb) stimulates the CO release from CORMs. Thus, we expected a decrease in CO release at 1:1 CO-RM:Mb ratio. As shown in Figure 3, after 210 min of incubation, compounds **1**, **5** and **6** released a lower amount of CO when compared to the one observed in 1:6 conditions (Figure 2). Moreover, all the compounds released the same CO units (0.11–0.12 CO units), regardless of their different chemical structure. Therefore, disfavoring the CO release seems to reduce the differences in CO releasing efficiencies observed when Mb is present in excess (1:6/CO-RM:Mb).

3.3. *Effects of Compounds* **1–5** *on Human Tenocytes*

Once the influence of electronic and steric properties on CO release has been established, a fine-tuning of the drug sphere should also focus on treating particular conditions and on the biological activity of the drug sphere itself [17,34]. Compounds **1-5** drug sphere belongs to a series of sulphone and sulfamoyl diarylpyrrole derivatives developed by our research group as COX-2 selective inhibitors [35–39]. This class of compounds showed promising in vitro and in vivo anti-nociceptive and anti-inflammatory properties and tolerates a wide range of substituents at position C3. As augmented PGE_2 levels are a marker of oxidative-stress inflammation, the modulation of PGE_2 secretion might be a valuable strategy for therapeutic intervention of tendon diseases [27,40]. We therefore speculated that the conjugation of a COX-2 inhibiting scaffold and a CO-releasing moiety could help to achieve promising CORM-candidates for the treatment of tendon inflammatory-based diseases, as COX-2 inhibition and CO release could act synergically to resolve inflammation and restore oxidative homeostasis. Moreover, the conjugation of structural fragments of anti-inflammatory drugs with metal carbonyl moieties is well documented in the literature [18,19,24], and the five selected compounds showed quite similar CO release profiles (Figure 2), allowing us to make a proper comparison and rationalization of the observed biological activities. To help to discriminate between COX-2 mediated and independent activities, the efficacies of these compounds against inflammation and oxidative cytotoxicity were studied through the analysis of different parameters: the metabolic activity of tenocytes before and after H_2O_2 stimulation and the quantification of PGE_2 secretion.

Tendinopathies are characterized by a higher level of tenocyte apoptosis and a decreased metabolic activity, which can reduce the resistance of tendon structures and lead to failure in healing [41,42]. Unstimulated tenocytes were therefore exposed to increasing concentrations of CORMs to evaluate their biocompatibility and effects under non-oxidative stress conditions (Figure 4). It is worth noting that Meloxicam exerted no significant effects on tenocytes when administered in the same experimental conditions [26].

On the other hand, compounds **1–5** significantly increased the metabolic activity of tendon-derived cells after 24 h, as observed for CAI-CORMs hybrids [26]. In more detail, all the tested compounds showed a dose-dependent rise up to 25–50 μM already after 24 h of exposure, which was particularly significant in the presence of compounds **3** and **5**. Interestingly, the tested compounds seemed less active after a 72 h exposure, as the metabolic activity was comparable to that of the control up to the concentration of 25 μM. This might be related to their slightly different CO release kinetics, but further studies are needed to corroborate this hypothesis.

3.4. *Establishment of the Inflammatory Cell Model and Effects of Compounds* **1–5** *on Human Tenocytes under Oxidative Stress Conditions*

These preliminary results highlighted that compounds **1–5** have good proliferative effects on tendon-derived cells and provided a proof-of-concept that the biological features of these compounds are not only COX-2 mediated but also rely on CO release. As reported elsewhere [43], increasing cell metabolism and proliferation are particularly important for tendon tissue repair after the acute inflammatory phase. With this rationale, compounds **1–5** were tested in sub-toxic oxidative stress conditions in vitro [26] to investigate their ability to counteract H_2O_2-induced oxidative stress. After 3 h of incubation with 100 μM H_2O_2, human tenocytes were exposed to increasing concentrations of CORMs. As reported in Figure 5, all the tested compounds were more active than Meloxicam in increasing the cell metabolism of tenocytes. Notably, compounds **3–5** were the most active of the series. In particular, compound **3** showed an outstanding efficacy, being able to increase metabolic activity up to 166.2% after 72 h when administered at 25 μM. Unlike under non-oxidative stress conditions, the percentage of metabolically active tenocytes increases after 72 h. This observation suggests a composite mechanism of action, which probably results from the combination of COX-2 inhibition and CO release. Consistent with previously obtained results, activities were maximum at 25 µM, then decreased at the higher concentrations tested.

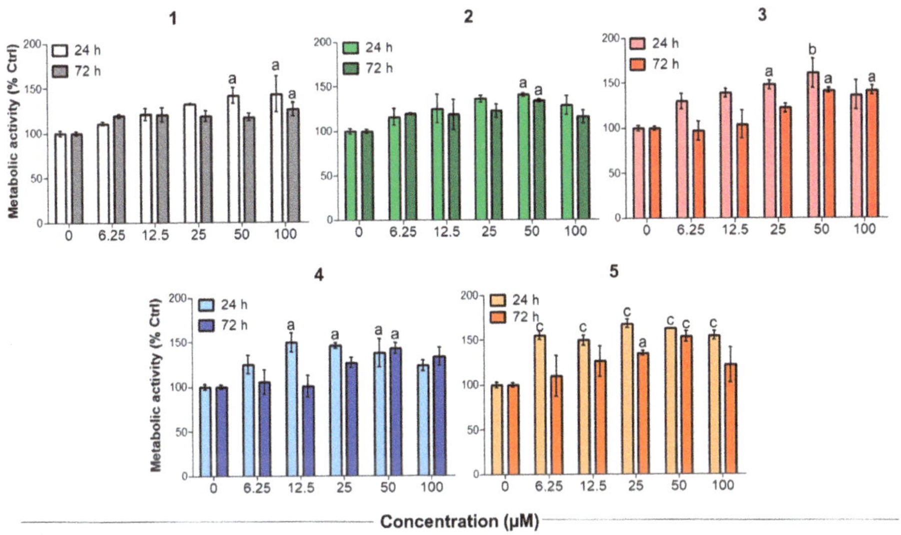

Figure 4. Metabolic activity of human primary tendon-derived cells exposed to increasing concentrations of CORMs (compounds **1–5**) after 24 and 72 h. The control sample (0 µM = cells treated with DMSO 0.1%) is set as 100%. a = $p < 0.01$; b = $p < 0.001$; c = $p < 0.0001$ between cells treated with CORMs and the control sample.

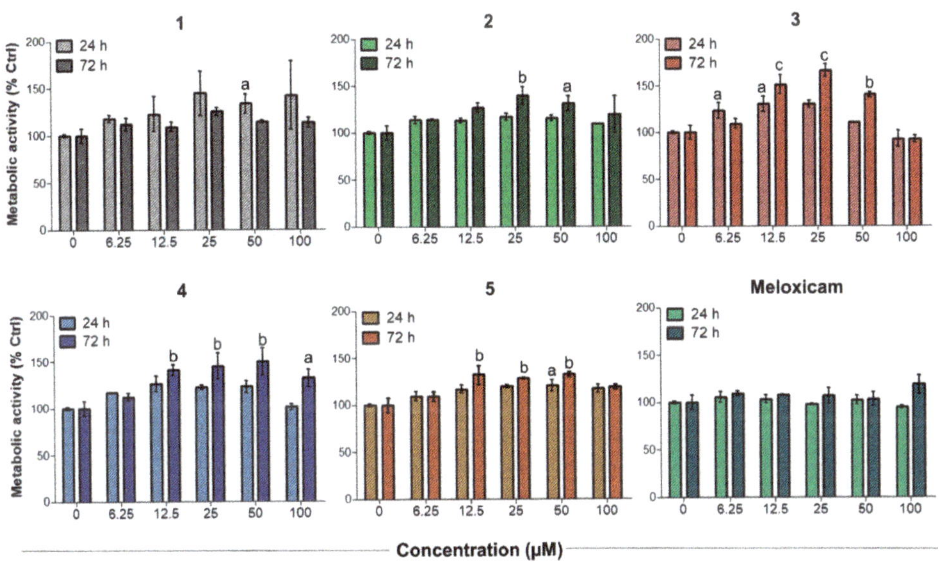

Figure 5. Metabolic activity of H_2O_2-pre-incubated human primary tendon-derived cells exposed to increasing concentrations of CORMs (compounds **1–5**) after 24 and 72 h. Cells were pre-incubated with H_2O_2 100 µM for 3 h. The control sample (0 µM = cells pre-incubated with H_2O_2 and treated with DMSO 0.1%) is set as 100%. a = $p < 0.01$; b = $p < 0.001$; c = $p < 0.0001$ between cells treated with CORMs and the control sample.

3.5. Effects of Compound **1–5** *on* PGE_2 *Secretion*

Anti-inflammatory COX-related activities of compounds **1–5** were evaluated by quantifying PGE_2 secretion in the established oxidative stress conditions in vitro (Figure 6).

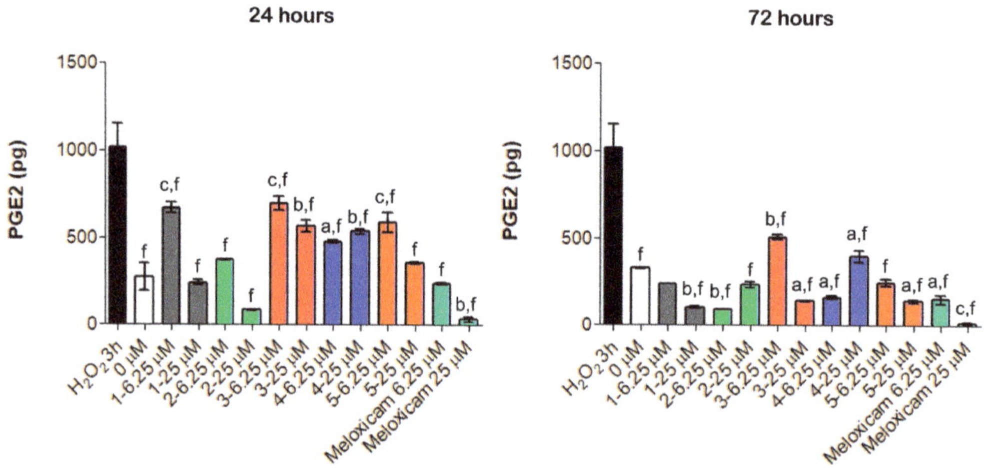

Figure 6. PGE_2 secretion from H_2O_2-pre-incubated human primary tendon-derived cells in the presence of increasing concentrations of CORMs after 24 and 72 h. Cells were pre-incubated with H_2O_2 100 μM for 3 h. 0 μM = cells pre-incubated with H_2O_2 and treated with DMSO 0.1%. The amount of PGE_2 secreted (pg/mL) was normalized on cell metabolic activity data, resulting in the PGE_2 secreted from each sample (total picograms). a = $p < 0.01$; b = $p < 0.001$; c = $p < 0.0001$ between cells treated with CORMs and the control sample. f = $p < 0.0001$ between cells exposed to CORMs and cells pre-treated with H_2O_2 for 3 h.

Consistent with literature data [27], even a short exposure to H_2O_2 for 3 h increased PGE_2 secretion up to 1014.2 pg (Figure 6). Notably, all the tested compounds lowered the amount of PGE_2 compared to the H_2O_2 pre-incubation but were less effective than Meloxicam (Figure 6). Compared to our previous experiments, the obtained data revealed a different trend of activity: compounds **1**, **2** and **5** showed remarkable anti-inflammatory effects and considerably reduced PGE_2 secretion already at 24 h and mainly at 25 μM (240.5 pg, 84.8 pg and 357.1 pg, respectively), whereas compounds **3** and **4** were less efficient. Moreover, the modulation of PGE_2 is time-dependent, being that this cytokine decreased over the time. Actually, the amount of PGE_2 was almost halved with all the tested compounds after 72 h of exposure compared to 24 h treatment (Figure 6). These findings support the hypothesis that the observed anti-inflammatory effects rely on a COX-2 mediated mechanism of action and that the organometallic complexes retain the ability to inhibit COX-2.

Collectively, HO-1 expression and enzymatic activity are confirmed to influence positively and negatively both innate and adaptive immune responses; this dual action seems to be related to the stage of the inflammatory response or disease. The therapeutic potential of HO-1 may rely on limiting early inflammation, hampering successive tissue damage and modulating key pathways in most cell types of the immune system, given the complexity of heme catabolism and the role of HO-1 as a critical mediator of innate immune response. Immunomodulation is mostly related to higher demolition of the pro-inflammatory heme group, macrophage activation towards an anti-inflammatory macrophage profile with reduced secretion of pro-inflammatory cytokines and iNOS and interferon production by macrophages and dendritic cells. Indeed, T cells constitutively express HO-1, and their expansion regulatory is positively influenced by a tolerogenic phenotype sustained through HO-1 induction in dendritic cells. HO-1 modulation or application of low concentrations of CO to LPS-challenged macrophages reduced TNF-α and IL-1β expression and simultaneously stimulated the anti-inflammatory IL-10 production through p38-MAPK activity [44–46].

This context was particularly evident in models of tendon-related diseases. In this light, the role of the macrophage is an area of emerging interest in tendinopathies and in general in the healing of tendons. In fact, inflammation appears to be driven by a high number of infiltrating macrophages at the inflamed tendon site [47]. Furthermore, damaged tendons from patients with tendinopathy show an abundance of $CD14^+$ and $CD68^+$ activated macrophages [48]. We have already reported that CORM hybrids exert their biological effects both on inflamed macrophages and tenocytes [22,26], disclosing a challenging field of application for molecules active on the HO-1/CO molecular axis not strictly related to the immune system. Finally, although once considered cells not involved in the immune-regulation and only related to tendon remodeling, tenocytes have been disclosed as active cells, secreting cytokines and expressing inflammation-related proteins [27].

4. Conclusions

We developed a novel series of DCH-CORMs based on a 1,5-diarylpyrrole scaffold. The screening of three small subsets of 1,5-diarylpyrroles and pyrazoles in a CO release assay allowed us to define the influence of the drug sphere electronic density on the kinetics of CO release. Based on these results, a series of 1,5-diarylpyrroles containing structural fragments of COX-2 selective inhibitors were selected for further biological studies on human primary tendon-derived cells. The observed results suggested the existence of different mechanisms of action and allowed us to conclude that the activities of these compounds result from the combination of COX-inhibition and CO release. Indeed, the obtained data suggest a multiple role for compounds **1–5** in tendon-derived diseases: a direct effect on tendon homeostasis and a cytoprotective effect in human tenocytes exposed to oxidative stress, reflecting their activity as CO-releasers, and a reduction of PGE_2 secretion, indicating a COX-2 mediated anti-inflammatory effect. Taken together, these findings indicate that these compounds could be potential double-acting therapeutic agents for the management of tendon-related diseases. Further studies are needed to better characterize their composite mechanism of action and the contribution of COX-2 inhibition to their biological activities.

Supplementary Materials: The following are available online at https://www.mdpi.com/article/10.3390/antiox10111828/s1, Scheme S1: Synthetic pathway for compounds **1–5**, Scheme S2: Synthetic pathway for compound **6**, Scheme S3: Synthetic pathway for compound **7**, Scheme S4: Synthetic pathway for compounds **8**,**9**.

Author Contributions: F.A., methodology, data curation and formal analysis; S.C. (Sara Consalvi), supervision, formal analysis, writing—original draft and writing—review and editing; S.C. (Simone Carradori), E.B., M.G. and A.G., methodology, data curation and formal analysis; C.C., supervision and formal analysis; S.C. (Simone Carradori) and M.B., conceptualization, funding acquisition, project administration and supervision; G.P., conceptualization, funding acquisition, project administration, supervision, data curation and writing—review and editing. All authors have read and agreed to the published version of the manuscript.

Funding: This work has been supported by the Italian Ministry of Education, Universities and Research—Dipartimenti di Eccellenza—L. 232/2016 and intramural grants to C.C. (FAR2019).

Institutional Review Board Statement: Not applicable.

Informed Consent Statement: Not applicable.

Data Availability Statement: Data are contained within the article.

Conflicts of Interest: The authors declare no conflict of interest.

References

1. Zobi, F. CO and CO-Releasing Molecules in Medicinal Chemistry. *Future Med. Chem.* **2013**, *5*, 175–188. [CrossRef]
2. Ryter, S.W.; Choi, A.M.K. Heme Oxygenase-1/Carbon Monoxide. *Am. J. Respir. Cell. Mol. Biol.* **2009**, *41*, 251–260. [CrossRef] [PubMed]
3. Constantin, M.; Choi, A.J.S.; Cloonan, S.M.; Ryter, S.W. Therapeutic Potential of Heme Oxygenase-1/Carbon Monoxide in Lung Disease. *Int. J. Hypertens* **2012**, *2012*, e859235. [CrossRef]
4. Motterlini, R.; Haas, B.; Foresti, R. Emerging Concepts on the Anti-Inflammatory Actions of Carbon Monoxide-Releasing Molecules (CO-RMs). *Med. Gas. Res.* **2012**, *2*, 28. [CrossRef]
5. Castruccio Castracani, C.; Longhitano, L.; Distefano, A.; Di Rosa, M.; Pittalà, V.; Lupo, G.; Caruso, M.; Corona, D.; Tibullo, D.; Li Volti, G. Heme Oxygenase-1 and Carbon Monoxide Regulate Growth and Progression in Glioblastoma Cells. *Mol. Neurobiol.* **2020**, *57*, 2436–2446. [CrossRef]
6. Di Pietro, C.; Öz, H.H.; Murray, T.S.; Bruscia, E.M. Targeting the Heme Oxygenase 1/Carbon Monoxide Pathway to Resolve Lung Hyper-Inflammation and Restore a Regulated Immune Response in Cystic Fibrosis. *Front. Pharmacol.* **2020**, *11*, 1059. [CrossRef]
7. Pol, O. The Role of Carbon Monoxide, Heme Oxygenase 1, and the Nrf2 Transcription Factor in the Modulation of Chronic Pain and Their Interactions with Opioids and Cannabinoids. *Med. Res. Rev.* **2021**, *41*, 136–155. [CrossRef]
8. Motterlini, R.; Otterbein, L.E. The Therapeutic Potential of Carbon Monoxide. *Nat. Rev. Drug. Discov.* **2010**, *9*, 728–743. [CrossRef]
9. Foresti, R.; Bani-Hani, M.G.; Motterlini, R. Use of Carbon Monoxide as a Therapeutic Agent: Promises and Challenges. *Intensive Care Med.* **2008**, *34*, 649–658. [CrossRef]
10. Knauert, M.; Vangala, S.; Haslip, M.; Lee, P.J. Therapeutic Applications of Carbon Monoxide. *Oxid. Med. Cell. Longev.* **2013**, *2013*, e360815. [CrossRef]
11. Hess, D.R. Inhaled Carbon Monoxide: From Toxin to Therapy. *Respir. Care* **2017**, *62*, 1333–1342. [CrossRef] [PubMed]
12. Adach, W.; Błaszczyk, M.; Olas, B. Carbon Monoxide and Its Donors—Chemical and Biological Properties. *Chem. Biol. Interact* **2020**, *318*, 108973. [CrossRef] [PubMed]
13. Goebel, U.; Wollborn, J. Carbon Monoxide in Intensive Care Medicine—Time to Start the Therapeutic Application?! *Intensive Care Med. Exp.* **2020**, *8*, 2. [CrossRef] [PubMed]
14. Ling, K.; Men, F.; Wang, W.-C.; Zhou, Y.-Q.; Zhang, H.-W.; Ye, D.-W. Carbon Monoxide and Its Controlled Release: Therapeutic Application, Detection, and Development of Carbon Monoxide Releasing Molecules (CORMs). *J. Med. Chem.* **2018**, *61*, 2611–2635. [CrossRef]
15. Cavicchioli, F.; Cesarotti, I.M.; Fangman, M.; Lua, J.; Hautamaki, R.; Doré, S. Carbon Monoxide Therapy Using Hybrid Carbon Monoxide-Releasing/Nrf2-Inducing Molecules through a Neuroprotective Lens. *Chemistry* **2021**, *3*, 800–817. [CrossRef]
16. Kautz, A.C.; Kunz, P.C.; Janiak, C. CO-Releasing Molecule (CORM) Conjugate Systems. *Dalton Trans.* **2016**, *45*, 18045–18063. [CrossRef]
17. Romão, C.C.; Blättler, W.A.; Seixas, J.D.; Bernardes, G.J.L. Developing Drug Molecules for Therapy with Carbon Monoxide. *Chem. Soc. Rev.* **2012**, *41*, 3571–3583. [CrossRef]
18. Ott, I.; Kircher, B.; Bagowski, C.P.; Vlecken, D.H.W.; Ott, E.B.; Will, J.; Bensdorf, K.; Sheldrick, W.S.; Gust, R. Modulation of the Biological Properties of Aspirin by Formation of a Bioorganometallic Derivative. *Angew. Chem. Int. Ed. Engl.* **2009**, *48*, 1160–1163. [CrossRef]
19. Zanellato, I.; Bonarrigo, I.; Ravera, M.; Gabano, E.; Gust, R.; Osella, D. The Hexacarbonyldicobalt Derivative of Aspirin Acts as a CO-Releasing NSAID on Malignant Mesothelioma Cells. *Metallomics* **2013**, *5*, 1604–1613. [CrossRef]
20. Heffern, M.C.; Yamamoto, N.; Holbrook, R.J.; Eckermann, A.L.; Meade, T.J. Cobalt Derivatives as Promising Therapeutic Agents. *Curr. Opin. Chem. Biol.* **2013**, *17*, 189–196. [CrossRef]
21. Gong, Y.; Zhang, T.; Liu, H.; Zheng, Y.; Li, N.; Zhao, Q.; Chen, Y.; Liu, B. Synthesis, Toxicities and Cell Proliferation Inhibition of CO-Releasing Molecules Containing Cobalt. *Transit. Met. Chem.* **2015**, *40*, 413–426. [CrossRef]
22. Berrino, E.; Carradori, S.; Angeli, A.; Carta, F.; Supuran, C.T.; Guglielmi, P.; Coletti, C.; Paciotti, R.; Schweikl, H.; Maestrelli, F.; et al. Dual Carbonic Anhydrase IX/XII Inhibitors and Carbon Monoxide Releasing Molecules Modulate LPS-Mediated Inflammation in Mouse Macrophages. *Antioxidants* **2021**, *10*, 56. [CrossRef]
23. Berrino, E.; Milazzo, L.; Micheli, L.; Vullo, D.; Angeli, A.; Bozdag, M.; Nocentini, A.; Menicatti, M.; Bartolucci, G.; di Cesare Mannelli, L.; et al. Synthesis and Evaluation of Carbonic Anhydrase Inhibitors with Carbon Monoxide Releasing Properties for the Management of Rheumatoid Arthritis. *J. Med. Chem.* **2019**, *62*, 7233–7249. [CrossRef]
24. Li, J.; Zhang, J.; Zhang, Q.; Bai, Z.; Zhao, Q.; He, D.; Wang, Z.; Chen, Y.; Liu, B. Syntheses and Anti-Cancer Activity of CO-Releasing Molecules with Targeting Galactose Receptors. *Org. Biomol. Chem.* **2018**, *16*, 8115–8129. [CrossRef] [PubMed]
25. Perontsis, S.; Dimitriou, A.; Fotiadou, P.; Hatzidimitriou, A.G.; Papadopoulos, A.N.; Psomas, G. Cobalt(II) Complexes with the Non-Steroidal Anti-Inflammatory Drug Diclofenac and Nitrogen-Donor Ligands. *J. Inorg. Biochem.* **2019**, *196*, 110688. [CrossRef] [PubMed]
26. Gallorini, M.; Berardi, A.C.; Ricci, A.; Antonetti Lamorgese Passeri, C.; Zara, S.; Oliva, F.; Cataldi, A.; Carta, F.; Carradori, S. Dual Acting Carbon Monoxide Releasing Molecules and Carbonic Anhydrase Inhibitors Differentially Modulate Inflammation in Human Tenocytes. *Biomedicines* **2021**, *9*, 141. [CrossRef]
27. Oliva, F.; Gallorini, M.; Antonetti Lamorgese Passeri, C.; Gissi, C.; Ricci, A.; Cataldi, A.; Colosimo, A.; Berardi, A.C. Conjugation with Methylsulfonylmethane Improves Hyaluronic Acid Anti-Inflammatory Activity in a Hydrogen Peroxide-Exposed Tenocyte Culture In Vitro Model. *Int. J. Mol. Sci.* **2020**, *21*, 7956. [CrossRef]
28. Darrieutort-Laffite, C.; Soslowsky, L.J.; Le Goff, B. Molecular and Structural Effects of Percutaneous Interventions in Chronic Achilles Tendinopathy. *Int. J. Mol. Sci.* **2020**, *21*, 7000. [CrossRef] [PubMed]

29. Zara, S.; De Colli, M.; di Giacomo, V.; Zizzari, V.L.; Di Nisio, C.; Di Tore, U.; Salini, V.; Gallorini, M.; Tetè, S.; Cataldi, A. Zoledronic acid at subtoxic dose extends osteoblastic stage span of primary human osteoblasts. *Clin. Oral. Investig.* **2015**, *19*, 601–611. [CrossRef] [PubMed]
30. Marconi, G.D.; Gallorini, M.; Carradori, S.; Guglielmi, P.; Cataldi, A.; Zara, S. The Up-Regulation of Oxidative Stress as a Potential Mechanism of Novel MAO-B Inhibitors for Glioblastoma Treatment. *Molecules* **2019**, *24*, 2005. [CrossRef] [PubMed]
31. Atkin, A.J.; Lynam, J.M.; Moulton, B.E.; Sawle, P.; Motterlini, R.; Boyle, N.M.; Pryce, M.T.; Fairlamb, I.J.S. Modification of the Deoxy-Myoglobin/Carbonmonoxy-Myoglobin UV-Vis Assay for Reliable Determination of CO-Release Rates from Organometallic Carbonyl Complexes. *Dalton Trans.* **2011**, *40*, 5755–5761. [CrossRef]
32. Smulevich, G.; Droghetti, E.; Focardi, C.; Coletta, M.; Ciaccio, C.; Nocentini, M. A Rapid Spectroscopic Method to Detect the Fraudulent Treatment of Tuna Fish with Carbon Monoxide. *Food Chem.* **2007**, *101*, 1071–1077. [CrossRef]
33. Wilson, J.L.; Fayad Kobeissi, S.; Oudir, S.; Haas, B.; Michel, B.; Dubois Randé, J.-L.; Ollivier, A.; Martens, T.; Rivard, M.; Motterlini, R.; et al. Design and Synthesis of New Hybrid Molecules That Activate the Transcription Factor Nrf2 and Simultaneously Release Carbon Monoxide. *Chem. Eur. J.* **2014**, *20*, 14698–14704. [CrossRef] [PubMed]
34. García-Gallego, S.; Bernardes, G.J.L. Carbon-Monoxide-Releasing Molecules for the Delivery of Therapeutic CO in Vivo. *Angew. Chem. Int. Ed. Engl.* **2014**, *53*, 9712–9721. [CrossRef] [PubMed]
35. Biava, M.; Porretta, G.C.; Poce, G.; Supino, S.; Forli, S.; Rovini, M.; Cappelli, A.; Manetti, F.; Botta, M.; Sautebin, L.; et al. Cyclooxygenase-2 Inhibitors. 1,5-Diarylpyrrol-3-Acetic Esters with Enhanced Inhibitory Activity toward Cyclooxygenase-2 and Improved Cyclooxygenase-2/Cyclooxygenase-1 Selectivity. *J. Med. Chem.* **2007**, *50*, 5403–5411. [CrossRef] [PubMed]
36. Biava, M.; Porretta, G.C.; Poce, G.; Supino, S.; Manetti, F.; Forli, S.; Botta, M.; Sautebin, L.; Rossi, A.; Pergola, C.; et al. Synthesis, in Vitro, and in Vivo Biological Evaluation and Molecular Docking Simulations of Chiral Alcohol and Ether Derivatives of the 1,5-Diarylpyrrole Scaffold as Novel Anti-Inflammatory and Analgesic Agents. *Bioorg. Med. Chem.* **2008**, *16*, 8072–8081. [CrossRef]
37. Biava, M.; Porretta, G.C.; Poce, G.; Battilocchio, C.; Manetti, F.; Botta, M.; Forli, S.; Sautebin, L.; Rossi, A.; Pergola, C.; et al. Novel Ester and Acid Derivatives of the 1,5-Diarylpyrrole Scaffold as Anti-Inflammatory and Analgesic Agents. Synthesis and in Vitro and in Vivo Biological Evaluation. *J. Med. Chem.* **2010**, *53*, 723–733. [CrossRef]
38. Battilocchio, C.; Poce, G.; Alfonso, S.; Porretta, G.C.; Consalvi, S.; Sautebin, L.; Pace, S.; Rossi, A.; Ghelardini, C.; Di Cesare Mannelli, L.; et al. A Class of Pyrrole Derivatives Endowed with Analgesic/Anti-Inflammatory Activity. *Bioorg. Med. Chem.* **2013**, *21*, 3695–3701. [CrossRef]
39. Consalvi, S.; Alfonso, S.; Di Capua, A.; Poce, G.; Pirolli, A.; Sabatino, M.; Ragno, R.; Anzini, M.; Sartini, S.; La Motta, C.; et al. Synthesis, Biological Evaluation and Docking Analysis of a New Series of Methylsulfonyl and Sulfamoyl Acetamides and Ethyl Acetates as Potent COX-2 Inhibitors. *Bioorg. Med. Chem.* **2015**, *23*, 810–820. [CrossRef]
40. Bergqvist, F.; Carr, A.J.; Wheway, K.; Watkins, B.; Oppermann, U.; Jakobsson, P.-J.; Dakin, S.G. Divergent Roles of Prostacyclin and PGE2 in Human Tendinopathy. *Arthritis Res. Ther.* **2019**, *21*, 74. [CrossRef]
41. Osti, L.; Berardocco, M.; di Giacomo, V.; Di Bernardo, G.; Oliva, F.; Berardi, A.C. Hyaluronic Acid Increases Tendon Derived Cell Viability and Collagen Type I Expression in Vitro: Comparative Study of Four Different Hyaluronic Acid Preparations by Molecular Weight. *BMC Musculoskelet Disord.* **2015**, *16*, 284. [CrossRef] [PubMed]
42. Gallorini, M.; Petzel, C.; Bolay, C.; Hiller, K.-A.; Cataldi, A.; Buchalla, W.; Krifka, S.; Schweikl, H. Activation of the Nrf2-Regulated Antioxidant Cell Response Inhibits HEMA-Induced Oxidative Stress and Supports Cell Viability. *Biomaterials* **2015**, *56*, 114–128. [CrossRef]
43. Sharma, P.; Maffulli, N. Biology of Tendon Injury: Healing, Modeling and Remodeling. *J. Musculoskelet. Neuronal. Interact.* **2006**, *6*, 181–190. [PubMed]
44. Canesin, G.; Hejazi, S.M.; Swanson, K.D.; Wegiel, B. Heme-Derived Metabolic Signals Dictate Immune Responses. *Front. Immunol.* **2020**, *11*, 66. [CrossRef] [PubMed]
45. Fernández-Fierro, A.; Funes, S.C.; Rios, M.; Covián, C.; González, J.; Kalergis, A.M. Immune Modulation by Inhibitors of the HO System. *Int. J. Mol. Sci.* **2020**, *22*, 294. [CrossRef] [PubMed]
46. Campbell, N.K.; Fitzgerald, H.K.; Dunne, A. Regulation of Inflammation by the Antioxidant Haem Oxygenase 1. *Nat. Rev. Immunol.* **2021**, *21*, 411–425. [CrossRef] [PubMed]
47. Frich, L.H.; Fernandes, L.R.; Schrøder, H.D.; Hejbøl, E.K.; Nielsen, P.V.; Jørgensen, P.H.; Stensballe, A.; Lambertsen, K.L. The Inflammatory Response of the Supraspinatus Muscle in Rotator Cuff Tear Conditions. *J. Shoulder. Elbow. Surg.* **2021**, *30*, e261–e275. [CrossRef]
48. Dakin, S.G.; Newton, J.; Martinez, F.O.; Hedley, R.; Gwilym, S.; Jones, N.; Reid, H.A.B.; Wood, S.; Wells, G.; Appleton, L.; et al. Chronic Inflammation Is a Feature of Achilles Tendinopathy and Rupture. *Br. J. Sports. Med.* **2018**, *52*, 359–367. [CrossRef] [PubMed]

antioxidants

Review

Reactive Oxygen Species (ROS) and Antioxidants as Immunomodulators in Exercise: Implications for Heme Oxygenase and Bilirubin

David Travis Thomas [1], Nicholas R. DelCimmuto [2], Kyle D. Flack [3], David E. Stec [4] and Terry D. Hinds, Jr. [5,6,7,*]

1 Department of Athletic Training and Clinical Nutrition, College of Health Sciences, University of Kentucky, Lexington, KY 40506, USA; dth225@uky.edu
2 Department of Neurosciences, University of Toledo College of Medicine, Toledo, OH 43614, USA; Nicholas.Delcimmuto@rockets.utoledo.edu
3 Department of Dietetics and Human Nutrition, University of Kentucky, Lexington, KY 40506, USA; kyle.flack@uky.edu
4 Center for Excellence in Cardiovascular-Renal Research, Department of Physiology & Biophysics, University of Mississippi Medical Center, 2500 North State St, Jackson, MS 392161, USA; dstec@umc.edu
5 Department of Pharmacology and Nutritional Sciences, University of Kentucky, Lexington, KY 40508, USA
6 Barnstable Brown Diabetes Center, University of Kentucky, Lexington, KY 40508, USA
7 Markey Cancer Center, University of Kentucky, Lexington, KY 40508, USA
* Correspondence: Terry.Hinds@uky.edu

Abstract: Exercise is commonly prescribed as a lifestyle treatment for chronic metabolic diseases as it functions as an insulin sensitizer, cardio-protectant, and essential lifestyle tool for effective weight maintenance. Exercise boosts the production of reactive oxygen species (ROS) and subsequent transient oxidative damage, which also upregulates counterbalancing endogenous antioxidants to protect from ROS-induced damage and inflammation. Exercise elevates heme oxygenase-1 (HO-1) and biliverdin reductase A (BVRA) expression as built-in protective mechanisms, which produce the most potent antioxidant, bilirubin. Together, these mitigate inflammation and adiposity. Moderately raising plasma bilirubin protects in two ways: (1) via its antioxidant capacity to reduce ROS and inflammation, and (2) its newly defined function as a hormone that activates the nuclear receptor transcription factor PPARα. It is now understood that increasing plasma bilirubin can also drive metabolic adaptions, which improve deleterious outcomes of weight gain and obesity, such as inflammation, type II diabetes, and cardiovascular diseases. The main objective of this review is to describe the function of bilirubin as an antioxidant and metabolic hormone and how the HO-1–BVRA–bilirubin–PPARα axis influences inflammation, metabolic function and interacts with exercise to improve outcomes of weight management.

Keywords: HO-1; biliverdin reductase; BVRA; PPARα; bilirubin; inflammation; metabolic disease; nutraceuticals; vitamin D; vitamin E; nitrate

Citation: Thomas, D.T.; DelCimmuto, N.R.; Flack, K.D.; Stec, D.E.; Hinds, T.D., Jr. Reactive Oxygen Species (ROS) and Antioxidants as Immunomodulators in Exercise: Implications for Heme Oxygenase and Bilirubin. *Antioxidants* **2022**, *11*, 179. https://doi.org/10.3390/antiox11020179

Academic Editors: Elias Lianos and Maria G. Detsika

Received: 9 December 2021
Accepted: 14 January 2022
Published: 18 January 2022

1. Introduction

Obesity and ectopic lipid accumulation are key contributing hallmarks of metabolic dysfunction, which is the cornerstone of pathogenesis for most comorbidities [1–3]. People with a BMI greater than 30 (>30) have an increased risk for obesity-associated comorbidities that include cardiovascular disease, hypertension, insulin-resistant diabetes, dyslipidemia, and certain cancers [4–7]. Alterations in lipid metabolism also contribute to ectopic lipid accumulation, exacerbating metabolic disorders, especially when combined with limited physical activity. Understanding and combating metabolic dysfunction is essential for improving clinical outcomes and quality of life.

Obesity treatment has been challenging, and exercise continues to be the foundation for obesity prevention and treatment. Despite the continued interest in exercise training in obesity, there are still challenges, including patients falling short on meeting exercise recommendations over time and limited effectiveness of exercise as a sole driver for stimulating weight loss [8–10].

A common theme found in the literature on the healthy athletic population is how different nutrients, hormones, dietary supplements, and other forms of ergogenic aids [referred to herein as "exercise enhancers" (EEs)] can improve exercise training outcomes to enhance athletic performance. There is also emerging evidence that EEs may augment the metabolic benefits of exercise and, in some cases, modulate inflammation. This review provides a brief overview of exercise in preventing metabolic dysfunction along with the potential role of select antioxidants (i.e., bilirubin and others), vitamin D, and nitrates on improving metabolic outcomes associated with exercise. The primary focus will then shift to describing bilirubin's emerging significance as a potential EE due to its role as a strong antioxidant and metabolic hormone.

This review describes how exercise interacts with bilirubin to further sensitize these newly defined antioxidant and protective metabolic functions as a hormone. The role of exercise and its influence on bilirubin catabolism will be discussed along with proposed theories on how bilirubin may influence physiological adaptations associated with exercise training and how this might impact inflammatory responses. A primary mechanism discussed postulates that as exercise increases reactive oxygen species (ROS) production, increased heme oxygenase (HO-1) activity raises plasma bilirubin levels, which can also directly bind and activate PPARα (peroxisome proliferator-activated receptor α) in metabolic tissues (e.g., adipose, liver, and muscle), which might explain some of the therapeutic benefits observed with exercise (Figure 1). Other important mediators such as HO-1 and PPARs and their impact on exercise and inflammation will be discussed.

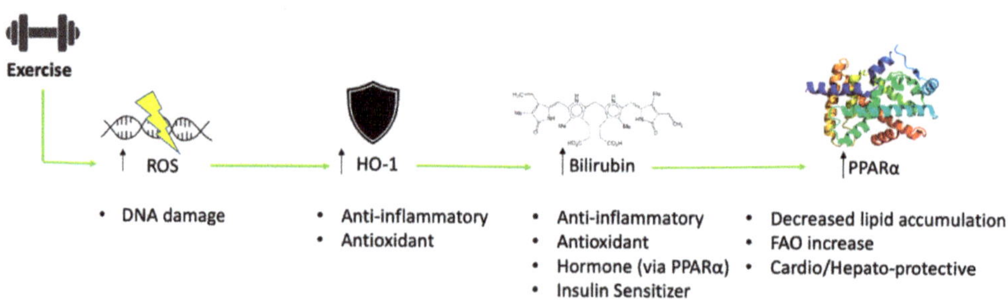

Figure 1. Overview of heme oxygenase and bilirubin interaction with exercise. Exercise increases reactive oxygen species (ROS) and potentiates oxidative DNA damage. The body compensates with oxidative stress by upregulating heme oxygenase-1 (HO-1), which generates the antioxidant bilirubin to help prevent excessive oxidative damage. Bilirubin also directly binds to the PPARα nuclear receptor to induce genes that suppress lipid accumulation and has cardiogenic and hepato-protective effects.

2. The Effect of Exercise on Weight Management and Inflammation

Exercise is regularly prescribed as a first-line treatment in preventing type 2 diabetes, coronary artery disease (CAD), and non-alcoholic fatty liver disease (NAFLD) [11–13]. It has strong therapeutic effects that usually meet or exceed expected improvements in metabolic function from pharmaceutical treatment [14]. Exercise training triggers significant metabolic adaptations that improve cardiorespiratory fitness, promoting greater capillary density and increases in HDL synthesis to protect from CAD [14,15]. Exercise also enhances glucose uptake through elevating translocation of GLUT4 in skeletal muscle and by increasing IRS-1 phosphorylation, an insulin receptor substrate that improves insulin sensitivity [16,17]. Therefore, exercise can be a reliable first-line and preventative therapeutic for type II diabetes by decreasing blood glucose [18] and CAD by reducing atherosclerotic plaque buildup and subsequent risk of stroke and myocardial infarction [19].

Although exercise training can improve blood glucose control, insulin sensitivity, and other aspects of metabolic syndrome without weight loss, these benefits are substantially greater when significant weight loss occurs [20–23]. Indeed, the American College of Sports Medicine issued separate recommendations to maintain health [24] or support weight loss through exercise [25]. Exercise is also one of the primary recommendations of the Diabetes Prevention Program (DPP) and a pivotal component to the classic Look AHEAD trial primarily due to the role that exercise is thought to play in weight loss and weight management [26,27]. Unfortunately, weight loss from exercise is often suboptimal due to compensatory mechanisms that resist the maintenance of an energy deficit [8–10]. For instance, an individual may exercise to expend 3000 kcal per week for ten weeks through exercise to expend a total of 30,000 kcal. However, this individual rarely loses 30,000 kcal of body mass. By comparing changes in bodily energy stores with the amount of total energy expended through exercise, we have demonstrated this compensatory response to equate to roughly 1000 kcal per week during a 12-week exercise intervention, and that energy expenditures of greater than 2700 kcal per week are needed to achieve significant weight loss after 12 weeks. [28,29]. Others have reported that greater amounts of exercise can evoke a proportionally greater compensatory response [30], potentially explaining why exercise interventions with large differences in daily and weekly exercise energy expenditures can promote similar weight loss [30–33]. Because of this, many have turned to various EE's to improve both weight loss and metabolic health outcomes with exercise. We will discuss some specific studies on EE's below and how they might impact exercise-induced ROS and inflammation.

Exercise-Induced Formation of Reactive Oxygen Species and Select Antioxidant Defense Mechanisms

ROS generation from exercise has a significant role in triggering and sustaining the healthy cellular, tissue, and organ level adaptations that help improve and maintain cardio-metabolic health. Acute ROS generation from exercise occurs via electron transport oxygen catabolism in the skeletal muscle. This is triggered by a substantial increase in mitochondrial oxygen uptake into the skeletal muscle cell, increasing ROS production [34]. Exercise-induced ROS production can also produce muscle injury, which sends inflammatory signals that attract polymorphonuclear neutrophils and macrophages and produce additional ROS in its defense mechanism of oxidative burst [35]. If there is a chronic imbalance of more ROS production than antioxidant activity, chronic oxidative stress may lead to apoptotic pathways in tissues [36]. A skeletal muscle oxidative stress imbalance is commonly seen in untrained individuals who begin a strenuous training program or "weekend-warriors" who perform a single bout of infrequent strenuous exercise (Figure 2). Although exercise is known to increase the abundance of ROS, progressive exercise training allows time for the upregulation of defense mechanisms that help protect the body from oxidative damage. This is known as redox balance (Figure 3), where free radicals are balanced by the adaptive antioxidants produced [37]. Exercise may stimulate the generation of antioxidants by triggering significant cell adaptations and upregulating antioxidant-producing enzymes [38]. Those who are exercise-trained and perform a single bout of exercise can leverage the benefits of endogenous antioxidant upregulation, along with mitochondrial expansion, cryoprotection, and insulin sensitivity [35]. The importance of ROS is highlighted in experimental models treated with allopurinol. This compound inhibits ROS production and protects muscle tissue from oxidative stress. However, because exercise-induced ROS was not produced, the important adaptive signaling pathways for oxidative protection were blunted. Thus, the formation of ROS in exercise can help activate these intrinsic protective pathways. Notwithstanding, allopurinol is a competitive inhibitor for xanthine oxidase that produces H_2O_2. H_2O_2 is a kind of ROS; thus, xanthine oxidase is an intrinsic prooxidant. This adaptive signaling response is an important body regulation, and if extrinsic, pharmaceutical dose antioxidants are administered, the body's natural adaptive ability to produce in trinsic antioxidants may be thwarted [38].

Figure 2. Relationship of HO-1 and ROS in habitually trained versus untrained subjects. Individuals who have performed a single bout of exercise (non-trained) versus individuals with exercise training experience (trained).

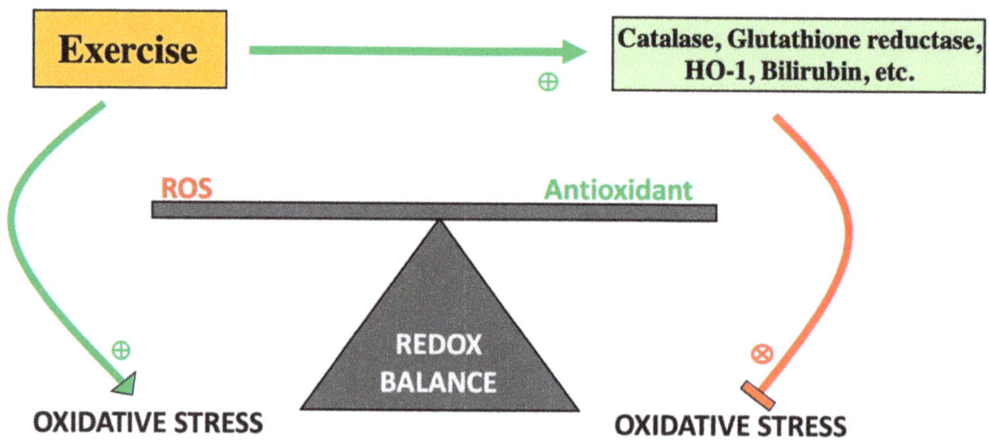

Figure 3. Redox Balance of ROS and Antioxidants. Exercise potentiates the release of reactive oxygen species due to increased oxidative exposure. However, exercise training also induces an adaptive response with the upregulation of antioxidant defense mechanisms that will help restore redox balance. The downregulation of endogenous antioxidant systems or the increased production of reactive oxygen species can precipitate an imbalance in redox balance and potentiate chronic oxidative damage.

Ultimately, exercise-related ROS adaptations improve oxygen transport and delivery that translate into better aerobic fitness that help explain many of the health benefits of exercise. Furthermore, upregulation of endogenous antioxidant systems can work in concert with exogenous dietary antioxidants to mitigate ROS-related tissue damage and support normal metabolic function and healthy aging. Conversely, the accumulation of ROS and inadequate ROS defense responses has been implicated as a key mechanism leading to significant atrophy in muscle tissue. Muscle atrophy due to chronic excessive ROS exposure progresses slowly as part of the normal aging process but is more pronounced and accelerated in severe underlying pathological processes such as in cancer wasting (cancer cachexia) [39], neurodegenerative diseases (Parkinson's, Alzheimer's), and immobilization (musculoskeletal injury) [40].

3. HO-1, BVRA, and Bilirubin as Inflammatory Mediators

3.1. Exercise-Induced HO-1 as a Mediator of Immune System Responses

HO-1 produces a known potent antioxidant and enzyme responsible for the cleavage of heme, yielding biliverdin, iron (Fe^{2+}), and carbon monoxide (CO) [41,42]. The HO-1 pathway also regulates some of the metabolic and inflammatory aspects of insulin resistance. While there is a connection between inflammation and the development of insulin resistance, it is unclear which development precedes and which is causative [43]. HO-1's role in inflammation and insulin resistance appears equivocal in the literature.

The presence of HO-1 mimics the same efficacious properties as bilirubin [44]. Bilirubin downregulates the M1 macrophages associated with the release of pro-inflammatory cytokines [45–47]. Future work to understand how HO-1 may affect M1 cells may shed light on potential underlying mechanisms to explain how bilirubin downregulates M1 cells. HO-1 also protects the liver from ischemia-reperfusion injury by modulating the macrophage phenotype into the anti-inflammatory M2 state in mouse livers [48,49]. This serves as evidence for an HO-1 role as a hepatic cryoprotective agent. In this same study, low HO-1 mRNA levels in human liver transplants correlated with increased expression of M1 pro-inflammatory markers [48,49]. Liver-specific biliverdin reductase A (BVRA) knockout animals with reduced hepatic bilirubin had worsened fatty liver on a high-fat diet compared to littermate controls [50], which was confirmed in global BVRA knockout animals [51]. Bilirubin reduces lipid content and inflammatory markers in mouse models of obesity-induced NAFLD [46,47].

Work by Gobert et al. found HO-1 to prevent an inflammatory response and has implicated HO-1 as a virulence factor in *H. Pylori* and other bacteria in order to evade the immune system [52]. Other work has described using a heme-inducing compound to effectively reduce obesity, insulin sensitivity and increase serum adiponectin levels. Inhibition of the HO-1 system decreased adiponectin and increased pro-inflammatory cytokines, TNFα, IL-6, and IL-1 [53–55]. Adiponectin, a known anti-inflammatory hormone, is thought to be working indirectly through HO-1-activation [56]. The complete mechanism of this anti-inflammatory activity is not fully understood, but some theories with convincing evidence reveal new insight on HO-1 and the importance of its catabolic products.

The HO-1 pathway can decrease inflammation by producing biliverdin/bilirubin, which has protective anti-inflammatory effects, especially in vascular endothelial tissue [57]. Another anti-inflammatory action of HO-1 is through carbon monoxide production, which is a known cryoprotectant and anti-apoptotic factor in endothelial cells that have the potential to crosstalk with nitric oxide, a known vasodilator. Although this work serves as further evidence that HO-1 has important underlying anti-inflammatory and insulin-sensitizing mechanisms that may augment bilirubin's therapeutic value, it is essential to note that the supporting evidence is not unequivocal. In contrast to these aforementioned findings, conflicting data suggest that HO-1 is implicated in driving inflammation and may even support insulin resistance in humans. Jais et al. demonstrated that HO-1 levels predict a strong positive prediction of metabolic disease in human subjects [58], while Ghio et al. reported HO-1 elevation due to cytokine stimulation in inflammatory disease [59]. Whether HO-1 is present in response to the inflammation or if it is the direct cause of inflammation is not completely clear.

Although HO-1's direct role on insulin resistance and inflammation has not been fully elucidated, the influence of exercise on the HO-1 pathway may shed light on these equivocal data. Niess et al. showed that HO-1 expression in leukocytes increased significantly after sustained endurance exercise in marathon runners. The authors interpreted this to be due to the excessive amount of free radical production, although the mechanism that causes this upregulation of HO-1 in exercise is not completely clear [60]. However, it can be postulated that since exercise promotes ROS generation, it would induce nuclear factor (erythroid-derived 2)-like-2 (Nrf2) expression, which is a key transcription factor in inducing HO-1 [61–63]. A potential explanation for the upregulation of HO-1 may be that acute exercise can also propagate a transient pro-inflammatory state to increase levels of HO-1 via increased cytokine activity. Others have suggested that ROS, themselves, can induce and upregulate HO-1. Kurata et al. found that the HO gene was induced by 12-O-tetradecanoylphorbol 13-acetate response element (TRE) in the presence of hydrogen peroxide, a ROS [64]. These oxidative free radical levels vary based on habitually trained versus untrained subjects. The trained individuals had a much more robust adaptive antioxidant defense system and thus a lower level of ROS production [65]. HO-1 levels at rest are significantly reduced in trained subjects compared to untrained subjects [60]. This suggests an adaptive regulatory feedback mechanism to which, at rest, basal ROS are downregulated in trained individuals and hence, a lower HO-1 level. The prevailing hypothesis surrounding this observation is that HO-1 is upregulated to offer protection from the free radicals that are produced with exercise (Figure 3) [59].

3.2. The Emerging Role of Biliverdin Reductase in Immune Response

BVRA plays a vital role in macrophage polarization and as a target for regulating responses to bacterial lipopolysaccharides and complement activation products. BVRA is expressed in macrophages where it is tyrosine phosphorylated. Phosphorylated BVRA then binds to phosphatidylinositol 3-kinase (PI3K) at the p85α subunit to activate downstream signaling to Akt [66,67]. Macrophage classification occurs according to activation state and function. M-1 macrophages are classically activated macrophages that express cytokines such as TNFα and interleukin-17A. M-2 macrophages are alternatively activated macrophages that express anti-inflammatory cytokines such as interleukin-10 (IL-10) and transforming growth factor-beta (TGFβ). Overexpression of BVRA in macrophages elevates expression of M-2 macrophage markers, while knockdown of BVRA increases M-1 macrophage markers [68]. Renal ischemia-reperfusion injury increases the levels of BVRA positive macrophages increasing the levels of IL-10, helping in the reparative process [68]. The recruitment of macrophages is an influential process in the inflammatory response. Release of chemokines that act on specific receptors such as the complement activation fragment 5a receptor one (C5aR1) recruits macrophages to sites of tissue injury. Loss of macrophage BVRA results in greater levels of C5aR1 increasing inflammation [69]. These studies demonstrate the critical role of BVRA in both macrophage chemotaxis and polarization. Augmentation of macrophage BVRA levels may be an effective treatment to bolster anti-inflammatory pathways in a number of inflammatory diseases. How they might affect metabolic adaptations to exercise is yet to be determined.

3.3. The Effect of Exercise on Bilirubin and Its Actions

Given that HO-1 expression is directly influenced by exercise training (Figure 4), it is logical to assume that exercise increases plasma bilirubin levels. Hinds et al. recently conducted a study where rats genetically selected for high capacity running (HCR) and low capacity running (LCR) were used to identify the metabolic pathways in the liver altering plasma bilirubin levels through exercise [70]. The investigators observed that HCR rats had significantly greater plasma bilirubin and hepatic BVRA expression while having a reduced expression of the glucuronyl hepatic enzyme UGT1A1. Significant increases in PPARα-target genes were also observed in HCR rats compared to the LCR. For the first time, these results suggest hepatic mechanisms involved in bilirubin synthesis and metabolism that may explain the positive effects of exercise on plasma bilirubin and metabolic health.

There are a limited number of articles published on this topic in humans [71,72]. In a controlled study that examined different levels of training intensity, researchers found that the high-intensity training group (defined as 12 kilocalories per kilogram per week (KKW) energy expenditure) presented a significant increase in total serum bilirubin in comparison to the sedentary control group. Those who trained at moderate intensity levels (defined as 4 and 8 KKW) experienced no significant differences in serum bilirubin levels [71]. Priest et al. observed an increase in bilirubin in male runners after a 13-mile run along with an increase in alkaline phosphatase. Bile acids and bilirubin have been shown to be elevated in these runners [73]. In both studies, bilirubin levels seem to be elevated in response to high-volume, exhaustive forms of exercise with high energy expenditure.

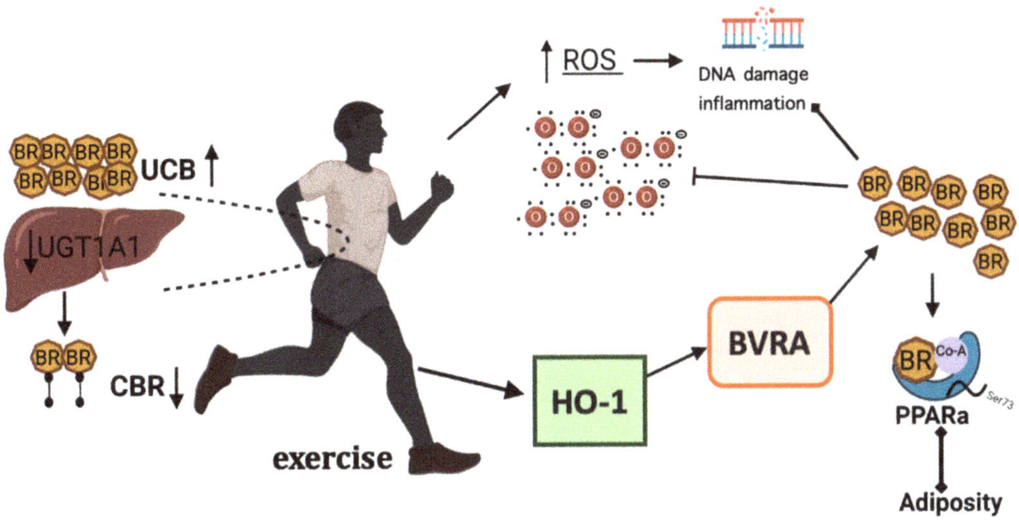

Figure 4. The heme oxygenase pathway signaling during exercise. Exercising (aerobic) raises plasma bilirubin levels by (1) suppression of the glucuronyl transferase enzyme UGT1A1 that conjugates bilirubin for removal from blood, and (2) activation of the heme oxygenase pathway (HO-1-BVRA-PPARα). The increased bilirubin combats reactive oxygen species (ROS) and ROS-induced inflammation and DNA damage. The bilirubin also activates the nuclear receptor transcription factor, PPARα, to reduce adiposity. Created with BioRender.com (accessed on 9 December 2021).

A subgroup analysis from Swift et al. revealed another interesting trend that showed those who were insulin resistant in the high-intensity exercise group had a significant increase in bilirubin compared to the insulin-sensitive group [71]. A more recent study confirmed these observations by examining why moderate-to-vigorous physical activity (MVPA) resulted in a significant increase in serum bilirubin in insulin-resistant subjects but not in insulin-sensitive subjects. The authors hypothesized that the observed increase in bilirubin in the insulin-resistant subjects could be due to a lower basal level of bilirubin, resulting in a more remarkable absolute change in bilirubin in response to MVPA [74]. The underlying rationale for this pattern of bilirubin change in response to different exercise volumes should be further explored to improve our understanding of the connection between insulin resistance and changing bilirubin levels.

Several studies in athletes have also reported a strong correlation between elevated bilirubin and the degree of exercise intensity and an associated increase in erythrocyte hemolysis [71,75–77]. Witek et al. reported normal bilirubin reference ranges for 339 male and female Polish athletes [72]. While approximately 45% of the samples had bilirubin levels in the range of 7–14 μM, 12% of the athletes had 21–28 μM. Nineteen percent of the total bilirubin values exceeded the established normal limit of 21 μM. These elevated concentrations appeared to be related to changes caused by regular exercise and were not directly related to increased hemolysis. The authors suggested that other exercise-induced mechanisms seem to affect bilirubin concentrations, such as altered liver function and upregulation of bilirubin production (to serve as an antioxidant) in response to increased oxidative stress (ROS). A study of young Polish athletes (aged 18–40 years) reported that bilirubin levels increased in response to both a ketogenic diet and short-term, high-intensity exercise (CrossFit) [78]. Study subjects increased their bilirubin concentrations in both diet groups in response to exercise (Customary diet: 10 ± 5 to 19 ± 8; Ketogenic diet: 14 ± 0 to 20 ± 8 μM; $p < 0.05$). These studies bring to question if bilirubin levels are being controlled by exercise to correct metabolic imbalances, mitigate oxidative stress, and reduce inflammation.

3.4. The Hormonal Function of Bilirubin in Exercise and the Impact of PPAR Signaling

The PPARs are a subfamily of ligand-activated nuclear receptor transcription factors with three distinct isoforms: α, β/δ, γ [79]. These isoforms are found in different tissues, each with a predominant isoform. PPARα is expressed in hepatocytes, enterocytes, and vascular endothelium and works to improve mitochondrial efficacy in FA oxidation in these tissue types. PPARβ/δ are expressed more ubiquitously in the body but predominate in skeletal muscle and macrophages and are important in fatty acid oxidation and macrophage immunosuppression through the reduction in NF-κB inflammatory cytokines [80,81]. PPARγ is found mainly in white and brown fat adipocytes and enhances genes involved in the metabolism of glucose and adipocyte differentiation [82–86]. PPAR's are activated in the presence of their specific corresponding natural or synthetic pharmacological ligands. All PPAR isoforms will activate in the presence of unsaturated fatty acid (PPAR pan agonist), which acts as a ligand to the PPAR isoforms [87]. It should be noted that all of the PPAR isoforms are considered to drive anti-inflammatory pathways. A hepatocyte-specific and adipocyte-specific knockout of PPARα in mice fed a high-fat diet showed greater fat content in each of the KOs, which both also exhibited significantly higher inflammation compared to control littermates [88,89]. Similarly, studies showing that overexpression of inflammatory meditator glucocorticoid receptor beta (GRβ) in the liver of C57/bl6 mice induced hepatic lipid accumulation in 5 days on a normal chow diet by suppression of hepatic PPARα [90].

We have shown that bilirubin (unconjugated form) binds directly to the PPARα nuclear receptor to induce transcription of genes (Figure 5) [91–94], which control adiposity and glucose sensitivity. Interestingly, competitive binding studies and transcriptional activity assays demonstrated that bilirubin's binding to the PPARs is specific to only PPARα, and it has no actions or binding to PPARγ or PPARβ/δ [91,92]. In looking more specifically at ligands for PPARα, a synthetic ligand such as fenofibrate (fibrates) is widely used in the treatment of hypertriglyceridemia in order to reduce serum triglyceride levels. Through the binding and subsequent activation of the PPARα nuclear receptor, fenofibrate reduces plasma triglycerides and VLDL/LDL concentrations [95]. An increased expression of PPARα offers significant induction of β-oxidation [46,47,63,92,94] and myocardial ATP production, which are markers for myocardial viability [96,97]. It can also reduce the oxidative stress that occurs after a high-fat meal [95]. As mentioned above, unconjugated bilirubin has been demonstrated to act as a novel endocrine ligand that activates the transcriptional activity of PPARα by direct interaction, which changes coregulator proteins bound to the nuclear receptor to control gene activity [92]. PPARα activation by bilirubin in obese mice with glucose intolerance leads to a decrease in fasting blood glucose, as well increase in lean body mass and an increased presence of FGF21 (fibroblast growth factor 21) [42]. FGF21 can act as a metabolic regulator by rapid reduction in blood glucose and insulin levels in obese models [42,98] (readers are referred to another review discussing modulation of metabolism by FGF21 for more information [99]). The impact that bilirubin has on exercise via FGF21 is unknown. More studies are needed to elucidate the protective properties of bilirubin that occur via it driving the PPARα-FGF21 pathway that reduces adiposity and improves insulin sensitivity.

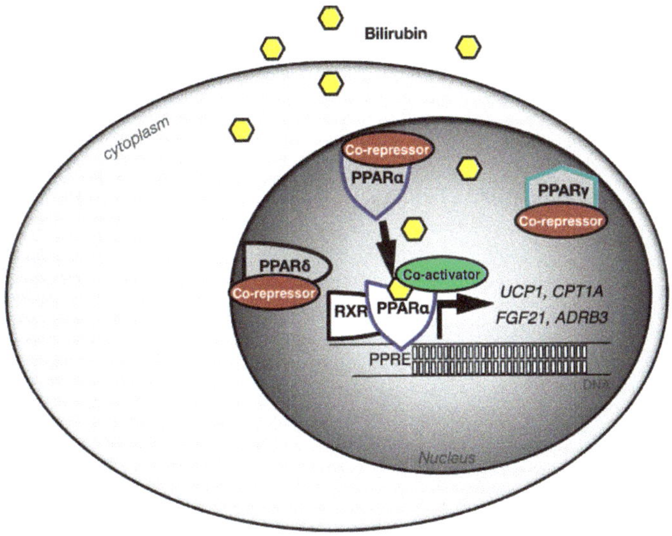

Figure 5. Selectivity of bilirubin for the PPAR isoforms and signaling mechanisms. The PPAR isoforms are bound by corepressors proteins until they are bound to the ligand, which induces a change from co-repressors to co-activators. Unconjugated bilirubin enters the cells and activates PPARα and not the PPARγ or PPARβ/δ isoforms. Bilirubin binding to PPARα induces a complex with RXR causing an exchange of corepressor proteins for co-activators. The bilirubin-induced PPARα-RXR complex controls specific genes for metabolic control of adiposity (*UCP1*, *CPT1A*, *FGF21*, *ADRB3*, and others), which might be based upon specific co-activators (PGC1α, NCOA1, NCOA2, MED1, etc.) bound in the complex.

Exercise plays a role in the activation of the PPAR systems. Exercise increases the levels of AMP-activated kinase (AMPK), ERK1/2-MAPK, and PKC, which are kinases in the skeletal muscle involved in increasing the expression of downstream transcription factors. These kinases are found to increase the transactivation of PPARα and thus an increase in FA oxidation and glucose production, which can be used as fuel during exercise [100]. PPARα, in particular, has strong actions on improving the efficacy of FA oxidation in the liver and adipose tissues [88,89]. PPARα mRNA upregulates in bouts of exercise and in times of starvation in order to metabolize fat and use it for an effective energy source [101,102]. Acute exercise also provides increases in liver and serum FGF21, which provides systemic insulin sensitization [103]. PPARα expression is necessary for optimized endurance exercise. PPARα knockout models had significantly less tolerance to endurance exercise than the control. This lack of tolerance is due to a rapid depletion of hepatic glycogen [104]. We have shown that reducing PPARα activity in the liver leads to lower hepatic glycogen content [88,90], and activation by bilirubin increases it [47,70]. Similarly, hepatocyte-specific BVRA knockout animals on a high-fat diet had reduced bilirubin-PPARα activity and lower glycogen levels [50]. Endurance athletes were found to have a specific polymorphism that produces an increased binding capacity of PPARα in skeletal muscle and more type I slow-twitch fibers [105]. This suggests that PPARα may have critical roles in exercise and is necessary to perform enhanced endurance activity [106]. Similar to PPARα, PPARγ and PPARβ/δ mRNA is also elevated as a result of an aerobic exercise training program [107,108]. PPARβ/δ are the least studied of the isoforms. There is evidence to support PPARβ/δ's ability to rectify metabolic disorders and enhance β-oxidation in muscle [109]. Many of its effects mimic the functionality of PPARα; however, the PPARβ/δ is more ubiquitously expressed than PPARα [110].

PPARγ is upregulated after sustained exercise programs and showed beneficial effects in skeletal muscle [111]. This skeletal muscle had signs of mitochondrial biogenesis and thus, improved aerobic respiration. The mitochondrial biogenesis is also seen in adipose tissue and is phenotypically evident by the increased conversion of white fat into brown fat in the presence of a highly induced PPARγ [112]. This exercise-induced PPARγ can provide antidiabetic effects through upregulation of monocyte PPARγ-control genes [111]. PPARγ is also in charge of controlling adipocyte differentiation [113,114]. In a PPARγ knockout model, severe lipoatrophy is observed, along with insulin resistance [115]. The PPARγ knockout mice have significantly decreased body mass; however, the liver showed a 1.5-fold increase in weight and increased lipid deposition in hepatic tissue. The increased lipid deposition in the liver is due to disrupted adipogenesis in white adipose tissue (WAT), causing increased plasma triglycerides that can deposit in the liver [116]. These PPAR systems have been correlated with decreased levels of atherosclerosis, insulin resistance, and inflammation in conjunction with metabolic syndrome and hypertriglyceridemia [114].

Another novel metabolic role designated to bilirubin is its natural ability to act as an insulin sensitizer [44,117]. PPARγ is elevated following bilirubin administration in mice with improved insulin sensitivity. This isomer of PPAR is implicated as a potent factor in adipocyte differentiation and adiponectin secretion [117]. Bilirubin administration has also improved obesity and hyperglycemia in rodent models. Bilirubin-treated obese mice increased phosphorylation of Akt (Thr309), an insulin-signaling molecule, in skeletal muscle and hepatocytes, indicating preservation of insulin sensitivity [44]. Bilirubin-treated mice also presented with greater adiponectin levels [117]. It should be noted that while bilirubin induced PPARγ expression in diabetic mice, it is not a ligand for this receptor as was previously demonstrated [91,92]. Because bilirubin levels rise with exercise more effectively in insulin-resistant subjects, there is therapeutic potential for bilirubin to control cholesterol metabolism and glucose tolerance in insulin-resistant patients. Therefore, exercise in pre-diabetic patients may offer metabolic benefits by raising HO-1, upregulating adiponectin and bilirubin levels, enhancing insulin signaling, activating PPARα pathways, and thus, decreasing insulin resistance (Figure 6).

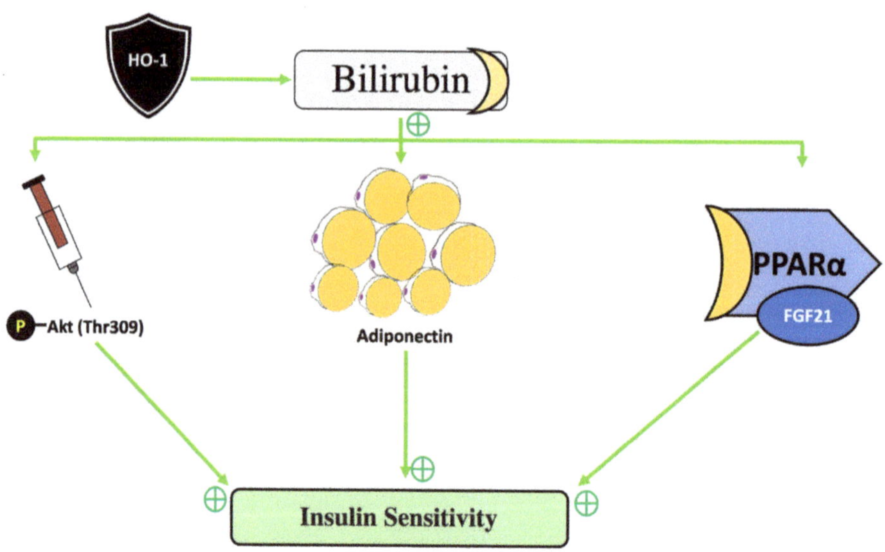

Figure 6. Model of how HO-1-bilirubin improves insulin sensitivity. Upregulation of HO-1-bilirubin is a multifactorial influencer of different metabolic processes such as induction phosphorylated Akt (Thr309), adiponectin production, and activation of PPARα-FGF21 pathways.

4. The Signaling Mechanisms of Heme Oxygenase and Bilirubin in Metabolism

4.1. Generation and Catabolism of Bilirubin

Bilirubin is a tetrapyrrole compound formed from the catabolism of heme to biliverdin that is converted to bilirubin by biliverdin reductase (BVR) [17,41,42,118–122]. Tetrapyrroles are seen as an orange-yellow pigment, which may indicate underlying disease processes if extremely elevated (>150 μM) in the skin (jaundice) or the urine [41]. When erythrocytes (red blood cells) are lysed, the hemoglobin is broken down into heme and protoporphyrin. The heme is oxidatively cleaved by the enzyme heme oxygenase (HO), yielding biliverdin, iron, and carbon monoxide (CO) [41]. This biliverdin can be converted to bilirubin through the cytosolic enzyme biliverdin reductase [123–125]. The conversion to bilirubin has been empirically shown to produce potent antioxidant effects that can regulate cellular redox reactions, decrease ROS, and decrease the activity of NADPH oxidase [3]. Bilirubin circulates bound to water-soluble albumin, where it is transferred to the hepatocyte as unconjugated bilirubin. Then, bilirubin is conjugated by the UDP glucuronosyltransferase 1A1 (UGT1A1) enzyme [121,126]. Once conjugated, bilirubin is then metabolized by colonic bacterial proteases and is either reabsorbed into the hepatobiliary system as urobilinogen or excreted in the feces and urine as stercobilin or urobilin, respectively [122]. The bilirubin pathway (illustrated and described in more detail elsewhere [17,41,42,119–122,126]) is increased with exercise [70], and a better understanding of the pathway regulation may identify areas that alter the bilirubin half-life that might lead to pathological consequences.

4.2. Biliverdin Reductase and Metabolism

While there are limited studies showing that BVRA is regulated by exercise [70], there have been supporting studies showing a role for the enzyme in metabolism [50,51,124,125,127]. Adipocyte-specific deletion of BVRA results in adipocyte hypertrophy and increased inflammation while decreasing mitochondrial number and markers of adipocyte browning such as PPARα and β3 adrenergic receptor (*Adrb3*) [127]. The loss of adipocyte BVRA also decreases insulin signaling in white adipose tissue contributing to increased fasting hyperglycemia in knockout mice [127]. These results agree with the finding from obese human patients who exhibit lower levels of BVRA, increased levels of inflammation, and increased adipocyte size [128]. CRISPR knockout of BVRA in hepatocytes and kidney proximal tubules cells induces oxidative stress and lipid accumulation [124,125]. Similarly, mice with a global knockout of BVRA have increased oxidative stress [41]. Deficiencies in BVRA also correlate with brain insulin resistance in Alzheimer's disease patients [129,130]. Administration of BVRA peptides improved intranasal insulin treatment in a mouse model of Alzheimer's disease, suggesting a potential therapeutic role for targeting BVRA for treatment [131].

BVRA also plays an essential role in the development of metabolic diseases associated with obesity, like NAFLD. Hepatocyte-specific BVRA knockout mice develop more severe dietary-induced NAFLD as compared to wild-type littermates [50]. The loss of hepatocyte BVRA increases activation of glycogen synthase kinase 3beta (GSK3β) via decreased levels of serine 9 (Ser9) phosphorylation which in turn increases serine 73 (Ser73) phosphorylation of PPARα, increasing protein turnover and decreasing its transcriptional activity [50]. Interestingly, reduced adipocyte levels in BVRA in obese human patients resulted in significantly more hepatic steatosis and NAFLD [128]. These results suggest that BVRA can have both direct and indirect effects to contribute to hepatic steatosis and the development of NAFLD. More studies are needed to determine factors that regulate BVRA expression and how these are affected by exercise.

4.3. Bilirubin and Metabolic Dysfunction

Bilirubin was once believed to act only as a toxic bile substance and end-product. However, more recent studies have uncovered potential metabolic benefits of greater yet subclinical bilirubin concentrations. These include its role in ROS scavenging, anti-inflammatory properties, and reduction in adipocyte size from increased fat oxidation [94,132,133]. While marked extreme hyperbilirubinemia (>150 μM) can be a sign of a more ominous clinical diagnosis, raised basal concentrations are also associated with protecting metabolic function (25–50 μM as discussed in [41]). The metabolic syndrome [134] is associated with increased insulin resistance and oxidative stress, which can also lend significant inflammatory and cardiovascular risk factors. Increased serum bilirubin concentration acts as a protective factor against the development of MetS. Subjects with increased basal bilirubin levels have a lower odds ratio to develop MetS [134]. It is thought this observation is due to the antioxidant, anti-inflammatory, and hormonal properties of bilirubin (discussed in more detail above). Conversely, in subjects diagnosed with metabolic syndrome, serum bilirubin is typically reduced (<10 μM, discussed further in [41]) [135]. Thus, the clinical assessment of serum bilirubin may have some future utility as a screening or prediction tool for those with high risk for metabolic dysfunction. In support of this, coronary artery disease severity was recently predicted with an odds ratio of 0.155 (95% confidence interval), revealing an inverse relationship between bilirubin and CAD severity [136]. NAFLD was also predicted in patients with an odds ratio of 0.88 (95% confidence interval), showing a strong inverse relationship between serum bilirubin and NAFLD [137]. A study of obese children showed that those with NAFLD had the lowest serum bilirubin [133]. Low bilirubin has also been associated with a greater risk of cerebral deep white matter lesions in healthy subjects [138], suggesting that low levels may impair cognitive function or lead to stroke [139–141]. These studies might suggest that increasing bilirubin levels could be therapeutic for improving metabolic dysfunction and reducing stroke risk. Factors that induce heme oxygenase production of bilirubin, such as nutraceuticals, may have several benefits [63]. These studies highlight a potential protective effect of bilirubin against metabolic disease and should be examined further to elucidate more of its positive benefits.

5. Strategies to Improve Metabolic Outcomes through Nutraceuticals

5.1. The Influence of Diet on Antioxidants

Fresh fruit and vegetable intake represent the largest source of dietary antioxidants that are essential in maintaining health [142]. A great deal of research has evaluated fruit and vegetable intake as a means to counter the inflammation that has been attributed to nearly all chronic diseases of modern society, with many studies demonstrating a strong inverse correlation between fruit and vegetable intake and inflammatory markers [143,144]. Much of the research concerning dietary antioxidants and inflammation has centered on the Mediterranean diet due to its emphasis on fresh fruit and vegetable intake. Both cross-sectional and longitudinal trials have demonstrated a substantial lowering effect for the Mediterranean diet on a wide variety of inflammatory markers, including IL-6, IL-7, IL-19, CRP, and TNFα [145–147]. This has prompted the use of the Mediterranean diet in hopes of managing metabolic and vascular diseases, endocrine disorders, and some cancers [148–150]. Elucidation of specific antioxidants, and the benefits of dietary supplementation, have been a focus for many current research studies, giving rise to various nutraceuticals. Below, we describe the effects of such nutraceuticals and their beneficial actions on the HO-1 pathway and inflammation.

5.2. The Benefits of Moderately Raising Plasma Bilirubin

Natural substances that raise plasma bilirubin have been of interest for reducing adiposity [63]. One herbal method that is gaining interest in elevating plasma bilirubin is the use of the milk thistle plant (*Silybum marianum*) [63]. The plant contains a mix of polyphenols such as p-coumaric, vanillic acid, silybin, and α-tocopherol [151,152]. The primary compound in milk thistle that is considered the active component that increases plasma bilirubin is the silymarin flavonoids that suppress hepatic UGT1A1 [153]. Silymarin may protect against liver injury and hepatic fat accumulation [154–156]. However, how milk thistle or silymarin might function combined with exercise in reducing adiposity is unknown.

Bilirubin is a potent endogenous antioxidant that the body uses to support oxidative balance [17,41,42,119–122]. Plasma bilirubin has been empirically correlated with decreased risk for oxidative disorders such as coronary artery disease (CAD) [42,119]. The theory of action stemmed from individuals with Gilbert Syndrome, who have a mutation in the UGT1A1 gene, which causes defective processing of bilirubin [47,126]. Hence, lower hepatic UGT1A1 causes higher plasma unconjugated bilirubin [47]. Individuals with Gilbert Syndrome were found to have decreased incidence of CAD [157]. Previous studies postulate that unconjugated bilirubin is fluxing back into cells and acting as a scavenging agent of oxidative radicals. However, researchers have also hypothesized that elevated serum bilirubin acts as a marker that could predict greater expression or inducibility of intracellular HO-1, which will increase the intracellular concentration of bilirubin [132]. Using HPLC-TLS, these researchers detected bilirubin levels within vascular endothelial cells. They also showed that bilirubin within these vascular endothelial cells could effectively modulate HO-1 upregulation [158]. These findings suggest strong potential for developing pharmacotherapeutics that can target and upregulate this intrinsic antioxidant system within the vascular endothelium through the induction of HO-1 and help prevent or follow the progression of cardiovascular disease. This has the potential to lend more focused antioxidant and anti-inflammatory therapeutic approaches.

5.3. *Vitamin D Repletion*

Vitamin D is an important secosteroid in understanding metabolic disease [159]. Vitamin D deficiency (defined as a 25(OH)D level less than 20 ng/dL) is common and is associated with decreased muscle endurance, function, and strength [160–167]. Vitamin D deficiency is connected to muscle metabolic perturbations, including insulin resistance [168–170], and is linked to mitochondrial dysfunction [171] in both young and aged adults. Vitamin D deficiency is highly prevalent in obesity without vitamin D supplementation [172,173]. Obese adults are commonly prescribed a high-dose vitamin D repletion protocol to combat vitamin D deficiency and obesity-associated vitamin D resistance. Aggressive vitamin D repletion to correct the deficiency is linked to improved muscle mitochondrial function [171,174]. Increasing vitamin D status is consistently associated with skeletal muscle lipid deposition and distribution [175–178]. There is also evidence that vitamin D may improve hepatic steatosis with just 4-weeks of supplementation [179].

Calcitriol, the active form of vitamin D [1,25$(OH)_2D_3$], is the only form that can bind to the Vitamin D receptor (VDR). The VDR is a nuclear receptor transcription factor that controls gene expression changes that improve mitochondrial function in myotubes [180], insulin sensitivity, and myocellular lipid partitioning in high fat-treated SkM cells [181]. In humans, we found that vitamin D combined with aerobic exercise potentiated the metabolic benefit of training by producing the most intramyocellular lipid (IMCL) loss and increasing skeletal muscle tissue-level VO_2 in older adults at risk for metabolic dysfunction [182]. These benefits were greater than when providing vitamin D repletion or exercising independently. These observations are consistent with reports that vitamin D coupled with exercise has positively affected muscle mitochondrial function [171,174]. In addition, VDR expression in SkM was increased by exercise [183]. Vitamin D supplementation has been associated with muscle regeneration and repair [123,145,146], suggesting an additive effect when combined with vitamin D repletion.

In addition to these findings, vitamin D has been described to have anti-inflammatory effects and is linked to insulin sensitivity and immuno-modulation [184–186]. Recent work has also highlighted a novel role of vitamin D in upregulating HO-1 expression in intestinal cells and reduced expression of macrophage HO-1 with an associated reduction in conjugated bilirubin [187]. Vitamin D has been shown to block the activation of M-1 macrophages, increase activation of M-2 macrophages, and impair monocyte/macrophage recruitment [187]. Collectively, these data suggest that vitamin D may ameliorate metabolic dysfunction by altering lipid availability for oxidation in response to exercise training and may help regulate inflammatory pathways. These observations require further exploration in obesity-inflammation studies.

Along with evidence that vitamin D repletion augments oxidative metabolism [171,174], we show in a muscle cell line that active calcitriol treatment altered total lipid, lipid species content, and increased gene expression of PLIN2, a lipid coatomer protein that facilitates IMCL availability for β-oxidation [188,189]. PLIN2-containing lipid droplets are also preferentially used during moderate-intensity exercise [190], suggesting that increased PLIN2 expression may increase lipolytic potential. In vitro findings from our group [180,181] indicate increased PLIN2-associated lipid accumulation and lipolysis after calcitriol treatment. These changes suggest an increased lipid flux—defined here as the rate at which lipids pass through SkM via IMCL accumulation and oxidation—and, by association, a decrease in lipid-mediated pathologies [191]. These cell culture results suggest that vitamin D is involved in muscle lipid packaging, partitioning, and mitochondrial lipid oxidation. Together with data showing that exercise improves muscle sensitivity to vitamin D storage and retention [192], evidence of muscle adaptations to the combination of vitamin D supplementation with exercise are tightly connected with improved mitochondrial function and may serve an integral role in delaying stagnant ectopic fat infiltration and metabolic dysfunction.

5.4. Nitrate from Foods and Dietary Supplements

Dietary nitrate is predominately found in green leafy vegetables and concentrated food sources (e.g., beetroot juice) and dietary supplements. Physically active individuals commonly use this EE to increase plasma nitrate concentrations and subsequently increase nitric oxide availability [193]. Increasing nitric oxide via the nitrate-nitrite-NO signaling pathway (with supplemental dietary nitrate) has been shown to decrease NADPH oxidase-derived oxidative stress via HO-1 induction and reduce p47phox expression [194]. Metabolically, nitrate has also been shown to reduce the oxygen cost of exercise [195] and improve exercise tolerance, economy, and performance. These benefits may also extend to those newly committed to exercise to lose fat and improve metabolic function. In addition to these observations, it seems reasonable that dietary nitrate may also work alongside exercise to preserve endothelial function [196]. Basaqr et al. recently found that four weeks of concentrated beetroot juice combined with vitamin C improved endothelial function and the lipid profile of overweight subjects with evidence of endothelial dysfunction [197]. The exact mechanism of action to explain these findings is unknown but is partially explained by the combined antioxidant effects of vitamin C and concentrated dietary nitrate supplements to decrease oxidative stress [198–201]. Improvements in blood lipids from others suggest that dietary nitrate supplementation with the addition of vitamin C (or other nutraceutical antioxidants) may be a valuable dietary approach alongside exercise to improve metabolic and cardiovascular health [60,61]. Future studies could determine how these impact exercise, inflammation, and metabolic outcomes.

5.5. Vitamin E Supplementation

Vitamin E (α/γ-tocopherol) is one of the most important dietary antioxidants that play a critical role as a radical savaging agent and mechanistic inducer [202]. Vitamin E acts as a potent antioxidant to neutralize free radicals and superoxide by using its free hydroxyl group to accept unpaired electrons [203]. Furthermore, unlike other dietary antioxidants (e.g., vitamin C, carotenoids, etc.), Vitamin E is uniquely connected with exercise-induced oxidative stress and insulin sensitivity. The regulation and distribution of Vitamin E are controlled by alpha-tocopherol transfer protein (α-TTP) in the liver. α-TTP secretes Vitamin E from the liver by releasing α-tocopherol into the circulation. However, this mechanism is still not clearly understood [204]. Data suggests that α-TTP is lower when the α-tocopherol levels are low and subjects with an α-TTP gene (TTPA) knockout presented with symptoms of vitamin E serum deficiency [202]. The α-TTP is also known to be induced by hypoxic states and stress-induced free radical production [205]. The administration of vitamin E in hypoxic states (similar to hypoxia observed with high-intensity exercise) has mitigated ROS-related biochemical changes in many tissues by preventing increases in malondialdehyde and myeloperoxidase and protecting against lipid peroxidation [206]. This hypoxia-based regulative mechanism has the potential to be evident during times of exercise; however, this hypothesis has yet to be tested. The rationale behind this hypothesis is that during exercise, hypoxia-induction of α-TTP will help increase serum α-tocopherol and protect the cells from free-radical damage during exercise-induced oxidative stress. This tocopherol can also provide non -antioxidant functions and induce mRNA levels of transcription factors, PPARγ, and the hormone, adiponectin [207,208]. Adiponectin and PPARγ are activated by vitamin E and are known to improve insulin sensitivity in diabetes. The vitamin E-induction of PPARγ is not through direct binding but through the increase in 15d-PGJ2, a commonly described ligand of PPARγ [208] that is also known to induce HO-1 through p38 MAP kinase and the Nrf-2 pathway [209]. This induction may further increase exercise-induced HO-1 activation and influence the BVRA-bilirubin-PPARαaxis. Both Vitamin E and PPARs are valuable targets in hepatic protection in non-alcoholic hepatic steatosis and fibrosis [210].

6. Conclusions

Exercise has clear benefits in reducing adiposity and inflammation while improving insulin sensitivity. A deeper understanding of the mechanisms of how exercise functions to improve these beneficial actions is needed. While bilirubin was once thought to be a harmful bile substance, current research argues otherwise and that slightly elevated levels have numerous health benefits against metabolic dysfunction. Studies on how exercise influences factors such as heme oxygenase, BVRA, and UGT1A1 that control bilirubin's turnover (half-life) are needed. Furthermore, nutraceuticals that activate and control these pathways might be beneficial in improving weight-loss regimens. Investigations in these areas might also benefit patients with inflammatory disorders as increasing plasma bilirubin has anti-inflammatory properties that probably originate from its antioxidant and hormone (as a ligand for PPARα) properties. Future work determining the interplay of exercise and nutraceuticals has many health benefits to help a broad spectrum of diseases.

Author Contributions: T.D.H.J. conceived the project. N.R.D. created graphic illustrations for Figures 1–3 and 6 using PowerPoint; D.E.S. developed Figure 4 using BioRender.com (accessed on 9 December 2021); T.D.H.J. generated Figure 5 using OmniGraffle. D.T.T., N.R.D., K.D.F., D.E.S. and T.D.H.J. wrote the manuscript and reviewed and revised the final manuscript. All authors have read and agreed to the published version of the manuscript.

Funding: This work was supported by the National Institutes of Health, National Institute of Diabetes and Digestive and Kidney Diseases 1R01DK121797-01A1 (T.D.H.J.) and 1R01DK126884-01A1 (D.E.S.), the National Heart, Lung and Blood Institute K01HL-125445 (T.D.H.J.) and P01 HL05197-11 (D.E.S.), and the National Institute of General Medical Sciences P20GM104357-02 (D.E.S.). The content is solely the responsibility of the authors and does not necessarily represent the official views of the National Institutes of Health.

Conflicts of Interest: T.D.H.J. and D.E.S. have submitted patents on bilirubin and obesity-related disorders.

References

1. Budui, S.L.; Rossi, A.P.; Zamboni, M. The pathogenetic bases of sarcopenia. *Clin. Cases Miner. Bone Metab.* **2015**, *12*, 22–26. [CrossRef]
2. Dewidar, B.; Kahl, S.; Pafili, K.; Roden, M. Metabolic liver disease in diabetes—From mechanisms to clinical trials. *Metabolism* **2020**, *111*, 154299. [CrossRef] [PubMed]
3. Takada, S.; Sabe, H.; Kinugawa, S. Abnormalities of Skeletal Muscle, Adipocyte Tissue, and Lipid Metabolism in Heart Failure: Practical Therapeutic Targets. *Front. Cardiovasc. Med.* **2020**, *12*, 79. [CrossRef] [PubMed]
4. Allison, D.B.; Fontaine, K.R.; Manson, J.E.; Stevens, J.; VanItallie, T.B. Annual deaths attributable to obesity in the United States. *JAMA* **1999**, *282*, 1530–1538. [CrossRef]
5. Williams, E.P.; Mesidor, M.; Winters, K.; Dubbert, P.M.; Wyatt, S.B. Overweight and Obesity: Prevalence, Consequences, and Causes of a Growing Public Health Problem. *Curr. Obes. Rep.* **2015**, *4*, 363–370. [CrossRef]

6. Ogden, C.L.; Carroll, M.D.; Kit, B.K.; Flegal, K.M. Prevalence of obesity among adults: United States, 2011–2012. *NCHS Data Brief* **2013**, *131*, 1–8.
7. National Center for Health Statistics. *Health, United States, 2016: With Chart Book on Long-Term Trends in Health*; National Center for Health Statistics: Hyattsville, MD, USA, 2017.
8. Thomas, D.M.; Bouchard, C.; Church, T.; Slentz, C.; Kraus, W.E.; Redman, L.M.; Martin, C.K.; Silva, A.M.; Vossen, M.; Westerterp, K.; et al. Why do individuals not lose more weight from an exercise intervention at a defined dose? An energy balance analysis. *Obes. Rev.* **2012**, *13*, 835–847. [CrossRef]
9. Thomas, D.M.; Kyle, T.K.; Stanford, F.C. The gap between expectations and reality of exercise-induced weight loss is associated with discouragement. *Prev. Med.* **2015**, *81*, 357–360. [CrossRef] [PubMed]
10. King, N.A.; Caudwell, P.; Hopkins, M.; Byrne, N.M.; Colley, R.; Hills, A.P.; Stubbs, J.R.; Blundell, J.E. Metabolic and behavioral compensatory responses to exercise interventions: Barriers to weight loss. *Obesity* **2007**, *15*, 1373–1383. [CrossRef]
11. Kirwan, J.P.; Sacks, J.; Nieuwoudt, S. The essential role of exercise in the management of type 2 diabetes. *Clevel. Clin. J. Med.* **2017**, *84*, S15–S21. [CrossRef]
12. Winzer, E.B.; Woitek, F.; Linke, A. Physical Activity in the Prevention and Treatment of Coronary Artery Disease. *J. Am. Heart Assoc.* **2018**, *7*, e007725. [CrossRef] [PubMed]
13. Whitsett, M.; VanWagner, L.B. Physical activity as a treatment of non-alcoholic fatty liver disease: A systematic review. *World J. Hepatol.* **2015**, *7*, 2041–2052. [CrossRef] [PubMed]
14. Vina, J.; Sanchis-Gomar, F.; Martinez-Bello, V.; Gomez-Cabrera, M.C. Exercise acts as a drug; the pharmacological benefits of exercise. *Br. J. Pharmacol.* **2012**, *167*, 1–12. [CrossRef] [PubMed]
15. Singh, R. The importance of exercise as a therapeutic agent. *Malays. J. Med. Sci.* **2002**, *9*, 7–16.
16. Bird, S.R.; Hawley, J.A. Update on the effects of physical activity on insulin sensitivity in humans. *BMJ. Open Sport Exerc. Med.* **2017**, *2*, e000143. [CrossRef] [PubMed]
17. O'Brien, L.; Hosick, P.A.; John, K.; Stec, D.E.; Hinds, T.D., Jr. Biliverdin reductase isozymes in metabolism. *Trends Endocrinol. Metab.* **2015**, *26*, 212–220. [CrossRef]
18. Colberg, S.R.; Sigal, R.J.; Fernhall, B.; Regensteiner, J.G.; Blissmer, B.J.; Rubin, R.R.; Chasan-Taber, L.; Albright, A.L.; Braun, B.; American College of Sports, M.; et al. Exercise and type 2 diabetes: The American College of Sports Medicine and the American Diabetes Association: Joint position statement. *Diabetes Care* **2010**, *33*, e147–e167. [CrossRef]
19. Bruning, R.S.; Sturek, M. Benefits of exercise training on coronary blood flow in coronary artery disease patients. *Prog. Cardiovasc. Dis.* **2015**, *57*, 443–453. [CrossRef]
20. Gilbertson, N.M.; Eichner, N.Z.M.; Heiston, E.M.; Gaitan, J.M.; Francois, M.E.; Mehaffey, J.H.; Hassinger, T.E.; Hallowell, P.T.; Weltman, A.; Malin, S.K. A low-calorie diet with or without interval exercise training improves adiposopathy in obese women. *Appl. Physiol. Nutr. Metab.* **2019**, *44*, 1057–1064. [CrossRef]
21. Sargeant, J.A.; Gray, L.J.; Bodicoat, D.H.; Willis, S.A.; Stensel, D.J.; Nimmo, M.A.; Aithal, G.P.; King, J.A. The effect of exercise training on intrahepatic triglyceride and hepatic insulin sensitivity: A systematic review and meta-analysis. *Obes. Rev.* **2018**, *19*, 1446–1459. [CrossRef]
22. Francois, M.E.; Gilbertson, N.M.; Eichner, N.Z.M.; Heiston, E.M.; Fabris, C.; Breton, M.; Mehaffey, J.H.; Hassinger, T.; Hallowell, P.T.; Malin, S.K. Combining Short-Term Interval Training with Caloric Restriction Improves ss-Cell Function in Obese Adults. *Nutrients* **2018**, *10*, 717. [CrossRef]
23. Mora-Rodriguez, R.; Ortega, J.F.; Ramirez-Jimenez, M.; Moreno-Cabanas, A.; Morales-Palomo, F. Insulin sensitivity improvement with exercise training is mediated by body weight loss in subjects with metabolic syndrome. *Diabetes Metab.* **2019**, *46*, 210–218. [CrossRef] [PubMed]
24. Haskell, W.L.; Lee, I.M.; Pate, R.R.; Powell, K.E.; Blair, S.N.; Franklin, B.A.; Macera, C.A.; Heath, G.W.; Thompson, P.D.; Bauman, A. Physical activity and public health: Updated recommendation for adults from the American College of Sports Medicine and the American Heart Association. *Med. Sci. Sports Exerc.* **2007**, *39*, 1423–1434. [CrossRef] [PubMed]
25. Donnelly, J.E.; Blair, S.N.; Jakicic, J.M.; Manore, M.M.; Rankin, J.W.; Smith, B.K.; American College of Sports Medicine. American College of Sports Medicine Position Stand. Appropriate physical activity intervention strategies for weight loss and prevention of weight regain for adults. *Med. Sci. Sports Exerc.* **2009**, *41*, 459–471. [CrossRef] [PubMed]
26. Diabetes Prevention Program Research, G. Long-term effects of lifestyle intervention or metformin on diabetes development and microvascular complications over 15-year follow-up: The Diabetes Prevention Program Outcomes Study. *Lancet Diabetes Endocrinol.* **2015**, *3*, 866–875. [CrossRef]
27. De Vries, T.I.; Dorresteijn, J.A.N.; van der Graaf, Y.; Visseren, F.L.J.; Westerink, J. Heterogeneity of Treatment Effects From an Intensive Lifestyle Weight Loss Intervention on Cardiovascular Events in Patients with Type 2 Diabetes: Data From the Look AHEAD Trial. *Diabetes Care* **2019**, *42*, 1988–1994. [CrossRef] [PubMed]
28. Flack, K.D.; Hays, H.M.; Moreland, J.; Long, D.E. Exercise for Weight Loss: Further Evaluating Energy Compensation with Exercise. *Med. Sci. Sports Exerc.* **2020**, *52*, 2466–2475. [CrossRef]
29. Flack, K.D.; Ufholz, K.; Johnson, L.; Fitzgerald, J.S.; Roemmich, J.N. Energy compensation in response to aerobic exercise training in overweight adults. *Am. J. Physiol. Regul. Integr. Comp. Physiol.* **2018**, *315*, R619–R626. [CrossRef]

30. Rosenkilde, M.; Auerbach, P.; Reichkendler, M.H.; Ploug, T.; Stallknecht, B.M.; Sjodin, A. Body fat loss and compensatory mechanisms in response to different doses of aerobic exercise—A randomized controlled trial in overweight sedentary males. *Am. J. Physiol. Regul. Integr. Comp. Physiol.* **2012**, *303*, R571–R579. [CrossRef]
31. Church, T.S.; Martin, C.K.; Thompson, A.M.; Earnest, C.P.; Mikus, C.R.; Blair, S.N. Changes in weight, waist circumference and compensatory responses with different doses of exercise among sedentary, overweight postmenopausal women. *PLoS ONE* **2009**, *4*, e4515. [CrossRef]
32. King, N.A.; Hopkins, M.; Caudwell, P.; Stubbs, R.J.; Blundell, J.E. Individual variability following 12 weeks of supervised exercise: Identification and characterization of compensation for exercise-induced weight loss. *Int. J. Obes.* **2008**, *32*, 177–184. [CrossRef] [PubMed]
33. King, N.A.; Tremblay, A.; Blundell, J.E. Effects of exercise on appetite control: Implications for energy balance. *Med. Sci. Sports Exerc.* **1997**, *29*, 1076–1089. [CrossRef] [PubMed]
34. Tryfidou, D.V.; McClean, C.; Nikolaidis, M.G.; Davison, G.W. NA Damage Following Acute Aerobic Exercise: A Systematic Review and Meta-analysis. *Sports Med.* **2020**, *50*, 103–127. [CrossRef] [PubMed]
35. Steinbacher, P.; Eckl, P. Impact of oxidative stress on exercising skeletal muscle. *Biomolecules* **2015**, *5*, 356–377. [CrossRef]
36. Fittipaldi, S.; Mercatelli, I.; Caporossi, D. Role of exercise-induced reactive oxygen species in the modulation of heat shock protein response. *Free Radic. Res.* **2014**, *48*, 52–70. [CrossRef]
37. Kawamura, T.; Muraoka, I. Exercise-Induced Oxidative Stress and the Effects of Antioxidant Intake from a Physiological Viewpoint. *Antioxidants* **2018**, *7*, 119. [CrossRef] [PubMed]
38. Gomez-Cabrera, M.C.; Domenech, E.; Vina, J. Moderate exercise is an antioxidant: Upregulation of antioxidant genes by training. *Free Radic. Biol. Med.* **2008**, *44*, 126–131. [CrossRef] [PubMed]
39. John, K.; Marino, J.S.; Sanchez, E.R.; Hinds, T.D., Jr. The glucocorticoid receptor: Cause of or cure for obesity? *Am. J. Physiol. Endocrinol. Metab.* **2016**, *310*, E249–E257. [CrossRef] [PubMed]
40. Muller, F.L.; Song, W.; Jang, Y.C.; Liu, Y.; Sabia, M.; Richardson, A.; Van Remmen, H. Denervation-induced skeletal muscle atrophy is associated with increased mitochondrial ROS production. *Am. J. Physiol. Regul. Integr. Comp. Physiol.* **2007**, *293*, R1159–R1168. [CrossRef] [PubMed]
41. Creeden, J.F.; Gordon, D.M.; Stec, D.E.; Hinds, T.D., Jr. Bilirubin as a metabolic hormone: The physiological relevance of low levels. *Am. J. Physiol. Endocrinol. Metab.* **2021**, *320*, E191–E207. [CrossRef]
42. Hinds, T.D., Jr.; Stec, D.E. Bilirubin, a Cardiometabolic Signaling Molecule. *Hypertension* **2018**, *72*, 788–795. [CrossRef] [PubMed]
43. Ndisang, J.F. Role of Heme Oxygenase in Inflammation, Insulin-Signalling, Diabetes and Obesity. *Mediat. Inflamm.* **2010**, *2010*, 359732. [CrossRef]
44. Dong, H.; Huang, H.; Yun, X.; Kim, D.-s.; Yue, Y.; Wu, H.; Sutter, A.; Chavin, K.D.; Otterbein, L.E.; Adams, D.B.; et al. Bilirubin increases insulin sensitivity in leptin-receptor deficient and diet-induced obese mice through suppression of ER stress and chronic inflammation. *Endocrinology* **2014**, *155*, 818–828. [CrossRef] [PubMed]
45. Takei, R.; Inoue, T.; Sonoda, N.; Kohjima, M.; Okamoto, M.; Sakamoto, R.; Inoguchi, T.; Ogawa, Y. Bilirubin reduces visceral obesity and insulin resistance by suppression of inflammatory cytokines. *PLoS ONE* **2019**, *14*, e0223302. [CrossRef]
46. Hinds, T.D., Jr.; Creeden, J.F.; Gordon, D.M.; Stec, D.F.; Donald, M.C.; Stec, D.E. Bilirubin Nanoparticles Reduce Diet-Induced Hepatic Steatosis, Improve Fat Utilization, and Increase Plasma beta-Hydroxybutyrate. *Front. Pharm.* **2020**, *11*, 594574. [CrossRef] [PubMed]
47. Hinds, T.D., Jr.; Hosick, P.A.; Hankins, M.W.; Nestor-Kalinoski, A.; Stec, D.E. Mice with hyperbilirubinemia due to Gilbert's Syndrome polymorphism are resistant to hepatic steatosis by decreased serine 73 phosphorylation of PPARalpha. *Am. J. Physiol. Endocrinol. Metab.* **2017**, *312*, E244–E252. [CrossRef]
48. Zhang, M.; Nakamura, K.; Kageyama, S.; Lawal, A.O.; Gong, K.W.; Bhetraratana, M.; Fujii, T.; Sulaiman, D.; Hirao, H.; Bolisetty, S.; et al. Myeloid HO-1 modulates macrophage polarization and protects against ischemia-reperfusion injury. *JCI Insight* **2018**, *3*, e120596. [CrossRef]
49. Kato, H.; Amersi, F.; Buelow, R.; Melinek, J.; Coito, A.J.; Ke, B.; Busuttil, R.W.; Kupiec-Weglinski, J.W. Heme Oxygenase-1 Overexpression Protects Rat Livers from Ischemia/Reperfusion Injury with Extended Cold Preservation. *Am. J. Transplant.* **2001**, *1*, 121–128. [CrossRef]
50. Hinds, T.D., Jr.; Burns, K.A.; Hosick, P.A.; McBeth, L.; Nestor-Kalinoski, A.; Drummond, H.A.; AlAmodi, A.A.; Hankins, M.W.; Vanden Heuvel, J.P.; Stec, D.E. Biliverdin reductase A attenuates hepatic steatosis by inhibition of glycogen synthase kinase (GSK) 3beta phosphorylation of serine 73 of peroxisome proliferator-activated receptor (PPAR) alpha. *J. Biol. Chem.* **2016**, *291*, 25179–25191. [CrossRef]
51. Chen, W.; Tumanov, S.; Fazakerley, D.J.; Cantley, J.; James, D.E.; Dunn, L.L.; Shaik, T.; Suarna, C.; Stocker, R. Bilirubin deficiency renders mice susceptible to hepatic steatosis in the absence of insulin resistance. *Redox Biol.* **2021**, *47*, 102152. [CrossRef]
52. Gobert, A.P.; Verriere, T.; Asim, M.; Barry, D.P.; Piazuelo, M.B.; de Sablet, T.; Delgado, A.G.; Bravo, L.E.; Correa, P.; Peek, R.M., Jr.; et al. Heme oxygenase-1 dysregulates macrophage polarization and the immune response to Helicobacter pylori. *J. Immunol.* **2014**, *193*, 3013–3022. [CrossRef] [PubMed]
53. Li, M.; Kim, D.H.; Tsenovoy, P.L.; Peterson, S.J.; Rezzani, R.; Rodella, L.F.; Aronow, W.S.; Ikehara, S.; Abraham, N.G. Treatment of obese diabetic mice with a heme oxygenase inducer reduces visceral and subcutaneous adiposity, increases adiponectin levels, and improves insulin sensitivity and glucose tolerance. *Diabetes* **2008**, *57*, 1526–1535. [CrossRef]

54. Ndisang, J.F.; Lane, N.; Syed, N.; Jadhav, A. Up-regulating the heme oxygenase system with hemin improves insulin sensitivity and glucose metabolism in adult spontaneously hypertensive rats. *Endocrinology* **2010**, *151*, 549–560. [CrossRef] [PubMed]
55. Lin, H.; Yu, C.-H.; Jen, C.-Y.; Cheng, C.-F.; Chou, Y.; Chang, C.-C.; Juan, S.-H. Adiponectin-mediated heme oxygenase-1 induction protects against iron-induced liver injury via a PPARα dependent mechanism. *Am. J. Pathol.* **2010**, *177*, 1697–1709. [CrossRef]
56. Yang, M.; Kimura, M.; Ng, C.; He, J.; Keshvari, S.; Rose, F.J.; Barclay, J.L.; Whitehead, J.P. Induction of heme-oxygenase-1 (HO-1) does not enhance adiponectin production in human adipocytes: Evidence against a direct HO-1—Adiponectin axis. *Mol. Cell. Endocrinol.* **2015**, *413*, 209–216. [CrossRef]
57. Araujo, J.A.; Zhang, M.; Yin, F. Heme oxygenase-1, oxidation, inflammation, and atherosclerosis. *Front. Pharm.* **2012**, *3*, 119. [CrossRef] [PubMed]
58. Jais, A.; Einwallner, E.; Sharif, O.; Gossens, K.; Lu, T.T.-H.; Soyal, S.M.; Medgyesi, D.; Neureiter, D.; Paier-Pourani, J.; Dalgaard, K.; et al. Heme oxygenase-1 drives metaflammation and insulin resistance in mouse and man. *Cell* **2014**, *158*, 25–40. [CrossRef]
59. Ghio, A.J.; Case, M.W.; Soukup, J.M. Heme oxygenase activity increases after exercise in healthy volunteers. *Free Radic. Res.* **2018**, *52*, 267–272. [CrossRef]
60. Niess, A.M.; Passek, F.; Lorenz, I.; Schneider, E.M.; Dickhuth, H.-H.; Northoff, H.; Fehrenbach, E. Expression of the antioxidant stress protein heme oxygenase-1 (HO-1) in human leukocytes: Acute and adaptational responses to endurance exercise. *Free Radic. Biol. Med.* **1999**, *26*, 184–192. [CrossRef]
61. Kitamuro, T.; Takahashi, K.; Ogawa, K.; Udono-Fujimori, R.; Takeda, K.; Furuyama, K.; Nakayama, M.; Sun, J.; Fujita, H.; Hida, W.; et al. Bach1 functions as a hypoxia-inducible repressor for the heme oxygenase-1 gene in human cells. *J. Biol. Chem.* **2003**, *278*, 9125–9133. [CrossRef] [PubMed]
62. Reichard, J.F.; Sartor, M.A.; Puga, A. BACH1 is a specific repressor of HMOX1 that is inactivated by arsenite. *J. Biol. Chem.* **2008**, *283*, 22363–22370. [CrossRef]
63. Stec, D.E.; Hinds, T.D., Jr. Natural Product Heme Oxygenase Inducers as Treatment for Nonalcoholic Fatty Liver Disease. *Int. J. Mol. Sci.* **2020**, *21*, 9493. [CrossRef]
64. Kurata, S.; Matsumoto, M.; Nakajima, H. Transcriptional control of the heme oxygenase gene in mouse M1 cells during their TPA-induced differentiation into macrophages. *J. Cell. Biochem.* **1996**, *62*, 314–324. [CrossRef]
65. Falone, S.; Mirabilio, A.; Pennelli, A.; Cacchio, M.; Di Baldassarre, A.; Gallina, S.; Passerini, A.; Amicarelli, F. Differential impact of acute bout of exercise on redox- and oxidative damage-related profiles between untrained subjects and amateur runners. *Physiol. Res.* **2010**, *59*, 953–961. [CrossRef] [PubMed]
66. Wegiel, B.; Baty, C.J.; Gallo, D.; Csizmadia, E.; Scott, J.R.; Akhavan, A.; Chin, B.Y.; Kaczmarek, E.; Alam, J.; Bach, F.H.; et al. Cell surface biliverdin reductase mediates biliverdin-induced anti-inflammatory effects via phosphatidylinositol 3-kinase and Akt. *J. Biol. Chem.* **2009**, *284*, 21369–21378. [CrossRef] [PubMed]
67. Stechschulte, L.A.; Wuescher, L.; Marino, J.S.; Hill, J.W.; Eng, C.; Hinds, T.D., Jr. Glucocorticoid receptor beta stimulates Akt1 growth pathway by attenuation of PTEN. *J. Biol. Chem.* **2014**, *289*, 17885–17894. [CrossRef] [PubMed]
68. Hu, Z.Z.; Pei, G.C.; Wang, P.G.; Yang, J.; Zhu, F.M.; Guo, Y.J.; Wang, M.; Yao, Y.; Zeng, R.; Liao, W.H.; et al. Biliverdin Reductase A (BVRA) Mediates Macrophage Expression of Interleukin-10 in Injured Kidney. *Int. J. Mol. Sci.* **2015**, *16*, 22621–22635. [CrossRef]
69. Bisht, K.; Canesin, G.; Cheytan, T.; Li, M.; Nemeth, Z.; Csizmadia, E.; Woodruff, T.M.; Stec, D.E.; Bulmer, A.C.; Otterbein, L.E.; et al. Deletion of Biliverdin Reductase A in Myeloid Cells Promotes Chemokine Expression and Chemotaxis in Part via a Complement C5a-C5aR1 Pathway. *J. Immunol.* **2019**, *202*, 2982–2990. [CrossRef] [PubMed]
70. Hinds, T.D., Jr.; Creeden, J.F.; Gordon, D.M.; Spegele, A.C.; Britton, S.L.; Koch, L.G.; Stec, D.E. Rats Genetically Selected for High Aerobic Exercise Capacity Have Elevated Plasma Bilirubin by Upregulation of Hepatic Biliverdin Reductase-A (BVRA) and Suppression of UGT1A1. *Antioxidants* **2020**, *9*, 889. [CrossRef]
71. Swift, D.L.; Johannsen, N.M.; Earnest, C.P.; Blair, S.N.; Church, T.S. Effect of different doses of aerobic exercise training on total bilirubin levels. *Med. Sci. Sports Exerc.* **2012**, *44*, 569–574. [CrossRef]
72. Witek, K.; Scislowska, J.; Turowski, D.; Lerczak, K.; Lewandowska-Pachecka, S.; Pokrywka, A. Total bilirubin in athletes, determination of reference range. *Biol. Sport* **2017**, *34*, 45–48. [CrossRef]
73. Priest, J.B.; Oei, T.O.; Moorehead, W.R. Exercise-induced changes in common laboratory tests. *Am. J. Clin. Pathol.* **1982**, *77*, 285–289. [CrossRef]
74. Loprinzi, P.D.; Abbott, K. Physical activity and total serum bilirubin levels among insulin sensitive and insulin resistant U.S. adults. *J. Diabetes Metab. Disord.* **2014**, *13*, 47. [CrossRef]
75. Fallon, K.E. The clinical utility of screening of biochemical parameters in elite athletes: Analysis of 100 cases. *Br. J. Sports Med.* **2008**, *42*, 334–337. [CrossRef] [PubMed]
76. Hammouda, O.; Chtouror, H.; Chaouachi, A.; Chahed, H.; Ferchichi, S.; Kallel, C.; Chamari, K.; Souissi, N. Effect of short-term maximal exercise on biochemical markers of muscle damage, total antioxidant status, and homocysteine levels in football players. *Asian J. Sports Med. Clin. N. Am.* **2012**, *3*, 239–246. [CrossRef]
77. Kratz, A.; Lewandowski, K.B.; Siegel, A.J.; Chun, K.Y.; Flood, J.G.; van Cott, E.M. Effect of marathon running on hematologic and biochemical laboratory parameters, including cardiac markers. *Am. J. Clin. Pathol.* **2002**, *188*, 856–863. [CrossRef]
78. Drukalec-Michalski, K.; Nowaczyk, P.; Glowka, N.; Ziobrowska, A.; Podgorski, T. Is a Four-Week Ketogenic Diet an Effective Nutritional Strategy in CrossFit-Trained Female and Male Athletes? *Nutrients* **2021**, *13*, 864. [CrossRef]

79. Schoonjans, K.; Martin, G.; Staels, B.; Auwerx, J. Peroxisome proliferator-activated receptors, orphans with ligands and functions. *Curr. Opin. Lipidol.* **1997**, *8*, 159–166. [CrossRef] [PubMed]
80. Sodhi, K.; Puri, N.; Kim, D.H.; Hinds, T.D.; Stechschulte, L.A.; Favero, G.; Rodella, L.; Shapiro, J.I.; Jude, D.; Abraham, N.G. PPARdelta binding to heme oxygenase 1 promoter prevents angiotensin II-induced adipocyte dysfunction in Goldblatt hypertensive rats. *Int. J. Obes.* **2014**, *38*, 456–465. [CrossRef]
81. Fedorova, L.V.; Sodhi, K.; Gatto-Weis, C.; Puri, N.; Hinds, T.D., Jr.; Shapiro, J.I.; Malhotra, D. Peroxisome proliferator-activated receptor delta agonist, HPP593, prevents renal necrosis under chronic ischemia. *PLoS ONE* **2013**, *8*, e64436. [CrossRef]
82. Tyagi, S.; Gupta, P.; Saini, A.S.; Kaushal, C.; Sharma, S. The peroxisome proliferator-activated receptor: A family of nuclear receptors role in various diseases. *J. Adv. Pharm. Technol. Res.* **2011**, *2*, 236–240. [CrossRef]
83. Smedlund, K.B.; Sanchez, E.R.; Hinds, T.D., Jr. FKBP51 and the molecular chaperoning of metabolism. *Trends Endocrinol. Metab.* **2021**, *32*, 862–874. [CrossRef] [PubMed]
84. Hinds, T.D.; John, K.; McBeth, L.; Trabbic, C.J.; Sanchez, E.R. Timcodar (VX-853) Is a Non-FKBP12 Binding Macrolide Derivative That Inhibits PPARγand Suppresses Adipogenesis. *PPAR Res.* **2016**, *2016*, 6218637. [CrossRef]
85. Hinds, T.D., Jr.; Stechschulte, L.A.; Cash, H.A.; Whisler, D.; Banerjee, A.; Yong, W.; Khuder, S.S.; Kaw, M.K.; Shou, W.; Najjar, S.M.; et al. Protein phosphatase 5 mediates lipid metabolism through reciprocal control of glucocorticoid receptor and peroxisome proliferator-activated receptor-gamma (PPARgamma). *J. Biol. Chem.* **2011**, *286*, 42911–42922. [CrossRef] [PubMed]
86. Chiu, M.; McBeth, L.; Sindhwani, P.; Hinds, T.D. Deciphering the Roles of Thiazolidinediones and PPARgamma in Bladder Cancer. *PPAR Res.* **2017**, *2017*, 4810672. [CrossRef]
87. Grygiel-Górniak, B. Peroxisome proliferator-activated receptors and their ligands: Nutritional and clinical implications—A review. *Nutr. J.* **2014**, *13*, 17. [CrossRef] [PubMed]
88. Stec, D.E.; Gordon, D.M.; Hipp, J.A.; Hong, S.; Mitchell, Z.L.; Franco, N.R.; Robison, J.W.; Anderson, C.D.; Stec, D.F.; Hinds, T.D., Jr. The loss of hepatic PPARalpha promotes inflammation and serum hyperlipidemia in diet-induced obesity. *Am. J. Physiol. Regul. Integr. Comp. Physiol.* **2019**, *317*, R733–R745. [CrossRef] [PubMed]
89. Hinds, T.D., Jr.; Kipp, Z.A.; Xu, M.; Yiannikouris, F.B.; Morris, A.J.; Stec, D.F.; Wahli, W.; Stec, D.E. Adipose-Specific PPARalpha Knockout Mice Have Increased Lipogenesis by PASK-SREBP1 Signaling and a Polarity Shift to Inflammatory Macrophages in White Adipose Tissue. *Cells* **2021**, *11*, 4. [CrossRef] [PubMed]
90. Marino, J.S.; Stechschulte, L.A.; Stec, D.E.; Nestor-Kalinoski, A.; Coleman, S.; Hinds, T.D., Jr. Glucocorticoid receptor beta induces hepatic steatosis by augmenting inflammation and inhibition of the peroxisome proliferator-activated receptor (PPAR) alpha. *J. Biol. Chem.* **2016**, *291*, 25776–25788. [CrossRef]
91. Gordon, D.M.; Hong, S.H.; Kipp, Z.A.; Hinds, T.D., Jr. Identification of Binding Regions of Bilirubin in the Ligand-Binding Pocket of the Peroxisome Proliferator-Activated Receptor-A (PPARalpha). *Molecules* **2021**, *26*, 2975. [CrossRef] [PubMed]
92. Gordon, D.M.; Neifer, K.L.; Hamoud, A.A.; Hawk, C.F.; Nestor-Kalinoski, A.L.; Miruzzi, S.A.; Morran, M.P.; Adeosun, S.O.; Sarver, J.G.; Erhardt, P.W.; et al. Bilirubin remodels murine white adipose tissue by reshaping mitochondrial activity and the coregulator profile of peroxisome proliferator-activated receptor alpha. *J. Biol. Chem.* **2020**, *295*, 9804–9822. [CrossRef]
93. Gordon, D.M.; Blomquist, T.M.; Miruzzi, S.A.; McCullumsmith, R.; Stec, D.E.; Hinds, T.D., Jr. RNA sequencing in human HepG2 hepatocytes reveals PPAR-alpha mediates transcriptome responsiveness of bilirubin. *Physiol. Genom.* **2019**, *51*, 234–240. [CrossRef] [PubMed]
94. Stec, D.E.; John, K.; Trabbic, C.J.; Luniwal, A.; Hankins, M.W.; Baum, J.; Hinds, T.D., Jr. Bilirubin Binding to PPARα Inhibits Lipid Accumulation. *PLoS ONE* **2016**, *11*, e0153427. [CrossRef] [PubMed]
95. Rosenson, R.S. Fenofibrate: Treatment of hyperlipidemia and beyond. *Expert Rev. Cardiovasc.* **2008**, *6*, 1319–1330. [CrossRef]
96. Iemitsu, M.; Miyauchi, T.; Maeda, S.; Tanabe, T.; Takanashi, M.; Irukayama-Tomobe, Y.; Sakai, S.; Ohmori, H.; Matsuda, M.; Yamaguchi, I. Aging-induced decrease in the PPAR-alpha level in hearts is improved by exercise training. *Am. J. Physiol. Heart Circ. Physiol.* **2002**, *283*, H1750–H1760. [CrossRef]
97. Zhang, S.; Liu, Y.; Li, Q.; Dong, X.; Hu, H.; Hu, R.; Ye, H.; Wu, Y.; Hu, R.; Li, Y. Exercise improved rat metabolism by raising PPAR-α. *Int. J. Sports Med.* **2011**, *32*, 568–573. [CrossRef] [PubMed]
98. Xu, J.; Stanislaus, S.; Chinookoswong, N.; Lau, Y.Y.; Hager, T.; Patel, J.; Ge, H.; Weiszmann, J.; Lu, S.-C.; Graham, M.; et al. Acute glucose-lowering and insulin-sensitizing action of FGF21 in insulin-resistant mouse models—Association with liver and adipose tissue effects. *Am. J. Physiol.-Endocrinol. Metab.* **2009**, *297*, E1105–E1114. [CrossRef]
99. Tezze, C.; Romanello, V.; Sandri, M. FGF21 as Modulator of Metabolism in Health and Disease. *Front. Physiol.* **2019**, *10*, 419. [CrossRef]
100. Bougarne, N.; Weyers, B.; Desmet, S.J.; Deckers, J.; Ray, D.W.; Staels, B.; De Bosscher, K. Molecular Actions of PPARα in Lipid Metabolism and Inflammation. *Endocr. Rev.* **2018**, *39*, 760–802. [CrossRef] [PubMed]
101. Contreras, A.V.; Torres, N.; Tovar, A.R. PPAR-α as a Key Nutritional and Environmental Sensor for Metabolic Adaptation. *Adv. Nutr.* **2013**, *4*, 439–452. [CrossRef] [PubMed]
102. Zhang, H.; He, Y.; Chung, P.K.; Tong, T.K.; Fu, F.H.; Chen, Y.; Jiao, G. Effects of 12 Weeks of Exercise on Hepatic TNF-α and PPARα in an Animal Model of High-Fat Diet-Induced Nonalcoholic Steatohepatitis. *J. Exerc. Sci. Fit.* **2009**, *7*, 18–23. [CrossRef]
103. Tanimura, Y.; Aoi, W.; Takanami, Y.; Kawai, Y.; Mizushima, K.; Naito, Y.; Yoshikawa, T. Acute exercise increases fibroblast growth factor 21 in metabolic organs and circulation. *Physiol. Rep.* **2016**, *4*, e12828. [CrossRef] [PubMed]

104. Muoio, D.M.; MacLean, P.S.; Lang, D.B.; Li, S.; Houmard, J.A.; Way, J.M.; Winegar, D.A.; Corton, J.C.; Dohm, G.L.; Kraus, W.E. Fatty acid homeostasis and induction of lipid regulatory genes in skeletal muscles of peroxisome proliferator-activated receptor (PPAR) alpha knock-out mice. Evidence for compensatory regulation by PPAR delta. *J. Biol. Chem.* **2002**, *277*, 26089–26097. [CrossRef]
105. Lopez-Leon, S.; Tuvblad, C.; Forero, D.A. Sports genetics: The PPARA gene and athletes' high ability in endurance sports. A systematic review and meta-analysis. *Biol. Sport* **2016**, *33*, 3–6. [CrossRef]
106. Russell, A.P.; Feilchenfeldt, J.; Schreiber, S.; Praz, M.; Crettenand, A.; Gobelet, C.; Meier, C.A.; Bell, D.R.; Kralli, A.; Giacobino, J.P.; et al. Endurance training in humans leads to fiber type-specific increases in levels of peroxisome proliferator-activated receptor-gamma coactivator-1 and peroxisome proliferator-activated receptor-alpha in skeletal muscle. *Diabetes* **2003**, *52*, 2874–2881. [CrossRef] [PubMed]
107. Fritz, T.; Krämer, D.K.; Karlsson, H.K.; Galuska, D.; Engfeldt, P.; Zierath, J.R.; Krook, A. Low-intensity exercise increases skeletal muscle protein expression of PPARdelta and UCP3 in type 2 diabetic patients. *Diabetes Metab. Res. Rev.* **2006**, *22*, 492–498. [CrossRef]
108. Kannisto, K.; Chibalin, A.; Glinghammar, B.; Zierath, J.R.; Hamsten, A.; Ehrenborg, E. Differential expression of peroxisomal proliferator activated receptors alpha and delta in skeletal muscle in response to changes in diet and exercise. *Int. J. Mol. Med.* **2006**, *17*, 45–52.
109. Rachid, T.L.; Silva-Veiga, F.M.; Graus-Nunes, F.; Bringhenti, I.; Mandarim-de-Lacerda, C.A.; Souza-Mello, V. Differential actions of PPAR-α and PPAR-β/δ on beige adipocyte formation: A study in the subcutaneous white adipose tissue of obese male mice. *PLoS ONE* **2018**, *13*, e0191365. [CrossRef] [PubMed]
110. Varga, T.; Czimmerer, Z.; Nagy, L. PPARs are a unique set of fatty acid regulated transcription factors controlling both lipid metabolism and inflammation. *Biochim. Biophys. Acta* **2011**, *1812*, 1007–1022. [CrossRef]
111. Thomas, A.W.; Davies, N.A.; Moir, H.; Watkeys, L.; Ruffino, J.S.; Isa, S.A.; Butcher, L.R.; Hughes, M.G.; Morris, K.; Webb, R. Exercise-associated generation of PPARγ ligands activates PPARγ signaling events and upregulates genes related to lipid metabolism. *J. Appl Physiol.* **2012**, *112*, 806–815. [CrossRef] [PubMed]
112. Ma, X.; Wang, D.; Zhao, W.; Xu, L. Deciphering the Roles of PPARγ in Adipocytes via Dynamic Change of Transcription Complex. *Front. Endocrinol.* **2018**, *9*, 473. [CrossRef]
113. Vidal-Puig, A.J.; Considine, R.V.; Jimenez-Liñan, M.; Werman, A.; Pories, W.J.; Caro, J.F.; Flier, J.S. Peroxisome proliferator-activated receptor gene expression in human tissues. Effects of obesity, weight loss, and regulation by insulin and glucocorticoids. *J. Clin. Investig.* **1997**, *99*, 2416–2422. [CrossRef] [PubMed]
114. Verreth, W.; Keyzer, D.D.; Pelat, M.; Verhamme, P.; Ganame, J.; Bielicki, J.K.; Mertens, A.; Quarck, R.; Benhabilès, N.; Marguerie, G.; et al. Weight Loss–Associated Induction of Peroxisome Proliferator–Activated Receptor-α;and Peroxisome Proliferator–Activated Receptor-γ Correlate with Reduced Atherosclerosis and Improved Cardiovascular Function in Obese Insulin-Resistant Mice. *Circulation* **2004**, *110*, 3259–3269. [CrossRef]
115. Wang, F.; Mullican, S.E.; DiSpirito, J.R.; Peed, L.C.; Lazar, M.A. Lipoatrophy and severe metabolic disturbance in mice with fat-specific deletion of PPARγ. *Proc. Natl. Acad. Sci. USA* **2013**, *110*, 18656–18661. [CrossRef]
116. Jones, J.R.; Barrick, C.; Kim, K.-A.; Lindner, J.; Blondeau, B.; Fujimoto, Y.; Shiota, M.; Kesterson, R.A.; Kahn, B.B.; Magnuson, M.A. Deletion of PPARgamma in adipose tissues of mice protects against high fat diet-induced obesity and insulin resistance. *Proc. Natl. Acad. Sci. USA* **2005**, *102*, 6207–6212. [CrossRef] [PubMed]
117. Liu, J.; Dong, H.; Zhang, Y.; Cao, M.; Song, L.; Pan, Q.; Bulmer, A.; Adams, D.B.; Dong, X.; Wang, H. Bilirubin Increases Insulin Sensitivity by Regulating Cholesterol Metabolism, Adipokines and PPARγ Levels. *Sci. Rep.* **2015**, *5*, 9886. [CrossRef]
118. Shiels, R.G.; Vidimce, J.; Pearson, A.G.; Matthews, B.; Wagner, K.-H.; Battle, A.R.; Sakellaris, H.; Bulmer, A.C. Unprecedented Microbial Conversion of Biliverdin into Bilirubin-10-sulfonate. *Sci. Rep.* **2019**, *9*, 2988. [CrossRef] [PubMed]
119. Hinds, T.D., Jr.; Stec, D.E. Bilirubin Safeguards Cardiorenal and Metabolic Diseases: A Protective Role in Health. *Curr. Hypertens. Rep.* **2019**, *21*, 87. [CrossRef]
120. Weaver, L.; Hamoud, A.R.; Stec, D.E.; Hinds, T.D., Jr. Biliverdin reductase and bilirubin in hepatic disease. *Am. J. Physiol. Gastrointest. Liver Physiol.* **2018**, *314*, G668–G676. [CrossRef]
121. Sundararaghavan, V.L.; Binepal, S.; Stec, D.E.; Sindhwani, P.; Hinds, T.D., Jr. Bilirubin, a new therapeutic for kidney transplant? *Transpl. Rev.* **2018**, *32*, 234–240. [CrossRef]
122. Hamoud, A.R.; Weaver, L.; Stec, D.E.; Hinds, T.D., Jr. Bilirubin in the Liver-Gut Signaling Axis. *Trends Endocrinol. Metab.* **2018**, *29*, 140–150. [CrossRef]
123. Adeosun, S.O.; Moore, K.H.; Lang, D.M.; Nwaneri, A.C.; Hinds, T.D., Jr.; Stec, D.E. A Novel Fluorescence-Based Assay for the Measurement of Biliverdin Reductase Activity. *React. Oxyg. Species* **2018**, *5*, 35–45. [CrossRef] [PubMed]
124. Gordon, D.M.; Adeosun, S.O.; Ngwudike, S.I.; Anderson, C.D.; Hall, J.E.; Hinds, T.D., Jr.; Stec, D.E. CRISPR Cas9-mediated deletion of biliverdin reductase A (BVRA) in mouse liver cells induces oxidative stress and lipid accumulation. *Arch. Biochem. Biophys.* **2019**, *672*, 108072. [CrossRef]
125. Adeosun, S.O.; Gordon, D.M.; Weeks, M.F.; Moore, K.H.; Hall, J.E.; Hinds, T.D., Jr.; Stec, D.E. Loss of biliverdin reductase-A promotes lipid accumulation and lipotoxicity in mouse proximal tubule cells. *Am. J. Physiol. Ren. Physiol.* **2018**, *315*, F323–F331. [CrossRef]

126. Sundararaghavan, V.L.; Sindhwani, P.; Hinds, T.D., Jr. Glucuronidation and UGT isozymes in bladder: New targets for the treatment of uroepithelial carcinomas? *Oncotarget* **2017**, *8*, 3640–3648. [CrossRef] [PubMed]
127. Stec, D.E.; Gordon, D.M.; Nestor-Kalinoski, A.L.; Donald, M.C.; Mitchell, Z.L.; Creeden, J.F.; Hinds, T.D., Jr. Biliverdin Reductase A (BVRA) Knockout in Adipocytes Induces Hypertrophy and Reduces Mitochondria in White Fat of Obese Mice. *Biomolecules* **2020**, *10*, 387. [CrossRef]
128. Ceccarelli, V.; Barchetta, I.; Cimini, F.A.; Bertoccini, L.; Chiappetta, C.; Capoccia, D.; Carletti, R.; Di Cristofano, C.; Silecchia, G.; Fontana, M.; et al. Reduced Biliverdin Reductase-A Expression in Visceral Adipose Tissue is Associated with Adipocyte Dysfunction and NAFLD in Human Obesity. *Int. J. Mol. Sci.* **2020**, *21*, 9091. [CrossRef] [PubMed]
129. Barone, E.; Di Domenico, F.; Cassano, T.; Arena, A.; Tramutola, A.; Lavecchia, M.A.; Coccia, R.; Butterfield, D.A.; Perluigi, M. Impairment of biliverdin reductase-A promotes brain insulin resistance in Alzheimer disease: A new paradigm. *Free Radic. Biol. Med.* **2016**, *91*, 127–142. [CrossRef]
130. Triani, F.; Tramutola, A.; Di Domenico, F.; Sharma, N.; Butterfield, D.A.; Head, E.; Perluigi, M.; Barone, E. Biliverdin reductase-A impairment links brain insulin resistance with increased Abeta production in an animal model of aging: Implications for Alzheimer disease. *Biochim. Biophys. Acta Mol. Basis Dis.* **2018**, *1864*, 3181–3194. [CrossRef] [PubMed]
131. Barone, E.; Tramutola, A.; Triani, F.; Calcagnini, S.; Di Domenico, F.; Ripoli, C.; Gaetani, S.; Grassi, C.; Butterfield, D.A.; Cassano, T.; et al. Biliverdin Reductase-A Mediates the Beneficial Effects of Intranasal Insulin in Alzheimer Disease. *Mol. Neurobiol.* **2019**, *56*, 2922–2943. [CrossRef]
132. McCarty, M.F. Serum bilirubin may serve as a marker for increased heme oxygenase activity and inducibility in tissues—A rationale for the versatile health protection associated with elevated plasma bilirubin. *Med. Hypotheses* **2013**, *81*, 607–610. [CrossRef] [PubMed]
133. Lin, Y.C.; Chang, P.F.; Hu, F.C.; Chang, M.H.; Ni, Y.H. Variants in the UGT1A1 gene and the risk of pediatric nonalcoholic fatty liver disease. *Pediatrics* **2009**, *124*, e1221–e1227. [CrossRef] [PubMed]
134. Verlaan, S.; Maier, A.B.; Bauer, J.M.; Bautmans, I.; Brandt, K.; Donini, L.M.; Maggio, M.; McMurdo, M.E.; Mets, T.; Seal, C.; et al. Sufficient levels of 25-hydroxyvitamin D and protein intake required to increase muscle mass in sarcopenic older adults—The PROVIDE study. *Clin. Nutr.* **2017**, *37*, 551–557. [CrossRef]
135. Lee, M.J.; Jung, C.H.; Kang, Y.M.; Hwang, J.Y.; Jang, J.E.; Leem, J.; Park, J.Y.; Kim, H.K.; Lee, W.J. Serum bilirubin as a predictor of incident metabolic syndrome: A 4-year retrospective longitudinal study of 6205 initially healthy Korean men. *Diabetes Metab.* **2014**, *40*, 305–309. [CrossRef] [PubMed]
136. Turfan, M.; Duran, M.; Poyraz, F.; Yayla, C.; Akboga, M.K.; Sahinarslan, A.; Tavil, Y.; Pasaoglu, H.; Boyaci, B. Inverse relationship between serum total bilirubin levels and severity of disease in patients with stable coronary artery disease. *Coron. Artery Dis.* **2013**, *24*, 29–32. [CrossRef] [PubMed]
137. Jang, B.K. Elevated serum bilirubin levels are inversely associated with nonalcoholic fatty liver disease. *Clin. Mol. Hepatol.* **2012**, *18*, 357–359. [CrossRef]
138. Higuchi, S.; Kabeya, Y.; Uchida, J.; Kato, K.; Tsukada, N. Low Bilirubin Levels Indicate a High Risk of Cerebral Deep White Matter Lesions in Apparently Healthy Subjects. *Sci. Rep.* **2018**, *8*, 6473. [CrossRef] [PubMed]
139. Garde, E.; Mortensen, E.L.; Krabbe, K.; Rostrup, E.; Larsson, H.B. Relation between age-related decline in intelligence and cerebral white-matter hyperintensities in healthy octogenarians: A longitudinal study. *Lancet* **2000**, *356*, 628–634. [CrossRef]
140. Verdelho, A.; Madureira, S.; Moleiro, C.; Ferro, J.M.; Santos, C.O.; Erkinjuntti, T.; Pantoni, L.; Fazekas, F.; Visser, M.; Waldemar, G.; et al. White matter changes and diabetes predict cognitive decline in the elderly: The LADIS study. *Neurology* **2010**, *75*, 160–167. [CrossRef] [PubMed]
141. Debette, S.; Markus, H.S. The clinical importance of white matter hyperintensities on brain magnetic resonance imaging: Systematic review and meta-analysis. *BMJ* **2010**, *341*, c3666. [CrossRef] [PubMed]
142. Benzie, I.F. Evolution of dietary antioxidants. *Comp. Biochem. Physiol. Part A Mol. Integr. Physiol.* **2003**, *136*, 113–126. [CrossRef]
143. Watzl, B.; Kulling, S.E.; Moseneder, J.; Barth, S.W.; Bub, A. A 4-wk intervention with high intake of carotenoid-rich vegetables and fruit reduces plasma C-reactive protein in healthy, nonsmoking men. *Am. J. Clin. Nutr.* **2005**, *82*, 1052–1058. [CrossRef] [PubMed]
144. Nanri, A.; Moore, M.A.; Kono, S. Impact of C-reactive protein on disease risk and its relation to dietary factors. *Asian Pac. J. Cancer Prev.* **2007**, *8*, 167–177.
145. Esposito, K.; Marfella, R.; Ciotola, M.; Di Palo, C.; Giugliano, F.; Giugliano, G.; D'Armiento, M.; D'Andrea, F.; Giugliano, D. Effect of a mediterranean-style diet on endothelial dysfunction and markers of vascular inflammation in the metabolic syndrome: A randomized trial. *JAMA* **2004**, *292*, 1440–1446. [CrossRef]
146. Salas-Salvado, J.; Garcia-Arellano, A.; Estruch, R.; Marquez-Sandoval, F.; Corella, D.; Fiol, M.; Gomez-Gracia, E.; Vinoles, E.; Aros, F.; Herrera, C.; et al. Components of the Mediterranean-type food pattern and serum inflammatory markers among patients at high risk for cardiovascular disease. *Eur. J. Clin. Nutr.* **2008**, *62*, 651–659. [CrossRef]
147. Chrysohoou, C.; Panagiotakos, D.B.; Pitsavos, C.; Das, U.N.; Stefanadis, C. Adherence to the Mediterranean diet attenuates inflammation and coagulation process in healthy adults: The ATTICA Study. *J. Am. Coll. Cardiol.* **2004**, *44*, 152–158. [CrossRef]
148. Tuttolomondo, A.; Simonetta, I.; Daidone, M.; Mogavero, A.; Ortello, A.; Pinto, A. Metabolic and Vascular Effect of the Mediterranean Diet. *Int. J. Mol. Sci.* **2019**, *20*, 4716. [CrossRef] [PubMed]
149. Mentella, M.C.; Scaldaferri, F.; Ricci, C.; Gasbarrini, A.; Miggiano, G.A.D. Cancer and Mediterranean Diet: A Review. *Nutrients* **2019**, *11*, 2059. [CrossRef]

150. Martín-Peláez, S.; Fito, M.; Castaner, O. Mediterranean Diet Effects on Type 2 Diabetes Prevention, Disease Progression, and Related Mechanisms. A Review. *Nutrients* **2020**, *12*, 2236. [CrossRef]
151. Meddeb, W.; Rezig, L.; Abderrabba, M.; Lizard, G.; Mejri, M. Tunisian Milk Thistle: An Investigation of the Chemical Composition and the Characterization of Its Cold-Pressed Seed Oils. *Int. J. Mol. Sci.* **2017**, *18*, 2582. [CrossRef]
152. Pittala, V.; Vanella, L.; Salerno, L.; Romeo, G.; Marrazzo, A.; Di Giacomo, C.; Sorrenti, V. Effects of Polyphenolic Derivatives on Heme Oxygenase-System in Metabolic Dysfunctions. *Curr. Med. Chem.* **2018**, *25*, 1577–1595. [CrossRef] [PubMed]
153. Suk, J.; Jasprova, J.; Biedermann, D.; Petraskova, L.; Valentova, K.; Kren, V.; Muchova, L.; Vitek, L. Isolated Silymarin Flavonoids Increase Systemic and Hepatic Bilirubin Concentrations and Lower Lipoperoxidation in Mice. *Oxid. Med. Cell. Longev.* **2019**, *2019*, 6026902. [CrossRef] [PubMed]
154. Bechynska, K.; Kosek, V.; Fenclova, M.; Muchova, L.; Smid, V.; Suk, J.; Chalupsky, K.; Sticova, E.; Hurkova, K.; Hajslova, J.; et al. The Effect of Mycotoxins and Silymarin on Liver Lipidome of Mice with Non-Alcoholic Fatty Liver Disease. *Biomolecules* **2021**, *11*, 1723. [CrossRef] [PubMed]
155. Ni, X.; Wang, H. Silymarin attenuated hepatic steatosis through regulation of lipid metabolism and oxidative stress in a mouse model of nonalcoholic fatty liver disease (NAFLD). *Am. J. Transl. Res.* **2016**, *8*, 1073–1081.
156. Clichici, S.; Olteanu, D.; Nagy, A.L.; Oros, A.; Filip, A.; Mircea, P.A. Silymarin inhibits the progression of fibrosis in the early stages of liver injury in CCl(4)-treated rats. *J. Med. Food* **2015**, *18*, 290–298. [CrossRef]
157. Gupta, N.; Singh, T.; Chaudhary, R.; Garg, S.K.; Sandhu, G.S.; Mittal, V.; Gupta, R.; Bodin, R.; Sule, S. Bilirubin in coronary artery disease: Cytotoxic or protective? *World J. Gastrointest. Pharm.* **2016**, *7*, 469–476. [CrossRef]
158. Ziberna, L.; Martelanc, M.; Franko, M.; Passamonti, S. Bilirubin is an Endogenous Antioxidant in Human Vascular Endothelial Cells. *Sci. Rep.* **2016**, *6*, 29240. [CrossRef]
159. Cesari, M.; Incalzi, R.A.; Zamboni, V.; Pahor, M. Vitamin D hormone: A multitude of actions potentially influencing the physical function decline in older persons. *Geriatr. Gerontol. Int.* **2011**, *11*, 133–142. [CrossRef]
160. Sato, Y.; Iwamoto, J.; Kanoko, T.; Satoh, K. Low-dose vitamin D prevents muscular atrophy and reduces falls and hip fractures in women after stroke: A randomized controlled trial. *Cerebrovasc. Dis.* **2005**, *20*, 187–192. [CrossRef]
161. Hamilton, B. Vitamin D and human skeletal muscle. *Scand. J. Med. Sci. Sports* **2010**, *20*, 182–190. [CrossRef]
162. Girgis, C.M. Integrated therapies for osteoporosis and sarcopenia: From signaling pathways to clinical trials. *Calcif. Tissue Int.* **2015**, *96*, 243–255. [CrossRef]
163. Bunout, D.; Barrera, G.; Leiva, L.; Gattas, V.; de la Maza, M.P.; Avendano, M.; Hirsch, S. Effects of vitamin D supplementation and exercise training on physical performance in Chilean vitamin D deficient elderly subjects. *Exp. Gerontol.* **2006**, *41*, 746–752. [CrossRef]
164. Houston, D.K.; Tooze, J.A.; Neiberg, R.H.; Hausman, D.B.; Johnson, M.A.; Cauley, J.A.; Bauer, D.C.; Cawthon, P.M.; Shea, M.K.; Schwartz, G.G.; et al. 25-hydroxyvitamin D status and change in physical performance and strength in older adults: The Health, Aging, and Body Composition Study. *Am. J. Epidemiol.* **2012**, *176*, 1025–1034. [CrossRef] [PubMed]
165. Bischoff-Ferrari, H.A.; Dietrich, T.; Orav, E.J.; Hu, F.B.; Zhang, Y.; Karlson, E.W.; Dawson-Hughes, B. Higher 25-hydroxyvitamin D concentrations are associated with better lower-extremity function in both active and inactive persons aged > or =60 y. *Am. J. Clin. Nutr.* **2004**, *80*, 752–758. [CrossRef]
166. Brunner, F.; Schmid, A.; Sheikhzadeh, A.; Nordin, M.; Yoon, J.; Frankel, V. Effects of aging on Type II muscle fibers: A systematic review of the literature. *J. Aging Phys. Act.* **2007**, *15*, 336–348. [CrossRef]
167. Muir, S.W.; Montero-Odasso, M. Effect of vitamin D supplementation on muscle strength, gait and balance in older adults: A systematic review and meta-analysis. *J. Am. Geriatr. Soc.* **2011**, *59*, 2291–2300. [CrossRef]
168. Ford, E.S.; Ajani, U.A.; McGuire, L.C.; Liu, S. Concentrations of serum vitamin D and the metabolic syndrome among U.S. adults. *Diabetes Care* **2005**, *28*, 1228–1230. [CrossRef]
169. Liu, E.; Meigs, J.B.; Pittas, A.G.; McKeown, N.M.; Economos, C.D.; Booth, S.L.; Jacques, P.F. Plasma 25-hydroxyvitamin d is associated with markers of the insulin resistant phenotype in nondiabetic adults. *J. Nutr.* **2009**, *139*, 329–334. [CrossRef]
170. Scragg, R.; Sowers, M.; Bell, C.; Third National, H.; Nutrition Examination, S. Serum 25-hydroxyvitamin D, diabetes, and ethnicity in the Third National Health and Nutrition Examination Survey. *Diabetes Care* **2004**, *27*, 2813–2818. [CrossRef] [PubMed]
171. Sinha, A.; Hollingsworth, K.G.; Ball, S.; Cheetham, T. Improving the vitamin D status of vitamin D deficient adults is associated with improved mitochondrial oxidative function in skeletal muscle. *J. Clin. Endocrinol. Metab.* **2013**, *98*, E509–E513. [CrossRef] [PubMed]
172. Daniel, D.; Hardigan, P.; Bray, N.; Penzell, D.; Savu, C. The incidence of vitamin D deficiency in the obese: A retrospective chart review. *J. Community Hosp. Intern. Med. Perspect* **2015**, *5*, 26069. [CrossRef] [PubMed]
173. Vranic, L.; Mikolasevic, I.; Milic, S. Vitamin D Deficiency: Consequence or Cause of Obesity? *Medicina* **2019**, *55*, 541. [CrossRef] [PubMed]
174. Ryan, Z.C.; Craig, T.A.; Folmes, C.D.; Wang, X.; Lanza, I.R.; Schaible, N.S.; Salisbury, J.L.; Nair, K.S.; Terzic, A.; Sieck, G.C.; et al. 1alpha,25-Dihydroxyvitamin D3 Regulates Mitochondrial Oxygen Consumption and Dynamics in Human Skeletal Muscle Cells. *J. Biol. Chem.* **2016**, *291*, 1514–1528. [CrossRef] [PubMed]
175. Tagliafico, A.S.; Ameri, P.; Bovio, M.; Puntoni, M.; Capaccio, E.; Murialdo, G.; Martinoli, C. Relationship between fatty degeneration of thigh muscles and vitamin D status in the elderly: A preliminary MRI study. *AJR Am. J. Roentgenol.* **2010**, *194*, 728–734. [CrossRef] [PubMed]

176. Gilsanz, V.; Kremer, A.; Mo, A.O.; Wren, T.A.; Kremer, R. Vitamin D status and its relation to muscle mass and muscle fat in young women. *J. Clin. Endocrinol. Metab.* **2010**, *95*, 1595–1601. [CrossRef]
177. Redzic, M.; Powell, D.K.; Thomas, D.T. Vitamin D status is related to intramyocellular lipid in older adults. *Endocrine* **2014**, *47*, 854–861. [CrossRef] [PubMed]
178. Scott, D.; Joham, A.; Teede, H.; Gibson-Helm, M.; Harrison, C.; Cassar, S.; Hutchison, S.; Ebeling, P.R.; Stepto, N.; de Courten, B. Associations of Vitamin D with Inter- and Intra-Muscular Adipose Tissue and Insulin Resistance in Women with and without Polycystic Ovary Syndrome. *Nutrients* **2016**, *8*, 774. [CrossRef] [PubMed]
179. Papapostoli, P.; Lammert, F.; Stokes, C.S. Effect of Short-Term Vitamin D Correction on Hepatic Steatosis as Quantified by Controlled Attenuation Parameter (CAP). *J. Gastrointest. Liver Dis.* **2016**, *25*, 175–181. [CrossRef] [PubMed]
180. Schnell, D.M.; Walton, R.G.; Vekaria, H.J.; Sullivan, P.G.; Bollinger, L.M.; Peterson, C.A.; Thomas, D.T. Vitamin D produces a perilipin 2-dependent increase in mitochondrial function in C2C12 myotubes. *J. Nutr. Biochem.* **2019**, *65*, 83–92. [CrossRef]
181. Jefferson, G.E.; Schnell, D.M.; Thomas, D.T.; Bollinger, L.M. Calcitriol concomitantly enhances insulin sensitivity and alters myocellular lipid partitioning in high fat-treated skeletal muscle cells. *J. Physiol. Biochem.* **2017**, *73*, 613–621. [CrossRef]
182. Thomas, D.T.; Schnell, D.M.; Redzic, M.; Zhao, M.; Abraha, H.; Jones, D.; Brim, H.; Yu, G. Local In Vivo Measures of Muscle Lipid and Oxygen Consumption Change in Response to Combined Vitamin D Repletion and Aerobic Training in Older Adults. *Nutrients* **2019**, *11*, 930. [CrossRef] [PubMed]
183. Makanae, Y.; Ogasawara, R.; Sato, K.; Takamura, Y.; Matsutani, K.; Kido, K.; Shiozawa, N.; Nakazato, K.; Fujita, S. Acute bout of resistance exercise increases vitamin D receptor protein expression in rat skeletal muscle. *Exp. Physiol.* **2015**, *100*, 1168–1176. [CrossRef] [PubMed]
184. Giulietti, A.; van Etten, E.; Overbergh, L.; Stoffels, K.; Bouillon, R.; Mathieu, C. Monocytes from type 2 diabetic patients have a pro-inflammatory profile. 1,25-Dihydroxyvitamin D(3) works as anti-inflammatory. *Diabetes Res. Clin. Pract.* **2007**, *77*, 47–57. [CrossRef]
185. Di Rosa, M.; Malaguarnera, M.; Nicoletti, F.; Malaguarnera, L. Vitamin D3: A helpful immuno-modulator. *Immunology* **2011**, *134*, 123–139. [CrossRef]
186. Jablonski, K.; Chonchol, M.; Pierce, G.; Walker, A.E.; Seals, D.R. 25-Hydroxyvitamin D deficiency is associated with inflammation-linked vascular endothelial dysfunction in middle-aged and older adults. *Hypertension* **2010**, *57*, 63–69. [CrossRef]
187. Yin, K.; Agrawal, D.K. Vitamin D and inflammatory diseases. *J. Inflamm. Res.* **2014**, *7*, 69–87. [CrossRef]
188. Wolins, N.E.; Quaynor, B.K.; Skinner, J.R.; Tzekov, A.; Croce, M.A.; Gropler, M.C.; Varma, V.; Yao-Borengasser, A.; Rasouli, N.; Kern, P.A.; et al. OXPAT/PAT-1 is a PPAR-induced lipid droplet protein that promotes fatty acid utilization. *Diabetes* **2006**, *55*, 3418–3428. [CrossRef]
189. Bosma, M.; Hesselink, M.K.C.; Sparks, L.M.; Timmers, S.; Ferraz, M.J.; Mattijssen, F.; van Beurden, D.; Schaart, G.; de Baets, M.H.; Verheyen, F.K.; et al. Perilipin 2 Improves Insulin Sensitivity in Skeletal Muscle Despite Elevated Intramuscular Lipid Levels. *Diabetes* **2012**, *61*, 2679–2690. [CrossRef]
190. Shepherd, S.O.; Cocks, M.; Tipton, K.D.; Ranasinghe, A.M.; Barker, T.A.; Burniston, J.G.; Wagenmakers, A.J.; Shaw, C.S. Preferential utilization of perilipin 2-associated intramuscular triglycerides during 1 h of moderate-intensity endurance-type exercise. *Exp. Physiol.* **2012**, *97*, 970–980. [CrossRef]
191. Funai, K.; Semenkovich, C.F. Skeletal muscle lipid flux: Running water carries no poison. *Am. J. Physiol. Endocrinol. Metab.* **2011**, *301*, E245–E251. [CrossRef]
192. Abboud, M.; Puglisi, D.A.; Davies, B.N.; Rybchyn, M.; Whitehead, N.P.; Brock, K.E.; Cole, L.; Gordon-Thomson, C.; Fraser, D.R.; Mason, R.S. Evidence for a specific uptake and retention mechanism for 25-hydroxyvitamin D (25OHD) in skeletal muscle cells. *Endocrinology* **2013**, *154*, 3022–3030. [CrossRef] [PubMed]
193. Thomas, D.T.; Erdman, K.A.; Burke, L.M. American College of Sports Medicine joint position statement: Nutrition and athletic performance. *Med. Sci. Sports Exerc.* **2016**, *48*, 543–568. [CrossRef] [PubMed]
194. Tian, R.; Peng, R.; Yang, Z.; Peng, Y.; Lu, N. Supplementation of dietary nitrate attenuated oxidative stress and endothelial dysfunction in diabetic vasculature through inhibition of NADPH oxidase. *Nitric Oxide* **2020**, *96*, 54–63. [CrossRef] [PubMed]
195. Jones, A.M. Influence of dietary nitrate on the physiological determinants of exercise performance: A critical review. *Appl. Physiol. Nutr. Metab.* **2014**, *39*, 1019–1028. [CrossRef]
196. Velmurugan, S.; Ming Gan, J.; Rathod, K.; Khambata, R.; Ghosh, M.; Hartley, A. Dietary nitrate improves vascular function in patients with hypercholesterolemia: A randomized, double-blind, placebo-controlled study. *Am. J. Clin. Nutr.* **2016**, *103*, 25–38. [CrossRef]
197. Basaqr, R.; Skleres, M.; Jayswal, R.; Thomas, D.T. The Effect of Dietary Nitrate and Vitamin C on Endothelial Function, Oxidative Stress, and Blood Lipids in Untreated Hypercholesterolemic Subjects: A Randomized Double-blind Crossover Study. *Clin. Nutr.* **2021**, *40*, 1851–1860. [CrossRef] [PubMed]
198. Ashor, A.W.; Siervo, M.; van der Velde, F.; Willis, N.D.; Mathers, J.C. Systematic review and meta-analysis of randomised controlled trials testing the effects of vitamin C supplementation on blood lipids. *Clin. Nutr.* **2016**, *35*, 626–637. [CrossRef]
199. Chambial, S.; Dwivedi, S.; Shukla, K.K.; John, P.J.; Sharma, P. Vitamin C in disease prevention and cure: An overview. *Indian J. Clin. Biochem.* **2013**, *28*, 314–328. [CrossRef]
200. Charlton-Menys, V.; Durrington, P. Human cholesterol metabolism and therapeutic molecules. *Exp. Physiol.* **2008**, *93*, 27–42. [CrossRef]

201. McRae, M. Vitamin C supplementation lowers serum low-density lipoprotein cholesterol and triglycerides: A meta-analysis of 13 randomized controlled trials. *J. Chiropr. Med.* **2008**, *7*, 48–58. [CrossRef]
202. Galmés, S.; Serra, F.; Palou, A. Vitamin E Metabolic Effects and Genetic Variants: A Challenge for Precision Nutrition in Obesity and Associated Disturbances. *Nutrients* **2018**, *10*, 1919. [CrossRef] [PubMed]
203. Nakamura, Y.K.; Omaye, S.T. Vitamin E-modulated gene expression associated with ROS generation. *J. Funct. Foods* **2009**, *1*, 241–252. [CrossRef]
204. Traber, M.G. Vitamin E regulatory mechanisms. *Annu. Rev. Nutr.* **2007**, *27*, 347–362. [CrossRef] [PubMed]
205. Ulatowski, L.; Dreussi, C.; Noy, N.; Barnholtz-Sloan, J.; Klein, E.; Manor, D. Expression of the α-tocopherol transfer protein gene is regulated by oxidative stress and common single-nucleotide polymorphisms. *Free Radic. Biol. Med.* **2012**, *53*, 2318–2326. [CrossRef] [PubMed]
206. Inan, C.; Kiliç, I.; Kilinç, K.; Kalayci, O.; Kotiloğlu, E. The effect of high dose antenatal vitamin E on hypoxia-induced changes in newborn rats. *Pediatr. Res.* **1995**, *38*, 685–689. [CrossRef]
207. Campbell, S.E.; Stone, W.L.; Whaley, S.G.; Qui, M.; Krishnan, K. Gamma (gamma) tocopherol upregulates peroxisome proliferator activated receptor (PPAR) gamma (gamma) expression in SW 480 human colon cancer cell lines. *BMC Cancer* **2003**, *3*, 25. [CrossRef] [PubMed]
208. Landrier, J.-F.o.; Gouranton, E.; El Yazidi, C.; Malezet, C.; Balaguer, P.; Borel, P.; Amiot, M.-J.p. Adiponectin Expression Is Induced by Vitamin E via a Peroxisome Proliferator-Activated Receptor γ-Dependent Mechanism. *Endocrinology* **2009**, *150*, 5318–5325. [CrossRef] [PubMed]
209. Lim, H.-J.; Lee, K.-S.; Lee, S.; Park, J.-H.; Choi, H.-E.; Go, S.H.; Kwak, H.-J.; Parka, H.-Y. 15d-PGJ2 stimulates HO-1 expression through p38 MAP kinase and Nrf-2 pathway in rat vascular smooth muscle cells. *Toxicol. Appl. Pharm.* **2007**, *223*, 20–27. [CrossRef]
210. Polyzos, S.A.; Kang, E.S.; Boutari, C.; Rhee, E.J.; Mantzoros, C.S. Current and emerging pharmacological options for the treatment of nonalcoholic steatohepatitis. *Metabolism* **2020**, *111*, 154203. [CrossRef] [PubMed]

antioxidants

MDPI

Article

Effect of Heme Oxygenase-1 Depletion on Complement Regulatory Proteins Expression in the Rat

Maria G. Detsika [1,*], Eirini Theochari [2], Kostas Palamaris [2], Harikleia Gakiopoulou [2] and Elias A. Lianos [3]

[1] GP Livanos and M Simou Laboratories, 1st Department of Critical Care Medicine & Pulmonary Services, Evangelismos Hospital, National and Kapodistrian University of Athens, 10675 Athens, Greece
[2] Department of Pathology, School of Medicine, University of Athens, 11527 Athens, Greece
[3] Veterans Affairs Medical Center and Virginia Tech, Carilion School of Medicine, Salem, VA 24153, USA
* Correspondence: mdetsika@med.uoa.gr

Abstract: Heme oxygenase has been implicated in the regulation of various immune responses including complement activation. Using a transgenic rat model of HO-1 depletion, the present study assessed the effect of HO-1 absence on the expression of complement regulatory proteins: decay accelerating factor (DAF), CR1-related gene/protein Y (Crry) and CD59, which act to attenuate complement activation. Protein expression was assessed by immunohistochemistry in kidney, liver, lung and spleen tissues. DAF protein was reduced in all tissues retrieved from rats lacking HO-1 ($Hmox1^{-/-}$) apart from spleen tissue sections. Crry protein was also reduced, but only in $Hmox1^{-/-}$ kidney and liver tissue. C3b staining was augmented in the kidney and spleen from $Hmox1^{-/-}$ rats, suggesting that the decrease of DAF and Crry was sufficient to increase C3b deposition. The observations support an important role of HO-1 as a regulator of the complement system.

Keywords: complement; heme oxygenase (HO)-1; complement regulatory proteins

Citation: Detsika, M.G.; Theochari, E.; Palamaris, K.; Gakiopoulou, H.; Lianos, E.A. Effect of Heme Oxygenase-1 Depletion on Complement Regulatory Proteins Expression in the Rat. *Antioxidants* **2023**, *12*, 61. https://doi.org/10.3390/antiox12010061

Academic Editor: Stanley Omaye

Received: 15 November 2022
Revised: 14 December 2022
Accepted: 22 December 2022
Published: 27 December 2022

1. Introduction

Heme oxygenase (HO)-1 is a well-known cytoprotective enzyme, mainly due to its ability to degrade heme into the established antioxidant and anti-apoptotic metabolites products bilirubin, biliverdin and carbon monoxide (CO). Furthermore, it has been implicated in the regulation of various cellular, biological and immunological responses. As an integral part of cell defense mechanisms against oxidative stress, HO-1 expression is induced in various tissues upon specific stimuli, including hypoxia, inflammatory cytokines or hemolysis, and is regulated by the transcription factor Nrf2 (Nuclear factor erythroid 2-related factor 2). HO-1 is also induced in many pathological conditions, while its expression levels vary greatly between different organs, as well as between the various organ cell types. In the lung, HO-1 induction is detected predominantly in pneumocytes II and in alveolar macrophages in both inflammatory, such as acute respiratory distress syndrome (ARDS), and vascular abnormalities [1], while in the liver it is found primarily in Kupffer cells in conditions of ischemia or toxic injury [2,3]. In the kidneys, it is primarily induced in tubules, both in glomerular and tubulointerstitial diseases of various etiologies [4].

Recently, it was demonstrated that HO-1 regulates complement activation, both by generation of heme degradation products (bilirubin and CO) [5] and by limiting levels of unbound heme, which activates the complement system [6]. This finding implies that HO-1 may possess an important role in limiting excessive complement activation. The complement system is a key mechanism of the innate immunity. Its activation involves mainly three pathways: (i) the classical pathway, in which antibodies bound to antigen lead to the recruitment of the C1 complex; (ii) the alternative pathway, which remains continuously active at a low level due to spontaneous hydrolysis of C3 to C3 (H2O); and (iii) the mannose-binding lectin (MBL) pathway, in which MBL protein binds to its target (for example, mannose on the surface of a bacterium) and MASP (Mannan-binding lectin serine protease) 1 cleaves C4 into C4a and C4b. Activation of any of the three pathways leads to the C3 step, at which C3 cleavage by the C3 convertases generates active C3b. Binding of the C3b-active fragment to the C3 convertase results in deactivation of the C3 convertase and formation and activation of the C5 convertase, which cleaves circulating C5 into active C5a and C5b fragments. C5b generation enables activation of the terminal step, in which the membrane attack complex (MAC, C5b-9), a pore-like structure causing lysis of targeted cells, is formed [7–9]. Moreover, C3a and C5a, known to act as anaphylatoxins, recruit and activate various immune cells.

Complement activation is a central mechanism in many renal immune-mediated conditions. Its important role has been demonstrated in experimental models of immune- mediated forms of glomerular injury, such as Heyman nephritis [10], resembling membranous nephropathy, and the anti-glomerular basement membrane (GBM) model [11], resembling anti-GBM disease, also known as Goodpasture's syndrome. It also plays a key role in mediating kidney injury and progression to end-stage renal disease in a plethora of human kidney diseases. C3 glomerulopathy, a recently characterized disease entity that includes dense deposit disease (DDD) and C3 glomerulonephritis (C3 GN), is driven by uncontrolled activation of the alternative complement pathway [12]. Apart from the established role of complement activation in renal diseases, a potential role for complement has also been reported in lung and injury. Increased levels of complement factors have been measured in various conditions leading to lung injury, such as ARDS, pulmonary arterial hypertension (PAH) and idiopathic pulmonary fibrosis (IPF) [13]. Complement activation has also been implicated in non-alcoholic fatty acid liver disease (NAFLD) leading to non-alcoholic steatohepatitis (NASH), with increased deposition of complement factors around steatotic hepatocytes and reported to be associated with disease progression and a proinflammatory profile [14]. Finally, a previous study utilizing complement factor H (CFH)-deficient mice reported the development of splenomegaly with distorted spleen architecture [15].

Activation of the complement system is controlled by complement regulatory proteins (CRPs), which ensure maintenance of balanced activation of all three complement pathways by acting at multiple steps of the cascade [16]. Given the fact that activation of C3 is the key step in complement activation, it is not surprising that several of the regulatory proteins act at the C3 convertase step. In the rat, cell membrane-bound CRPs include decay accelerating factor (DAF), CD59 and CR1-related gene/protein Y (Crry) [17], which is considered as the murine analogue of the human membrane cofactor protein (MCP) [18,19]. CRPs are an important group of cell surface glycoproteins and share similar structural and functional characteristics. CRPs are all transmembrane proteins attached on the cell surface, allowing for the remaining part of the protein structure to perform its function. Specifically, DAF and CD59 are attached on the cell membrane via a glycosylphosphatidylinositol (gpi)-anchor, while Crry is attached via a transmembrane domain [13]. They are all initially synthesized in the cell nucleus and subsequently undergo heavy N- and O- glycosylation until they finally reach the cell membrane [14,15]. Soluble forms of DAF and CD59 have also been identified, both in humans and rats [7]. Upon anchoring on the cell membrane, DAF and Crry promote the decay of C3 and C5 convertases, thereby limiting the extent of complement cascade activation, whilst CD59 acts solely on the final step of C5b-9 (membrane attack com-plex MAC) formation and activation, thereby limiting cell lysis. Therefore, the balanced activation of the complement cascade in healthy systems largely depends on the functional integrity of these CRPs.

We previously reported a reduction in glomerular DAF expression in HO-1 knock out ($Hmox1^{-/-}$) rats, which was sufficient to minimize C3b deposition upon injury caused by administration of a complement fixing antibody against the GBM [20]. However, the effect of HO-1 depletion on other complement regulatory proteins, both in renal and extra-renal tissues, was not assessed. The present study examined protein levels of CRP (DAF, Crry, CD59), levels of the cleavage product of complement component C3 (C3b), and levels of the complement component 3a receptor (C3aR) in the kidney, lung, liver and spleen of $Hmox1^{-/-}$ rats.

2. Materials and Methods

2.1. Animals

Sprague Dawley rats were reared in accordance to the European Union Directive for care and use of laboratory animals, and all procedures were approved by the Hellenic Veterinary Administration and the ethical committee of 'Evangelismos' Hospital. For generation of transgenic rats, all procedures described were conducted according to the U.S. Department of Health and Human Services Guide for the Care and Use of Laboratory Animals.

HO-1 knock out animals (homozygotes, *Hmox1*$^{-/-}$) were obtained through breeding of *Hmox1*$^{+/-}$ rats, generated as previously described [21]. All invasive experimental procedures were carried out under ketamine:xylazine (1: 3, 0.1 mL/100 g body weight) anesthesia.

2.2. Tissue Retrieval and Immunohistochemistry

Wild type (WT) or *Hmox1*$^{-/-}$ kidney, lung, liver and spleen tissue samples were fixed in formalin (30%)-buffered saline for 24 h. Tissues were dehydrated with graded ethanol solutions (80–90%) and embedded in paraffin prior to sectioning (4-μm-thick sections). Sections were applied to Poly-L-Lysine-coated slides and left at 55 °C for paraffin excess removal. Sections were incubated with either of anti-rat DAF (clone: RDIII-7), Crry (clone: TLD-1C11), CD59 (clone:TH9), C3/C3b (clone: 2B10B9B2) (Hycult, Uden, Netherlands) and C3aR (clone: D-12), (DTX, Santa Cruz, CA, USA) primary antibodies, followed by incubation with species-specific secondary antibodies, using established techniques. Sections incubated in the absence of primary antibody served as negative controls. For the staining process, deparaffinization, rehydration and antigen retrieval were performed by heating the slides in a PT module (PT Link, DAKO, CA, USA) for 20 min at 96 °C, at high pH (EnVision FLEX Target Retrieval Solution, High pH (50X), DAKO, Santa Clara, CA, USA). Endogenous peroxidase was blocked by incubating the slides with hydrogen peroxide 3% for 15 min at room temperature. Sections were incubated with the primary antibodies overnight at 4 °C, at a dilution of 1:50 for each marker. After rinsing with tris buffer solution, slides were incubated with EnVision FLEX HRP (DAKO, Santa Clara, CA, USA), for 30 min at room temperature. DAB was used as chromogen and was provided by EnVision FLEX KIT (DAKO, Santa Clara, CA, USA). Sections were incubated with DAB for 6 min at room temperature.

3. Results

3.1. Effect of HO-1 Depletion on DAF Expression

Immunohistochemistry for assessment of DAF expression in *Hmox1*$^{-/-}$ tissues revealed reduced DAF staining in kidney, liver, lung and spleen tissue. DAF staining did not differ in *Hmox1*$^{-/-}$ spleen tissue compared to WT (Figure 1) (Table 1). Staining for DAF in kidney sections was observed in glomerular epithelial cells (podocytes). This is in line with previous reports demonstrating constitutive expression of DAF exclusively in renal podocytes [22]. DAF staining in the liver was observed in hepatocytes (Figure 1) with a reduction of expression in liver tissue from *Hmox1*$^{-/-}$ rats. A reduction of DAF staining was also observed in the lung, in which DAF was mainly detected in the lung parenchyma.

Table 1. CRP and C3aR expression in WT and *Hmox1*$^{-/-}$ tissue samples and C3b deposition.

	Kidney		Liver		Lung		Spleen	
	WT	*Hmox1*$^{-/-}$	WT	*Hmox1*$^{-/-}$	WT	*Hmox1*$^{-/-}$	WT	*Hmox1*$^{-/-}$
DAF	+	–	+	– (↓)	+	– (↓)	+	+
Crry	+	– (↓)	+	– (↓)	++	+ (↓)	–	–
CD59	–	–	–	–	–	–	–	–
C3b	–	+ (↑)	–	–	++	–	–	+
C3aR	++	+	+	++ (↑)	++	+ (↓)	++	+ (↓)

Scale: –, +, ++ indicate no staining, variable staining and marked staining, respectively. ↓ reduced protein expression, ↑ increased protein expression.

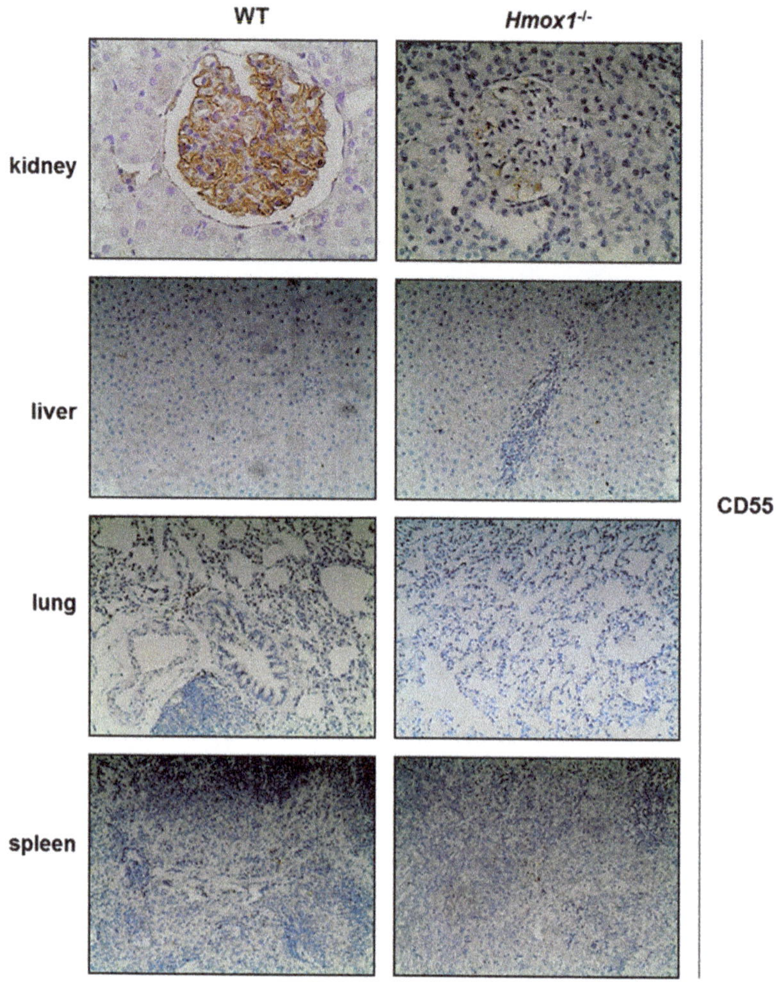

Figure 1. Immunostaining of DAF in tissue sections obtained from WT and *Hmox1*$^{-/-}$ rats. Reduced DAF staining was observed in kidney, liver and lung tissue. No apparent differences in DAF staining in spleen tissue sections were shown in *Hmox1*$^{-/-}$ rats. Magnification ×200.

3.2. *Effect of HO-1 Depletion on Crry Expression*

Reduced Crry staining was observed in kidney, liver and lung tissue sections from *Hmox1*$^{-/-}$ rats (Figure 2, Table 1). On the contrary, Crry could not be detected in spleen sections (Figure 2). In the kidney, reduced Crry expression was observed mainly in renal tubular cells and in glomerular endothelial and epithelial cells (Figure 2). A segmental staining along the capillary walls was also observed in glomeruli. Crry expression was substantially diminished in *Hmox1*$^{-/-}$ rats in both glomerular and tubular compartments. In the liver, Crry was located primarily in hepatocytes, with almost complete absence in liver tissue section obtained from HO-1 depleted rats. In WT liver sections, Crry immunostaining was also observed in sinusoid areas and was absent in *Hmox1*$^{-/-}$ rats. In the lung, Crry was localized primarily in the alveolar walls in lung tissue sections of WT rats, with no change of immunochemical staining in the lung tissue of *Hmox1*$^{-/-}$ compared to WT rats (Figure 2, Table 1).

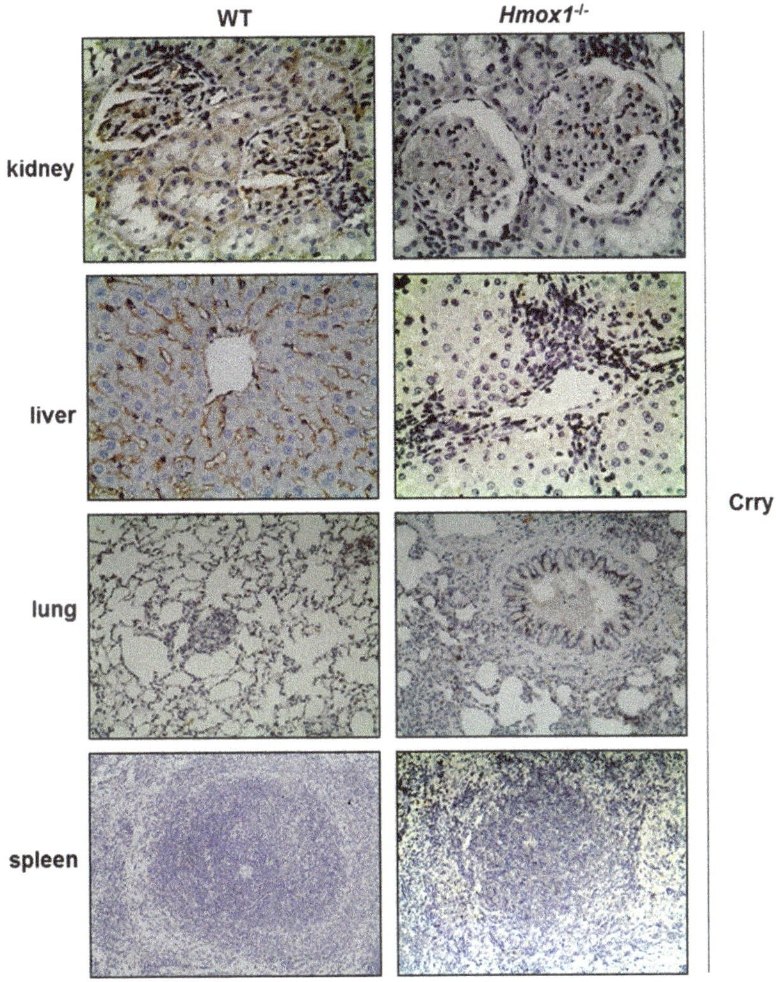

Figure 2. Crry immunostaining of tissues obtained from WT and *Hmox1*$^{-/-}$ rats. Crry was detected in all tissues apart from spleen. Reduced Crry protein staining was demonstrated in kidney, liver and lung tissue section with no apparent changes in spleen tissue sections. Magnification ×200.

3.3. *Effect of HO-1 Absence on CD59 Expression*

CD59 staining was not detected in any of the tissues examined (Figure 3, Table 1). Lung *Hmox1*$^{-/-}$ tissue exhibited some areas of staining, but this was nonspecific.

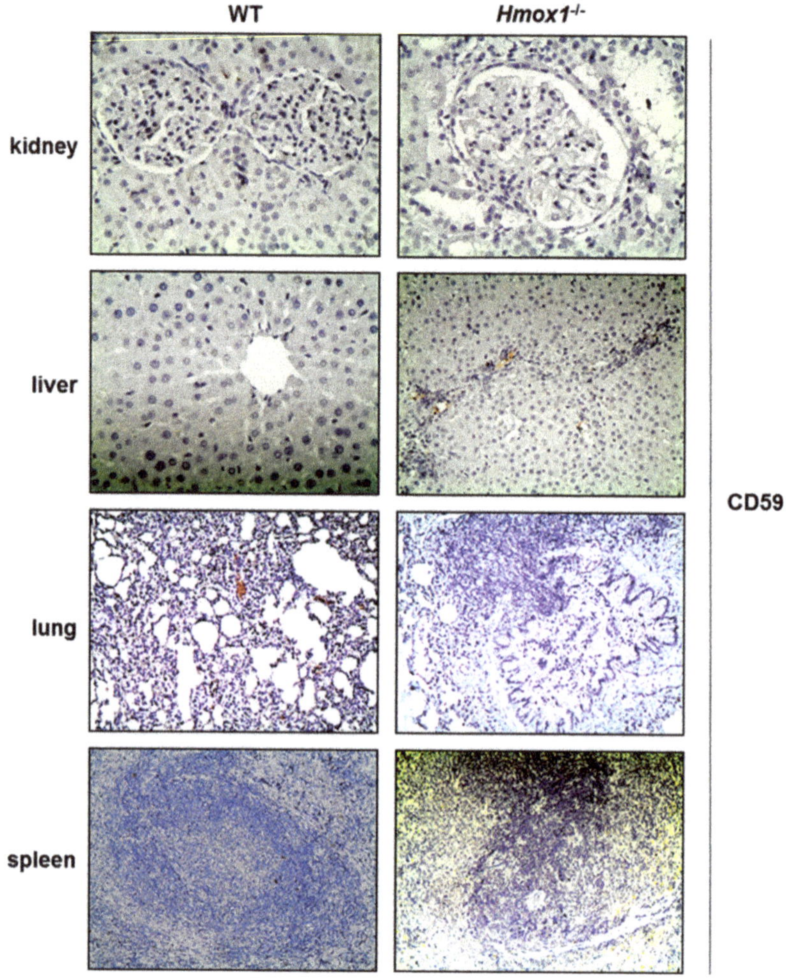

Figure 3. CD59 immunostaining of WT and *Hmox1*$^{-/-}$ rat tissue sections. CD59 was undetectable in all tissues examined. Magnification ×200.

3.4. *C3b Deposition in Hmox1*$^{-/-}$ *Tissue Samples*

C3b staining in kidney sections obtained from *Hmox1*$^{-/-}$ rats revealed focal (present in some glomeruli) glomerular deposition (Figure 4, Table 1) compared to WT kidney sections, in which no C3b expression was detected. Specifically, segmental glomerular staining, located in mesangial and endothelial cells of *Hmox1*$^{-/-}$ kidney tissue sections, was observed. In liver sections from *Hmox1*$^{-/-}$ rats, there was bridging consisting of inflammatory cells between central veins and biliary ducts with no C3b deposition. In the lung parenchyma of WT rats, C3b immunolocalized in alveolar walls and in cells surrounding bronchioles. On the contrary, in lung parenchyma of *Hmox1*$^{-/-}$ rats, C3b deposition was barely detectable. Increased C3b deposition, mainly in the red pulp and, to a lesser extent, in the marginal zones, periarteriolar sheath (PALS) and germinal centers, was detected in *Hmox1*$^{-/-}$ spleen sections (Figure 4, Table 1).

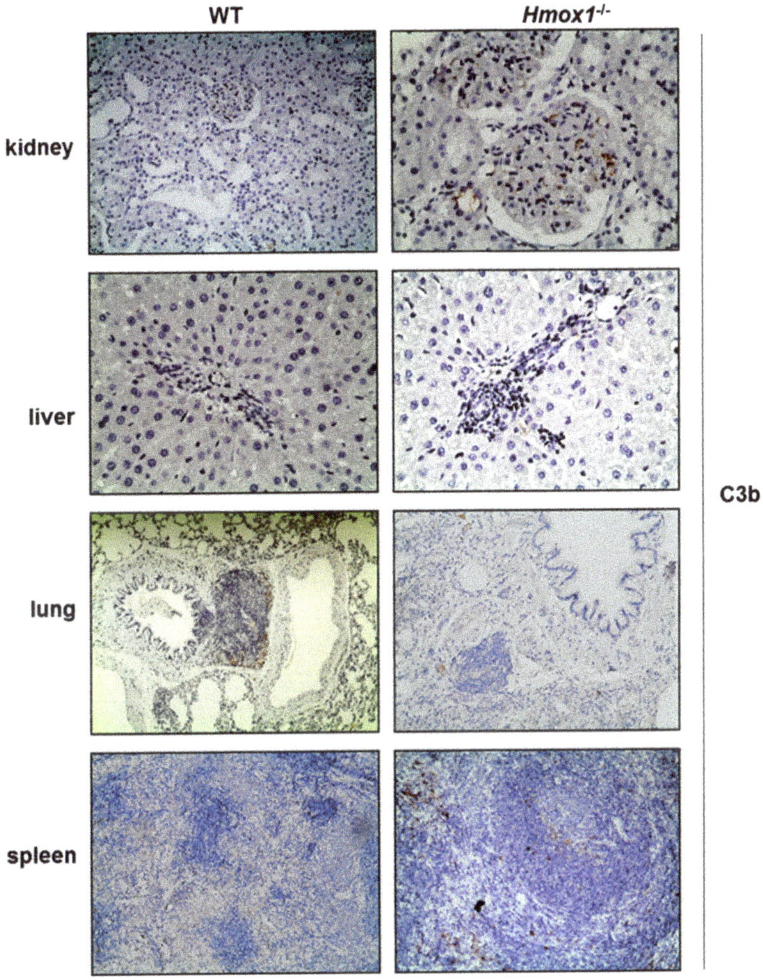

Figure 4. C3b staining of WT and *Hmox1*$^{-/-}$ rat tissue sections. Staining of C3b was detected in kidney, lung and spleen tissues sections. Magnification ×200.

3.5. Effect of HO-1 Depletion on Complement Component 3a Receptor (C3aR)

C3aR staining was detected in both tubules and glomeruli of kidney tissue sections (Figure 5, Table 1). In glomeruli, C3aR immunolocalized predominantly in podocytes and was reduced in those of *Hmox1*$^{-/-}$ rats. Faint cytoplasmic C3aR staining was observed in hepatocytes of WT rats (Figure 5, Table 1). This was substantially augmented in hepatocytes of *Hmox1*$^{-/-}$ rats. In lung tissue sections, cytoplasmic C3aR staining was observed in bronchial epithelium of WT and *Hmox1*$^{-/-}$ rats, with a significant reduction of expression in the latter. In spleen tissue, C3aR staining was detected in red and white pulp of both WT and *Hmox1*$^{-/-}$ rats, with weaker staining in *Hmox1*$^{-/-}$ rats (Figure 5, Table 1).

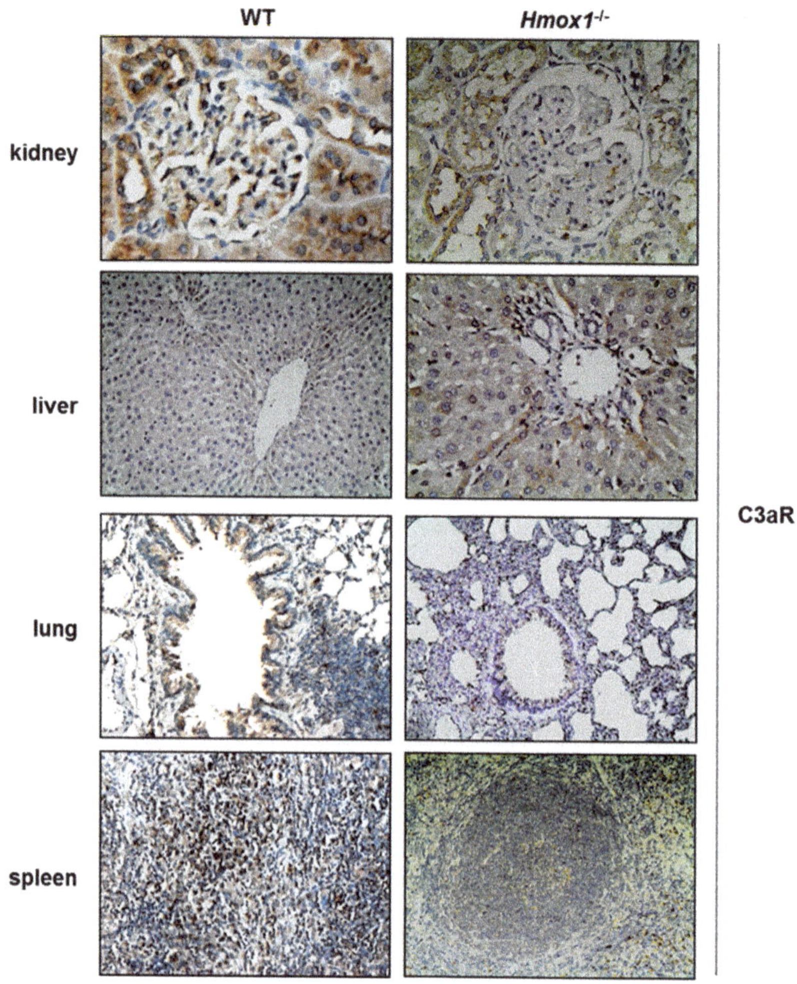

Figure 5. C3aR staining of WT and *Hmox1*$^{-/-}$ rat tissue sections. Staining of C3b was detected in all tissue sections examined. Magnification ×200.

3.6. Effect of HO-1 Depletion on Tissue Morphology

A mild mesangial hypercellularity in kidney tissue was noticed in *Hmox1*$^{-/-}$ rats (Figure 2). Lung tissue displayed focal inflammation, mainly with peribronchial distribution (Figure 3). Moreover, absence of HO-1 in the liver resulted in periportal inflammatory infiltrates (Figure 2) and "bridging" consisting of inflammatory cells. Specifically, *Hmox1*$^{-/-}$ rats showed increased portal inflammation consisting mainly of lymphocytes in the liver.

4. Discussion

In the current study, we used a previously generated and phenotypically characterized HO-1 deficient rat model to identify potential interactions of HO-1 with CRPs and unravel additional HO-1 complement-related functions. In the transgenic model used, complete deletion of *Hmox1* locus via zinc finger nuclease (ZFN)-mediated gene disruption was achieved. Efficiency of *Hmox1* ablation was validated by both western blotting and Real-time PCR amplification, revealing the complete absence of HO-1 protein and mRNA in renal and extra-renal (lung, liver, spleen) tissues of *Hmox1*$^{-/-}$ rats, respectively [23]. Moreover,

HO-1 deficiency was associated with a renal dysfunction phenotype, characterized by both albuminuria and increased serum creatinine, which accompanied specific glomerular histological alterations, defined by mild mesangial expansion and segmental scleroses. Electron microscopy study of glomeruli also revealed segmental effacement of podocytes foot processes. The generation of this transgenic model confirmed that HO-1 regulates DAF in glomeruli, with functional consequences in terms of C3b deposition.

In the kidney, constitutive expression of DAF is restricted exclusively in glomerular epithelial cells (podocytes) [22]. This was verified in the present study, which demonstrated a reduction in DAF staining in glomeruli of *Hmox1*$^{-/-}$ rats (Figure 1). Our previous work has also demonstrated that HO-1 ablation in transgenic rats resulted in a reduction of glomerular DAF expression [7], while podocyte-specific HO-1 overexpression increased DAF expression and reduced glomerular C3b deposition following administration of the complement fixing antibody anti-GBM.

The assessment of changes in tissue detection of CRPs other that DAF in WT and *Hmox1*$^{-/-}$ kidneys demonstrated that the most apparent differences were observed for Crry protein, which restricts complement activation via both alternative and classical pathways. A reduction of Crry was observed in *Hmox1*$^{-/-}$ kidney, liver and lung tissue. In contrast, CD59, which limits C5b-9-mediated cytotoxicity and specifically collaborates with Crry to control complement activation [24], could not be detected in either WT or *Hmox1*$^{-/-}$ rat tissues examined. Crry has been reported to functionally resemble MCP, while at the structural level, it shares more similarities with CR1, as it is attached on the cell via a transmembrane domain and does not possess any serine-threonine-rich structures [25]. The reduction of DAF and Crry protein staining observed in kidney tissue was associated with increased deposition of C3b (Figure 4), supporting a functional significance of this regulatory effect of HO-1 on these two CRPs.

Regarding the pathological consequences of HO-1 defective activity in renal parenchyma, previous results suggest that HO-1 absence results in lesions resembling focal segmental glomerular sclerosis (FSGS) in the kidney [23], a finding confirmed in the present study. Moreover, it was recently reported that reduced DAF expression may promote development of FSGS [26]. This was verified in DAF knock out mice and a model of adriamycin (ADR) induced nephropathy. The same study also confirmed reduced levels of DAF staining in human tissue specimens from patients with FSGS, with a simultaneous increase in C3b deposition, suggesting that lack of DAF results in FSGS lesions and that injury is primarily mediated by the C3a–C3aR axis [26]. In the present study, we also observed a significant decrease in DAF expression coupled with increased C3b staining in *Hmox1*$^{-/-}$ rat kidney sections, albeit C3aR staining was reduced compared to WT. This finding suggests that reduced DAF expression may also contribute to the development of FSGS lesions previously reported in HO-1 deficient rats [23]. Reduced immunostaining of DAF and Crry was also observed in *Hmox1*$^{-/-}$ liver tissue. However, there was no change in C3b detection, suggesting that the reduced expression of DAF and Crry, consequent to HO-1 depletion, in liver tissue was not sufficient to increase complement deposition and that lesions observed (periportal inflammatory infiltrates and bridging) are not mechanistically linked to reduced DAF or Crry levels.

In our study, it was not possible to assess changes in CD59, as this was not detected in any of the tissues. This is in contrast to previous studies, in which CD59 expression has been verified in kidney and stomach tissue samples [27,28]. The reason for this discrepancy could be related to detection methods used.

An effect of HO-1 ablation on C3aR expression was also observed in kidney, lung and spleen tissue, in which this receptor was reduced, and in liver tissue, in which it was augmented. C3aR is a G-protein-coupled receptor [29], via which C3a modulates immunity and certain cancers [30]. It also promotes podocyte autophagy and apoptosis [31]. Although C3aR function is not related to the function of CRPs, it is also attached on the cell membrane due its receptor function. A differential effect of HO-1 depletion on C3aR expression was identified in the present study, depending on the specific tissue, and it indicates a plausible effect of HO-1 on the C3-C3aR axis, thus extending its role as a contributor to the balanced complement cascade activation.

In a previous study assessing expression of complement factors in rat podocytes, increased C3aR expression was observed in glomeruli in puromycin (PAN)-induced renal injury, and this was accompanied by increased glomerular DAF expression [28]. In the present study, we observed decreased DAF immunostaining accompanied by a decrease in that of C3aR in the absence of injury. The reduction in DAF and Crry expression would be expected to increase C3aR expression, as the C3a-C3aR axis mode of action would be stimulated. However, our results indicate that the reduced DAF and Crry expression observed in *Hmox1*$^{-/-}$ rats was not sufficient to augment C3aR expression.

A mechanism by which HO-1 can regulate expression of CRPs could involve effects on the transcription factor Sp1, which is activated by the heme degradation product, CO [32]. Constitutive CD59 and DAF expression was shown to depend on Sp1 binding to specific promoter regions [33]. Further, the C3aR gene was shown to also include recognition sequences for Sp1 [34].

5. Conclusions

The present study provides an immunohistochemical characterization of the effect of HO-1 on CRP expression in various tissues. The study assesses the tissue-specific effect of HO-1 on CRP expression and shows a more prominent effect in the kidney, supporting the important role of HO-1 in mediating immune-mediated kidney injury. The observations further support the role of HO-1 as a regulator of complement activation.

Author Contributions: Conceptualization, M.G.D.; methodology, M.G.D. and E.T.; formal analysis, M.G.D., K.P., H.G. and E.A.L.; resources, M.G.D., H.G. and E.A.L.; data curation, M.G.D., E.T., K.P. and E.A.L.; writing—original draft preparation, M.G.D.; writing—review and editing, E.A.L. All authors have read and agreed to the published version of the manuscript.

Funding: This work was supported by a Merit Review Award # BX004333-02 from the United States (U.S.) Department of Veterans Affairs (Biomedical Laboratory Research and Development Service) to E.A.L.

Institutional Review Board Statement: The study was approved by the Project Evaluation Committee of the Agricultural and Veterinary Policy Sector of the Greek Ministry of Agricultural Development and Food. Project approval code: 4201. Date of approval: 2 July 2014.

Informed Consent Statement: Not applicable.

Data Availability Statement: The data is contained within the manuscript.

Conflicts of Interest: The authors declare no conflict of interest.

References

1. Fredenburgh, L.E.; Perrella, M.A.; Mitsialis, S.A. The Role of Heme Oxygenase-1 in Pulmonary Disease. *Am. J. Respir. Cell Mol. Biol.* **2007**, *36*, 158–165. [CrossRef] [PubMed]
2. Origassa, C.S.T.; Câmara, N.O.S. Cytoprotective role of heme oxygenase-1 and heme degradation derived end products in liver injury. *World J. Hepatol.* **2013**, *5*, 541–549. [CrossRef] [PubMed]
3. Fernandez, M.; Bonkovsky, H.L. Increased heme oxygenase-1 gene expression in liver cells and splanchnic organs from portal hypertensive rats. *Hepatology* **1999**, *29*, 1672–1679. [CrossRef] [PubMed]
4. Nath, K.A. Heme oxygenase-1 and acute kidney injury. *Curr. Opin. Nephrol. Hypertens.* **2014**, *23*, 17–24. [CrossRef] [PubMed]
5. Marilena, G. New Physiological Importance of Two Classic Residual Products: Carbon Monoxide and Bilirubin. *Biochem. Mol. Med.* **1997**, *61*, 136–142. [CrossRef]
6. Merle, N.S.; Grunenwald, A.; Rajaratnam, H.; Gnemmi, V.; Frimat, M.; Figueres, M.-L.; Knockaert, S.; Bouzekri, S.; Charue, D.; Noe, R.; et al. Intravascular hemolysis activates complement via cell-free heme and heme-loaded microvesicles. *J. Clin. Investig.* **2018**, *3*, e96910. [CrossRef]
7. Detsika, M.G.; Lianos, E.A. Regulation of Complement Activation by Heme Oxygenase-1 (HO-1) in Kidney Injury. *Antioxidants* **2021**, *10*, 60. [CrossRef]
8. Merle, N.S.; Church, S.E.; Fremeaux-Bacchi, V.; Roumenina, L.T. Complement System Part I—Molecular Mechanisms of Activation and Regulation. *Front. Immunol.* **2015**, *6*, 262. [CrossRef]

9. Mamidi, S.; Höne, S.; Kirschfink, M. The complement system in cancer: Ambivalence between tumour destruction and promotion. *Immunobiology* **2017**, *222*, 45–54. [CrossRef]
10. Farquhar, M.G.; Saito, A.; Kerjaschki, D.; Orlando, R.A. The Heymann nephritis antigenic complex: Megalin (gp330) and RAP. *J. Am. Soc. Nephrol.* **1995**, *6*, 35–47. [CrossRef]
11. Hatanaka, Y.; Yuzawa, Y.; Nishikawa, K.; Fukatsu, A.; Okada, N.; Okada, H.; Mizuno, M.; Matsuo, S. Role of a rat membrane inhibitor of complement in anti-basement membrane antibody-induced renal injury. *Kidney Int.* **1995**, *48*, 1728–1737. [CrossRef]
12. Barbour, T.D.; Ruseva, M.M.; Pickering, M.C. Update on C3 glomerulopathy. *Nephrol. Dial. Transplant.* **2016**, *31*, 717–725. [CrossRef]
13. Pandya, P.H.; Wilkes, D.S. Complement System in Lung Disease. *Am. J. Respir. Cell Mol. Biol.* **2014**, *51*, 467–473. [CrossRef]
14. Thorgersen, E.B.; Barratt-Due, A.; Haugaa, H.; Harboe, M.; Pischke, S.E.; Nilsson, P.; Mollnes, T.E. The Role of Complement in Liver Injury, Regeneration, and Transplantation. *Hepatology* **2019**, *70*, 725–736. [CrossRef]
15. Kiss, M.G.; Ozsvár-Kozma, M.; Porsch, F.; Göderle, L.; Papac-Miličević, N.; Bartolini-Gritti, B.; Tsiantoulas, D.; Pickering, M.C.; Binder, C.J. Complement Factor H Modulates Splenic B Cell Development and Limits Autoantibody Production. *Front. Immunol.* **2019**, *10*, 1607. [CrossRef]
16. Walport, M.J. Complement. *N. Engl. J. Med.* **2001**, *344*, 1058–1066. [CrossRef]
17. Harris, C.L.; Spiller, B.; Morgan, P. Human and rodent decay-accelerating factors (CD55) are not species restricted in their complement-inhibiting activities. *Immunology* **2000**, *100*, 462–470. [CrossRef]
18. Stankovic, J.A.; Vucevic, D.; Majstorovic, I.; Vasilijic, S.; Colic, M. The role of rat Crry, a complement regulatory protein, in proliferation of thymocytes. *Life Sci.* **2004**, *75*, 3053–3062. [CrossRef]
19. Caragine, T.A.; Imai, M.; Frey, A.; Tomlinson, S. Expression of rat complement control protein Crry on tumor cells inhibits rat natural killer cell–mediated cytotoxicity. *Blood* **2002**, *100*, 3304–3310. [CrossRef]
20. Detsika, M.G.; Duann, P.; Atsaves, V.; Papalois, A.; Lianos, E.A. Heme Oxygenase 1 Up-Regulates Glomerular Decay Accelerating Factor Expression and Minimizes Complement Deposition and Injury. *Am. J. Pathol.* **2016**, *186*, 2833–2845. [CrossRef]
21. Detsika, M.G.; Lygirou, V.; Frantzis, V.; Zoidakis, J.; Atsaves, V.; Poulaki, E.; Gakiopoulou, H.; Vlahou, A.; Lianos, E.A. Effect of Heme Oxygenase-1 Deficiency on Glomerular Proteomics. *Am. J. Nephrol.* **2016**, *43*, 441–450. [CrossRef] [PubMed]
22. Bao, L.; Spiller, O.B.; John, P.L.S.; Haas, M.; Hack, B.K.; Ren, G.; Cunningham, P.N.; Doshi, M.; Abrahamson, D.R.; Morgan, B.P.; et al. Decay-accelerating factor expression in the rat kidney is restricted to the apical surface of podocytes. *Kidney Int.* **2002**, *62*, 2010–2021. [CrossRef] [PubMed]
23. Atsaves, V.; Detsika, M.G.; Poulaki, E.; Gakiopoulou, H.; Lianos, E.A. Phenotypic characterization of a novel HO-1 depletion model in the rat. *Transgenic Res.* **2017**, *26*, 51–64. [CrossRef] [PubMed]
24. Mizuno, T.; Mizuno, M.; Morgan, B.P.; Noda, Y.; Yamada, K.; Okada, N.; Yuzawa, Y.; Matsuo, S.; Ito, Y. Specific collaboration between rat membrane complement regulators Crry and CD59 protects peritoneum from damage by autologous complement activation. *Nephrol. Dial. Transplant.* **2011**, *26*, 1821–1830. [CrossRef] [PubMed]
25. Miwa, T.; Song, W.-C. Membrane complement regulatory proteins: Insight from animal studies and relevance to human diseases. *Int. Immunopharmacol.* **2001**, *1*, 445–459. [CrossRef]
26. Angeletti, A.; Cantarelli, C.; Petrosyan, A.; Andrighetto, S.; Budge, K.; D'Agati, V.D.; Hartzell, S.; Malvi, D.; Donadei, C.; Thurman, J.M.; et al. Loss of decay-accelerating factor triggers podocyte injury and glomerulosclerosis. *J. Exp. Med.* **2020**, *217*. [CrossRef]
27. Yao, X.; Verkman, A.S. Complement regulator CD59 prevents peripheral organ injury in rats made seropositive for neuromyelitis optica immunoglobulin G. *Acta Neuropathol. Commun.* **2017**, *5*, 1–10. [CrossRef]
28. Li, X.; Ding, F.; Zhang, X.; Li, B.; Ding, J. The Expression Profile of Complement Components in Podocytes. *Int. J. Mol. Sci.* **2016**, *17*, 471. [CrossRef]
29. Ramos, T.N.; Bullard, D.C.; Barnum, S.R. ICAM-1: Isoforms and Phenotypes. *J. Immunol.* **2014**, *192*, 4469–4474. [CrossRef]
30. Sayegh, E.T.; Bloch, O.; Parsa, A.T. Complement anaphylatoxins as immune regulators in cancer. *Cancer Med.* **2014**, *3*, 747–758. [CrossRef]
31. Zhang, L.; Li, W.; Gong, M.; Zhang, Z.; Xue, X.; Mao, J.; Zhang, H.; Li, S.; Liu, X.; Wu, F.; et al. C-reactive protein inhibits C3a/C3aR-dependent podocyte autophagy in favor of diabetic kidney disease. *FASEB J.* **2022**, *36*, e22332. [CrossRef]
32. Lin, H.-H.; Lai, S.-C.; Chau, L.-Y. Heme Oxygenase-1/Carbon Monoxide Induces Vascular Endothelial Growth Factor Expression via p38 Kinase-dependent Activation of Sp1. *J. Biol. Chem.* **2011**, *286*, 3829–3838. [CrossRef]
33. Cauvi, D.M.; Cauvi, G.; Pollard, K.M. Constitutive Expression of Murine Decay-Accelerating Factor 1 Is Controlled by the Transcription Factor Sp1. *J. Immunol.* **2006**, *177*, 3837–3847. [CrossRef]
34. Hollma, T.J.; Haviland, D.L.; Kildsgaard, J.; Watts, K.; Wetsela, R.A. Cloning, expression, sequence determination, and chromosome localization of the mouse complement C3a anaphylatoxin receptor gene. *Mol. Immunol.* **1998**, *35*, 137–148. [CrossRef]

antioxidants

Review

The Role of Heme Oxygenase-1 as an Immunomodulator in Kidney Disease

Virginia Athanassiadou [1], Stella Plavoukou [1], Eirini Grapsa [1] and Maria G. Detsika [2,*]

[1] Department of Nephrology, School of Medicine, National and Kapodistrian University of Athens, Aretaieion University Hospital, 11528 Athens, Greece
[2] 1st Department of Critical Care Medicine & Pulmonary Services, GP Livanos and M Simou Laboratories, Evangelismos Hospital, National and Kapodistrian University of Athens, 10675 Athens, Greece
* Correspondence: mdetsika@med.uoa.gr

Abstract: The protein heme oxygenase (HO)-1 has been implicated in the regulations of multiple immunological processes. It is well known that kidney injury is affected by immune mechanisms and that various kidney-disease forms may be a result of autoimmune disease. The current study describes in detail the role of HO-1 in kidney disease and provides the most recent observations of the effect of HO-1 on immune pathways and responses both in animal models of immune-mediated disease forms and in patient studies.

Keywords: heme oxygenase; kidney; immune injury

Citation: Athanassiadou, V.; Plavoukou, S.; Grapsa, E.; Detsika, M.G. The Role of Heme Oxygenase-1 as an Immunomodulator in Kidney Disease. *Antioxidants* **2022**, *11*, 2454. https://doi.org/10.3390/antiox11122454

Academic Editor: Stanley Omaye

Received: 15 November 2022
Accepted: 8 December 2022
Published: 13 December 2022

1. Introduction

Heme oxygenase (HO)-1 has been linked to the regulations of immunological and pathophysiological processes, such as inflammation, apoptosis and cytoprotection, mainly through its downstream effector molecules, carbon monoxide, bilirubin and biliverdin. Inflammation and autoimmune diseases are key factors for kidney disease. In the kidney, glomerular capillaries may become targets of inflammation, ultimately leading to severe and irreversible tissue injury. The involvement of HO-1 in immune-mediated forms of kidney injury has been studied extensively. The current study describes in detail the potential role of HO-1, in various forms of kidney injury, as a mediator of immunological mechanisms that drive disease manifestation and progression.

The various forms of kidney injury usually occur due to injury of the renal glomerulus. Due to their highly specialized structure and function, renal glomeruli are known to be particularly vulnerable to injury. Briefly, the glomerulus (Figure 1) is a tri-cellular structure surrounded by the glomerular (Bowman's) capsule. Endothelial cells line the luminal side of the glomerular basement membrane (GBM); epithelial cells, also known as podocytes, are anchored on the outer surface of the GBM; and mesangial cells support the capillary loops. Glomeruli form a complex microvascular bed, the glomerular tuft, that functions as a highly selective plasma filter while retaining high-molecular-weight molecules and cells in circulation (Figure 1).

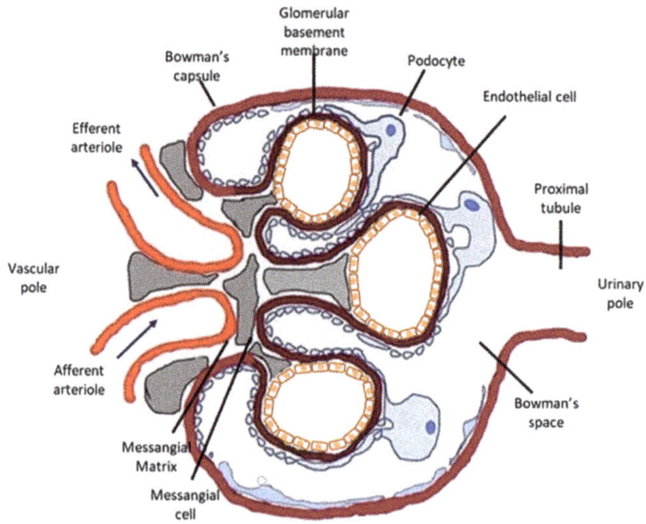

Figure 1. Schematic representation of the glomerulus. The glomerulus is a cluster of capillaries within the Bowman's capsule. Three different cell types comprise the glomerulus: endothelial cells, epithelial cells (podocytes) and mesangial cells. Endothelial cells are fenestrated, while mesangial cells and the mesangial matrix support the glomerular capillaries. Podocytes are located on the urinary side of the glomerular basement membrane, and they have long foot processes that wrap around the glomerular capillaries.

2. HO-1 and IgA Nephropathy

IgA nephropathy is a form of chronic glomerulonephritis characterized by the deposition of IgA immune complexes in glomeruli. It is the most common form of glomerulonephritis worldwide [1]. The majority of cases are idiopathic, but in recent years, secondary forms of the disease have appeared after various infections (Haemophilus parainfluenzae, HIV, cytomegalovirus) [2].

Symptoms include the onset of macroscopic hematuria usually one or two days after a febrile infectious episode, thus mimicking infectious glomerulonephritis. Urine analysis of patients with IgA has detected the presence of deformed red blood cells and sometimes red blood cell casts. Mild proteinuria (<1 g/day) is also typical and may occur without hematuria, while serum creatinine levels are usually normal at diagnosis [3].

The pathogenic mechanism that causes IgA nephropathy remains unknown, but accumulated evidence has led to the "four-hit hypothesis", starting with an abnormal glycosylation pattern of IgA (galactose-deficient IgA1) manifested through increased levels of poorly O-galactosylated IgA1 (gd-IgA1) in blood circulation, which causes the production of circulating auto-antibodies and consequentially the formation and deposition of immune complexes in the mesangium [4,5].

IgA nephropathy diagnosis is confirmed with renal biopsy and immunofluorescence methodology, which reveal granular IgA and complement factor 3 (C3) deposits located in the mesangium with foci of proliferative or necrotic segmental lesions. However, mesangial IgA deposits are considered non-specific and may be detected in many other disorders, such as immunoglobulin A-related vasculitis, HIV infection, psoriasis, lung cancer and several other disorders of the connective tissue. Examination of kidney tissue sections, obtained from IgA patients, under an electron microscope showed increased cellularity and an increased matrix in the mesangium, endocapillary proliferation of neutrophils and subendothelial deposits. Finally, normal levels of complement factors are detected with immunoassays, while an elevated IgA plasma concentration may sometimes be detected with serum electrophoresis [6].

Nitric oxide (NO) and advance oxidation protein products (AOPP) are strong markers of oxidative stress, and their elevated concentrations have been determined in serum samples of patients with severe IgA glomerulopathy [7]. Nakamura et al. compared patients who suffered from IgA nephritis with healthy controls, revealing that exposure to oxidative stress in IgA was of detrimental importance to the progression of renal injury [8]. This could be due to under-expression of superoxide dismutase (SOD) and the consequent exacerbation of tissue injury due to suppression of reactive oxygen species (ROS) scavenging ability (10). Furthermore, studies have identified a dinucleotide guanosine thymine (GT) repeat polymorphism of the HO-1 gene promoter that results in increased HO-1 expression when the GT length is shorter (S-allele) rather than when it is longer (L-allele). These two different alleles may influence the onsets and progressions of many different renal diseases [9], and several studies have shown a direct association between short (GT)n repeats and a higher induction rate of HO-1, which promotes the progression of IgA nephropathy [10].

Following intravascular hemolysis, free heme, the natural substrate of HO-1 and a powerful activator the complement cascade, is released. Heme has been reported to activate the alternative complement pathway [11] and may therefore be implicated in complement-mediated renal injury [12]. Free heme influences innate immune responses through the activation of Toll-like receptor 4 and ROS-dependent pathways, which in turn, through complex signaling pathways, promote the expression of proinflammatory cytokines. In addition, heme degradation byproducts (CO, biliverdin, bilirubin) and HO-1 constitute key molecules that upregulate the secretion of anti-inflammatory cytokines, such as IL-10 [13] (Figure 2). On the other hand, many pro-inflammatory enzymes, such as cyclooxygenase-2 (COX-2) and inducible nitric oxide synthase (iNOS), are in fact hemoproteins whose functions may be impaired due to insufficient free heme once that heme has been catalyzed by HO-1 [14]. Micro and/or macroscopic hematuria is a typical symptom of IgA nephropathy that suggests potential induction of HO-1 in the glomeruli [15]. A previous study used various concentrations of hemin to induce glomerular HO-1 expression in vitro and reported that when the concentration of hemin reached a critical level of 200 μM, HO-1 expression started to diminish. This could partly be explained by the fact that various forms of glomerular disease display increased levels of free radicals (OH^-), hydrogen peroxide (H_2O_2) and active Fe^{2+}. These may be further augmented by heme degradation of HO; therefore, a HO-1 expression-limiting mechanism may be necessary [16].

The clinical course of IgAN is highly variable. In many cases, the disease goes unnoticed and requires no treatment. The most important parameters that influence disease progression include the degree of proteinuria, uncontrollable hypertension and histopathological lesions on biopsies [17].

Patient assessment is usually performed with the "IgA Nephropathy Prediction Tool" (IIgAN-PT) that uses parameters such as histological findings, eGFR, the degree of proteinuria, the blood pressure value, medication prior to biopsy and demographics (sex, age, ethnicity) in order to estimate the individual five-year risk of renal progression toward end-stage renal disease (ESRD). Independently of IIgAN-PT results, all patients should receive supportive hypertension and proteinuria treatment. Supportive therapy consists of medications that inhibit the renin–angiotensin axon, such as angiotensin receptor blockers (ARB) or angiotensin-converting enzyme inhibitors (ACEi) [18]. These reduce both systemic and intraglomerular blood pressure—and thus glomerular injury due to hypertension—and lower the degree of proteinuria. If a patient's eGFR is greater than 30 mL/min/1.73 m^2, the addition of sodium–glucose co-transporter 2 (SGLT2) to their therapy could further reduce proteinuria [19]. Finally, due to toxicity effects, the use of steroids is only supported in cases of rapid declines in renal function and uncontrolled increases in proteinuria [20].

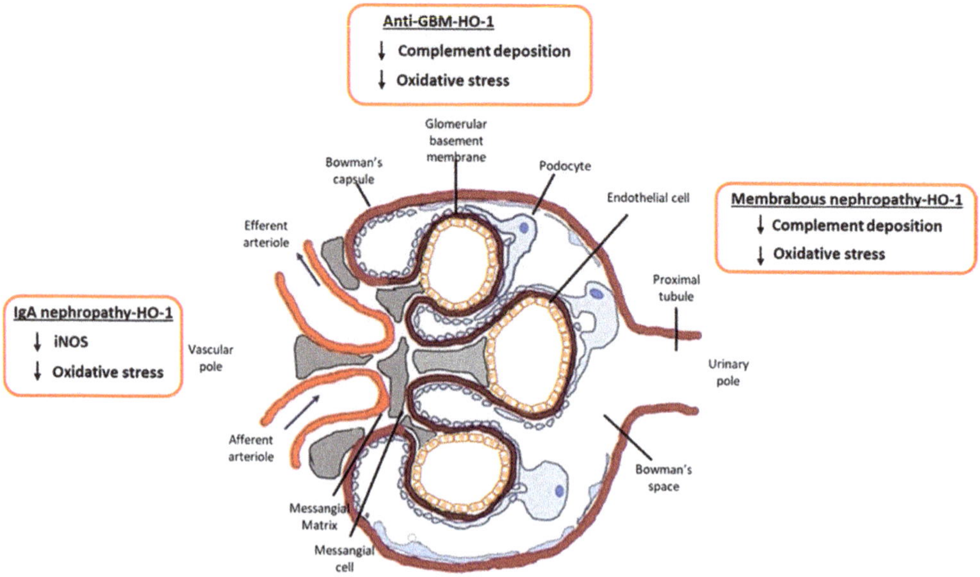

Figure 2. HO-1 immunomodulatory effects in glomerular diseases. Schematic representation of the effect of HO-1 in diseases that affect the glomerulus.

3. HO-1 and Membranous Nephropathy (MN)

MN is an autoimmune glomerular disease and a major cause of nephrotic syndrome in the adult population. Symptoms include insidious onset of edema, prominent proteinuria with mild urinary sediment, normal or deteriorating renal function and normal or elevated blood pressure [21]. Spontaneous remission can be seen in about 30% of patients. In most cases (about 70%), MN is reported as idiopathic. In these cases, antibodies against the M-type phospholipase A2 receptor (PLA2R) found in podocytes are linked to that specific locus and form immune complexes in situ that activate the complement membrane attack complex (MAC, C5b-9). However, MN can also develop as secondary MN due to a specific etiological factor. The main causes are solid tumors (lung, colon, rectum, kidney, breast, stomach), autoimmune diseases and microbial infections. In fact, in areas that are endemic of infections, such as malaria or schistosomiasis, the latter will consist of the main cause of MN [21].

The diagnosis of MN is also based on renal biopsy. Optical microscopy reveals the diffuse thickening of the GBM as a reaction to the subepithelial formation of immune complexes. No cellular hyperplasia or inflammatory cells are detected. The glomerular basement membrane thickens in a homogeneous way but gradually creates protrusions into the subepithelial space that eventually embrace and integrate the immune complexes into it. Immunofluorescence detects granular deposits of IgG and C3 along the basement membrane, while electron microscopy uncovers diffuse effacement of podocyte foot processes, as well as dense deposits in the subepithelial area [21].

Heymann et. al. described for the first time an experimental rat model of MN in which the pathogenic role of immune complexes was confirmed [22]. Heymann nephritis (HN) was induced through injections of proximal tubule brush border antigens (active HN) or their corresponding antibodies (passive HN) into rats. The autoantigen target in HN is megalin, a transmembrane protein located in the brush border of the proximal tubule and on podocyte foot processes in rats [23]. Studies of active and passive HN proved that subepithelial deposits are formed in situ against a constitutional endogenous antigen rather than through a circulating immune complex [24]. The HN model has allowed researchers to identify the important role of complement-mediated cytotoxicity in podocyte injury and proteinuria in this model. Early studies in HN showed co-localization of C3 and C5b-9 to the immune deposits. Furthermore, it was shown that the podocytes that were present in the urine of passive HN rats were coated with C5b-9 [25]. HN progression correlated with persisting urinary excretion of C5b-9, indicating continuous complement activation at the GBM.

Membrane-bound proteins, known as complement regulatory proteins (CRPs), act to eliminate complement activation and therefore MAC formation and lysis. A well-known CRP is the decay-accelerating factor (DAF, CD55). The DAF accelerates the decay of C3 and C5 convertases and thus restricts MAC formation [26]. In rats, other CRPs are Crry and CD59 [26]. Studies of DAF expression in rats have revealed constitutive DAF expression exclusively in podocytes [27]. In order to study the protective role of the DAF in complement-induced podocyte injury, a previous study generated a transgenic rat model of DAF depletion ($Daf^{-/-}$) [28]. Histological, clinical, and biochemical examinations (creatinine levels, albuminuria, urine albumin to urine creatinine ratio) in $Daf^{+/+}$ and $Daf^{-/-}$ rats demonstrated no significant differences prior to administration of complement-fixing antibody anti-Fx1A. Following anti-Fx1A administration, proteinuria levels were significantly elevated in $Daf^{-/-}$ rats. Immunofluorescence staining in rats that received anti-Fx1A evidenced greater C3 depositions in $Daf^{-/-}$ rats than in $Daf^{+/+}$ rats, suggesting a protective role of the DAF in podocyte complement-induced injury [28]. Several studies have demonstrated that HO-1 upregulates the DAF, which in turn reduces complement activation and complement-mediated injury [16]. In that context, HO-1 induction could be a useful tool as a potential treatment strategy against complement-mediated glomerulonephritis via its immunomodulatory effects, including DAF induction.

To explore the previously mentioned properties of HO-1, Wu et al. used an experimental animal model of induced MN in BALB/c mice by introducing intravenously cationic bovine serum albumin and dividing the animals into three groups. The first group was treated with a weekly intraperitoneal administration of cobalt protoporphyrin (CoPP), a HO-1 inducer; the second with tin protoporphyrin (SnPP), an HO-1 inhibitor; and the third with saline. The MN-CoPP group exhibited an HO-1 upregulation and presented a clear improvement of symptoms (a decrease in proteinuria and normalization of the serum-albumin and cholesterol levels). CoPP treatment also significantly reduced production of serum anti-cBSA antibodies. Although immunofluorescence staining remained positive for all three groups, the MN-CoPP group exhibited lesser intensity in the glomerular membrane and reduced C3 depositions in respect to the other two groups, especially the MN-SnPP group. The concentrations of markers of oxidative stress, such as thiobarbituric acid reactive substances (TBARSs), were evaluated both in the serum and the kidneys and found to be notably higher in respect to non-MN mice, but significant differences were assessed between the MN-CoPP group and the MN-SnPP group. CoPP treatment decreased oxidative stress markers both in the serum and the kidneys, suggesting that HO-1 may have an antioxidative effect on a local and on a systemic level [29].

Complement activation seems to induce ROS in podocytes that undergo mitochondrial dysregulation in MN [24]. Pyroptosis is a recently identified type of regulated cell death that follows bacterial or viral infections (40). Complex signaling pathways, including activation of inflammasomes and pro-inflammatory cytokines, lead to activation of caspase-1 and consequently formation of pores on the membranes, resulting in cell death and excess of pro-inflammatory cytokines and ROS [30]. Wang et al. described complement mediated pyroptosis in podocytes with a concurrent mitochondrial depolarization and ROS production. Blocking ROS production reversed complement mediated pyroptosis. Immunohistochemistry of MN glomeruli confirmed the co-localization of pyroptosis-related proteins, such as caspase-1 and gasdermin D (GSDMD), as well as synaptopodin, an actin-associated protein found in podocytes. Furthermore, C3a and C5a promoted overexpression of caspase-1 and GSDMD in the podocytes in vitro and influenced the integrity of cellular membranes and the depolarization of the podocyte mitochondrial membranes. When the podocytes of MN patients were incubated with inhibitors of key pyroptosis molecules, C3a and C5a did not affect podocyte membrane integrity [31].

Recent studies have assessed the role of HO-1 in blocking pyroptosis. Carbon monoxide (CO) as a byproduct of the HO-1 catalysis of hemin seems to block activation of caspase-1 through direct inhibition of its inducer molecule, NLRP3-ASC [32,33]. The NF-E2-related factor (Nrf2) is a transcriptional factor that regulates the cellular antioxidant response to oxidative stress by inducing the expressions of antioxidant and cytoprotective molecules, one of which is HO-1 [34]. Sirtuin is a deacetilase protein essential to the integrity of podocytes' cytoskeletons [35]. Under stress conditions, sirtuin seems to induce an overexpression of Nrf2 that leads to upregulation of HO-1 expression in podocytes [36]. In a murine renal ischemia/reperfusion (I/R) injury model, Diao et al. investigated the role of the NF-E2-related factor/heme oxygenase-1 (Nrf2/HO-1) as a protective factor against pyroptosis [37]. Protein arginine methylation transferase 5 (PRMT5) is involved in a vast number of physiological and pathological conditions, among them embryonic development, tissue homeostasis and malignancies [38]. PRMT5 is implicated in I/R-induced ROS production. Its inhibition resulted in an upregulation of Nrf2/HO-1, a reduction of oxidative stress markers and a decrease in tissue injury [37].

The highly variable course of MN renders its personalized treatment essential according to the risk of renal impairment progression. Spontaneous remission may occur in about 30% of patients; however, all patients presented with proteinuria should be treated either with an angiotensin receptor blocker (ARB) or an ACEi for three to six months. According to the latest MN treatment guidelines, all patients should be assessed for anti-M-type phospholipase A2 receptor (PLA2R) antibody levels in the blood prior to and during treatment, as an absence of anti PLA2R antibodies in a patient with an initially positive test indicates remission. Corticosteroids plus cyclophosphamide administration, along with supportive therapy, are recommended for patients with severe nephrotic syndrome and a rapid decline of renal function at the onset of the disease [39]. In moderate- to high-risk patients, if remission is not achieved within six months of supportive therapy alone, immunosuppressive treatment with rituximab, calcineurin inhibitors (CNIs) and corticosteroids plus cyclophosphamide may be used unless contraindicated due to severe renal impairment, diffuse interstitial fibrosis or recurrent infections [18].

4. HO-1 and Anti-GBM Disease

Anti-GBM disease, formerly known as Goodpasture's disease, is an infrequent autoimmune type of vasculitis that involves small vessels. It mainly affects the kidneys and the lungs, resulting in complement-induced, rapidly progressive crescentic glomerulonephritis and/or diffuse pulmonary bleeding if not treated promptly [40].

Epidemiological data suggest that anti-GBM affects both genders but at different age spectra. Specifically, there is a clear predominance of anti-GBM in male patients aged between 20 and 40 years and in female patients over 60 years old [40].

Anti-GBM is caused by autoantibodies that are directed against a specific epitope of type IV collagen present in the glomerular basement membrane as well as in the alveolar membrane of the lungs (autoantibodies against the NC1 domain of an α-3 chain of type IV collagen) Genetic predisposition and exposure to certain environmental agents have been postulated to trigger its onset. Various exogenous factors, such as infections; smoking; and drugs, such as alemtuzumab, a monoclonal anti-CD52 antibody used in the therapies of B cell leukemia and relapsing forms of multiple sclerosis, have been implicated. Regarding the genetic background of patients, the related literature demonstrates that there is a correlation between the HLA phenotype and the predisposition to develop the disease when exposed to a causal exogenous factor. In particular, it has been shown that people who present the types HLA DR15 and HLA DR4 show greater predisposition [41].

Anti-GBM diagnosis is confirmed with renal biopsy and immunoassays of anti-GBM antibodies in patients' serum samples. Characteristic histopathological findings in light microscopy of anti-GBM-diseased kidney sections include intense inflammation with focal segmental and necrotic lesions and the presence of diffuse lunar formations called crescents. Immunofluorescence staining reveals a linear GBM deposition of IgG, mostly of type 1 (IgG1), and, in almost 40% of cases, C3 [40]. Serum creatinine levels at diagnosis are directly correlated with the percentage of crescentic formations in biopsy.

In order to investigate these complicated immune mechanisms, anti-GBM-induced animal models have been vastly developed. In 1962, Steblay et al. were the first to acknowledge that inoculation of human GBM, along with Freund's adjuvant in sheep, induced crescentic anti-GBM disease [42]. Ryan et al. managed to induce anti-GBM in rats via the administration of a recombinant rat anti-α3 (IV) NC1 antibody, proving that murine experimental models of anti-GBM can be used to assess immune responses that could mirror the immunological responses in humans [43].

All three pathways of the complement cascade (classical, alternative and lectin) seem to be involved in complement-mediated characteristic histopathological lesions, as seen in immunofluorescence of pathological glomeruli both in rats and in humans, leading to the assembly of the complement MAC [44].

As mentioned above, the main immunomodulatory effect of HO-1 is indirect inhibition of MAC via an upregulation of the DAF [16]. Sogabe et al. used glycosylphosphatidylinositol (GPI)-DAF knockout mice to assess the correlation between the complement-regulator DAF and glomerular lesions in experimental anti-GBM nephritis. Renal-tissue biopsy samples from GPI-DAF knockout/anti-GBM-induced and wild-type/anti-GBM-induced mice were examined under optical microscopy. Glomeruli from GPI-DAF knockout mice presented increased mesangium cellularity as well as focal and segmental glomerulosclerosis (FSGS). On the contrary, glomeruli from wild-type mice showed minimal pathological signs at day eight post immunization. Immunofluorescence staining demonstrated linear depositions of IgG along the GBMs in both groups, but only in knockout mice were C3 and fibrinogen deposition observed [45].

Experimentally induced anti-GBM nephritis has provided plenty of evidence of a co-stimulation of cytotoxic enzymes, such as inducible nitric oxide synthase (iNOS), and cytoprotective molecules, such as HO-1 [46]. iNOS and HO-1 are both hemoproteins that, along with other heme-containing molecules, are upregulated by oxidative stress and inflammation. HO-1 activation, as mentioned above, catalyzes free heme, rendering it less available for synthesis of and tampering with the functionality of heme-containing enzymes, thus blocking the formation of oxidative byproducts [47]. On the other hand, NO seems to upregulate production of HO-1 in mesangial cells [48] and in renal tubular epithelial cells [49] suggesting complicated regulatory interactions between the two systems (iNOS and HO-1) and supporting the potential of HO-1 as a target for innovative future therapeutic strategies [50].

Due to the severity of the disease and the poor outcomes if it is left untreated, early treatment should be considered for all patients suspected to be positive for anti-GBM disease and concomitant rapidly progressive glomerulonephritis and/or a pulmonary hemorrhage, even if definite diagnosis through serological tests for anti-GBM antibodies and immunofluorescence is pending. KDIGO guidelines for treatment of anti-GBM suggest concurrent use of immunosuppression with corticosteroids and cyclophosphamide in alternative months, as well as plasmapheresis, except for in patients who need dialysis prior to therapy, are negative for a pulmonary hemorrhage and present 100% cellular crescents in the biopsy. [20]. A recently published retrospective multicenter observational study evaluated the risk for ESRD in patients diagnosed with anti-GBM disease over a period of 20 years and concluded that histopathological findings such as cellular crescents >50% and high creatinine (>4.7 mg/dL) at the onset of the disease are detrimental to renal survival, underlining the need for a more targeted and effective treatment of the disease [51].

5. HO-1 and Lupus Nephritis

Systemic lupus erythematosus (SLE) is a complex multisystemic autoimmune disease of unknown etiology that predominantly affects young women. There are a wide range of clinical manifestations and an involute pathogenesis as a result of interactions between genetic, epigenetic, ethnic, immunoregulatory and environmental factors [52]. Lupus nephritis (LN) occurs in about 20–40% of patients with SLE and remains the major risk factor for increased morbidity and mortality, despite advances in its diagnosis and treatment [53].

LN pathogenesis is characterized by multiple interactions between activated immune cells (extra- and intra-renal), the production of autoantibodies and the release of inflammatory mediators. Deposition of immune complexes (ICs) activates the complement cascade within the glomeruli or in the intratubular space, resulting in tissue inflammation [54].

Progress has been made in decoding the roles of innate and adaptive immune cells (particularly neutrophils, monocytes/macrophages, T and B cells) in the pathogenesis of SLE [54]. In particular, the role of monocytes/macrophages in the SLE pathogenesis has been widely studied, yet their exact inflammatory role remains uncertain due to their participation in multiple levels of disease development (phagocytosis, recruitment of other immune cells, cytokine secretion, tissue repair and fibrosis) [55].

Monocytes from patients with SLE—independently of disease activity—displayed significantly reduced HO-1 levels compared to those of healthy controls, suggesting that this low HO-1 expression and action could contribute to altered monocyte function in SLE and lupus nephritis. [13]. Kishimoto et al. have also demonstrated that glomerular M2-like macrophages from LN patients exhibit lower levels of HO-1 expression. This study showed that a transcriptional HO-1 repressor named Bach-1 can be induced by interferon type I. Bach-1-deficient MRL/lpr mice exhibited high HO-1 expression in kidneys with improved clinical biomarkers and unaltered anti-dsDNA antibody levels [56]. Thus, Bach-1 suggests a potential therapeutic target that could restore the M2-like macrophage function that is linked to increased HO-1 expression and activity [56].

Further studies of LN patients by Cuitino et al. confirm low HO-1 expression in pro-inflammatory monocytes and activated neutrophils with unbalanced function, such as increased phagocytosis and ROS production [57]. Interestingly, cobalt protoporphyrin (Co-PP) seems to induce HO-1 expression with a subsequent modification of LN monocyte phagocytic activity to a level similar to that of healthy controls. Thus, we can speculate that the impaired LN monocyte and neutrophil activity could be in part explained by reduced levels of HO-1 [57]. However, further studies are needed to confirm this hypothesis.

In LN, the primary events are the production of autoantibodies and the glomerular deposition of immune complexes (ICs) that activate complement cascade and immune cells bearing FCγ receptors (FCγRs) [58]. Dendritic cells, T helper cells, B cells and plasma cells all contribute to irregular polyclonal autoimmunity that is mediated with cell-to-cell interactions, immune tolerance and apoptotic mechanisms [58]. Tolerogenic dendritic cells (tolDCs) that are specialized to suppress the immune response may be a promising strategy in SLE treatment [59]. Funes et al. evaluated the therapeutic effect of tolDCs generated with the HO-1 inducer, CoPP, as well as, with dexamethasone and rosiglitazone, in two SLE mouse models [60]. Generation of tolDCs by use of the above agents, showed an efficient tolerogenic profile in vitro but did not improve LN severity or progression although it ameliorated other disease symptoms such as skin lesions [60]. In another study with SLE mice, the administration of hemin, the natural HO-1 substrate and inducer, attenuated disease progression. A marked reduction in proteinuria and glomerular immune complex deposition was observed and a concurrent reduction of inducible NOS expression in the kidney and spleen [61]. Furthermore, a reduction of autoantibodies was also identified. These findings demonstrate a double, anti-inflammatory and immunomodulatory, role of HO-1 induction, highlighting its potential as a novel therapeutic target in LN [61].

Furthermore, CO (a product of heme degradation catalyzed by HO-1) administration can attenuate autoimmunity and prevent clinical disease manifestation in FcγRIIb-deficient mice which is another SLE model [62]. The effects of CO exposure in SLE mice included decreased activated $B220^{+}CD3^{+}CD4^{-}$T cells in lungs and kidneys, together with low autoantibody levels [63].

In patients with SLE, LN affects both sexes equally, is more severe in children and men and is less severe in older adults. Approximately 10% of patients with LN will progress to ESRD [64], however, this depends on disease classification according to histologic evaluation. The risk for ESRD over 15 years was found to be as high as 44% in class IV LN [65]. SLE patients have a shorter life expectancy compared to those without nephritis and have a standardized mortality ratio of 6 to 6.8, versus the 2.4 ratio in lupus without renal manifestations [66]. An increase of this ratio to 14 is reported for patients with CKD and to 63 for patients with ESRD [67]. However, if LN remission is achieved through treatment, 10 year survival doubles to 95% [68].

Although nephritis may be characterized by clinical symptoms and laboratory markers, renal biopsy is required for confirmation, subclassification, prognosis and management options. The degree and type of glomerular involvement correlate directly with the clinical presentation and guide treatment decisions. LN is currently classified by the International Society of Nephrology (ISN)/Renal Pathology Society (RPS) system, which is based on glomerular histology using light and immunofluorescence microscopy [69]. Treatment recommendations are based on the ISN/RPS biopsy classification. All patients with SLE should be treated with hydroxychloroquine or an equivalent antimalarial unless contraindicated. In general, immunosuppressive therapy of extra renal lupus manifestations is sufficient for class I and II LN. A combination of high-dose corticosteroids plus an immunosuppressive agent is mainly used for patients with active focal proliferative LN (classes IIIA and IIIA/C), active diffuse proliferative LN (classes IVA and IVA/C) or membranous lupus (class V). The treatment of focal or diffuse LN has two phases. The initial or induction therapy with anti-inflammatory and immunosuppressive agent, which addresses acute life- or organ-threatening disease, and the long-term immunosuppressive treatment, which prevents relapses and consolidate remissions. According to the latest guidelines for the management of lupus nephritis, patients with active Class III or IV LN, with or without a membranous component, should be treated initially with glucocorticoids plus either a low-dose intravenous cyclophosphamide or a mycophenolic acid analogue (MPAA). An alternative immunosuppressive regimen that includes a calcineurin inhibitor (CNI) (usually tacrolimus or cyclosporine) with reduced-dose MPAA and glucocorticoids is reserved for patients who cannot tolerate standard-dose MPAA or are unfit for cyclophosphamide-based regimens. There is also an emerging role for B-lymphocyte-targeting biologics in the treatment of LN. Belimumab can be added to standard therapy in treatment of active LN and rituximab may be considered for patients with persistent disease activity or repeated flares. Maintenance therapy would be based on an MPAA regimen or, alternatively, azathioprine. Glucocorticoids should be tapered to the lowest possible dose [18].

6. HO-1 and Acute Kidney Injury (AKI)

AKI, characterized by a rapid increase in serum creatinine and/or a decrease in urine output, is common in critically ill patients and is associated with increased morbidity and mortality. The pathophysiology of AKI is complex, involving activation and crosstalk between multiple pathways, including inflammation and oxidative stress. HO-1 regulates oxidative stress, autophagy and inflammation. Furthermore, it controls cell-cycle progression directly and indirectly [70]. Recent studies have revealed that HO-1 expression in monocytes/macrophages may be beneficial, as it was shown to ease the inflammatory response in AKI [13]. HO-1-expressing macrophages were shown to possess a tendency towards an M2 phenotype polarization that contributes to upregulation of anti-inflammatory cytokine (IL-10) expression, suppression of pro-inflammatory cytokine (TNFα) secretion and expression of reparative genes that are beneficial to tissue recovery following AKI [71,72]. Furthermore, the cytoprotective effects of HO-1 attributed to heme degradation byproducts may also exert modulatory effects on AKI [73]. CO was shown to exhibit strong anti-proliferative effects on T cells via IL-2 downregulation, which diminishes inflammation [74]. HO-1 cytoprotective properties were first identified in the kidney in a model of heme-protein-induced AKI [75]. Various studies have confirmed the great potential of HO-1 induction, both pharmacologically and genetically, to regulate immune responses against AKI [71]. However, well-established modes for HO-1 upregulation, in terms of tolerance and efficacy, are essential to initiate the translation of its therapeutic potential into effective patient therapies for AKI.

7. HO-1 in Renal Ischemia/Reperfusion Injury (IRI)

IRI is a pathological condition characterized by an initial restriction of blood supply to an organ and a subsequent restoration of perfusion and re-oxygenation [76]. It involves activation of cell death programs, endothelial dysfunction, transcriptional reprogramming and activation of the immune system [77]. IRI is one of the most common causes of AKI. Current knowledge of the role of HO-1 in IRI-induced renal disease is largely based on experience with animal models of kidney disease. Chemical inhibition of HO-2 and HO-1 activity in the healthy kidney results in reduced medullary blood flow, thus supporting the role of HO-1 in the maintenance of medullary perfusion under physiological conditions. A reduction of IRI severity via HO-1 has also been demonstrated with the use of HO-1 chemical inducers [78]. However, the exact underlying mechanisms by which HO-1 exerts its protective effect against IRI remain unknown. In an attempt to elucidate the molecular mechanisms of the cytoprotective effect of HO-1 in IRI, a previous study utilized a mouse model of IRI in HO-$1^{+/-}$ mice and then measured the levels of blood urea nitrogen (BUN) and serum creatinine (SCr). Furthermore, that study investigated the severity of histological changes, as well as HO-1 and vascular cell adhesion molecule 1 (VCAM-1) protein expression levels, inflammatory factor expression and the effects of VCAM-1 blockades. That study reported elevated levels of VCAM-1 expression in HO-$1^{+/-}$ mice during IRI and an increase in the extent of renal-tissue damage and activation of the inflammatory response [79]. Another study imported an experimental model of repeated episodes of IR with 10-day intervals and found that it induced long-term renal protection accompanied by HO-1 overexpression and an increase in M2-macrophages. The aforementioned study investigated the transition between AKI and CKD and involved AKI induction via a single bilateral IR episode (1IR) or three episodes of IR separated by 10 day intervals (3IR) of mild (20 min) or severe (45 min) ischemia [80].

IRI also plays an important role in kidney transplantation and affects delayed renal function after transplantation. The protective role of HO-1 in IRI-mediated injury during kidney transplantation was shown in a recent study that utilized a myeloid-restricted mouse model of deletion of HO-1 (HO-$1^{M\text{-}KO}$). IRI in the HO-$1^{M\text{-}KO}$ mice resulted in significant renal histological damage, pro-inflammatory responses, and oxidative stress 24 h after reperfusion. An assessment of the animals at a following time point of seven days afterward revealed that the HO-$1^{M\text{-}KO}$ mice displayed impaired tubular repair and increased renal fibrosis [81]. Furthermore, the same study showed that hemin mediated HO-1 induction in WT mice, resulting in HO-1 upregulation within the CD11b^{+} F4/80^{lo} subset of the renal myeloid cells [81]. The findings supported the increasing potential of HO-1 as a target of therapeutic strategies in the field of kidney transplantation.

8. HO-1 Polymorphisms and Kidney Disease

As mentioned before, there is a direct association between short (GT)n repeats and a higher induction rate of HO-1 and progression of IgA nephropathy [10]. The HO-1 genotype is a risk factor for renal impairment of IgA nephropathy at diagnosis, which is a strong predictor of mortality [10]. Another study investigated the (GT)n short repeat genotype, which promotes induction of HO-1, in renal transplantation and allograft rejection. The study reported that the beneficial effect of the HO-1 genotype was attributed to the donor genotype and not to the recipient when an allograft was exposed to prolonged cold ischemia. Furthermore, allografts from L-donors resulted in more rejections, whereas kidneys that bared the S-allele were less susceptible to injury, thus resulting in reduced rates of allograft rejection [82].

9. Clinical Applications

Manipulation of HO-1 expression/activity for potential therapeutic strategies has already been explored. HO-1 may be activated through a wide range of both naturally occurring and chemically synthesized compounds. The most widely used HO-1 inducers/inducers are metalloporphyrins (MPs). These are all heme analogues that differ mainly in the metal moiety of the porphyrin structure. However, numerous other naturally occurring compounds have been reported to induce HO-1, including curcumin, resveratrol, quercentin, carnosic acid, carnosol and anthocyanin [83]. Both MPs, as well as all of the other naturally occurring compounds, have been proposed for their use in the treatments of various immune-mediated diseases, including kidney disease [84] as well as other immune-mediated diseases, such as multiple sclerosis (MS), type 1 diabetes, rheumatoid arthritis, lupus and inflammatory bowel disease [83]. The therapeutic potential of HO-1 has also been highlighted by other studies that have proposed the potential use of MPs with non-immune-mediated forms of disease, such as non-alcoholic fatty liver disease [85]. Finally, manipulation of the HO reaction has also enabled research towards potential therapeutic strategies via CO release via CO-releasing molecule (CORM) administration in preclinical models of type 1 diabetes [86] and in MS [87,88], as well as in models of autoimmune hepatitis [89].

10. Conclusions

Apart from its established role as a strong antioxidant and anti-apoptotic enzyme, HO-1 is by now also recognized as an important modulator of various immune pathways and responses. Furthermore, its increased inducibility through multiple varied types of inducers renders it a highly interesting target for novel therapeutic strategies against immune-mediated diseases, including kidney disease. Further research is needed, in order to unravel the exact mechanisms of immunomodulation in kidney disease, which will enable translation into innovative treatment strategies against immune-mediated kidney disease.

Funding: This research received no external funding.

Conflicts of Interest: The authors declare no conflict of interest.

References

1. Barratt, J.; Feehally, J. IgA nephropathy. *J. Am. Soc. Nephrol. JASN* **2005**, *16*, 2088–2097. [CrossRef]
2. Saha, M.K.; Julian, B.A.; Novak, J.; Rizk, D.V. Secondary IgA nephropathy. *Kidney Int.* **2018**, *94*, 674–681. [CrossRef]
3. Penfold, R.S.; Prendecki, M.; McAdoo, S.; Tam, F.W. Primary IgA nephropathy: Current challenges and future prospects. *Int. J. Nephrol. Renov. Dis.* **2018**, *11*, 137–148. [CrossRef] [PubMed]
4. Suzuki, H. Biomarkers for IgA nephropathy on the basis of multi-hit pathogenesis. *Clin. Exp. Nephrol.* **2019**, *23*, 26–31. [CrossRef] [PubMed]
5. Dotz, V.; Visconti, A.; Lomax-Browne, H.J.; Clerc, F.; Hipgrave Ederveen, A.L.; Medjeral-Thomas, N.R.; Cook, H.T.; Pickering, M.C.; Wuhrer, M.; Falchi, M. O- and N-Glycosylation of Serum Immunoglobulin A is Associated with IgA Nephropathy and Glomerular Function. *J. Am. Soc. Nephrol. JASN* **2021**, *32*, 2455–2465. [CrossRef]
6. Yau, T.; Korbet, S.M.; Schwartz, M.M.; Cimbaluk, D.J. The Oxford classification of IgA nephropathy: A retrospective analysis. *Am. J. Nephrol.* **2011**, *34*, 435–444. [CrossRef]
7. Descamps-Latscha, B.; Witko-Sarsat, V.; Nguyen-Khoa, T.; Nguyen, A.T.; Gausson, V.; Mothu, N.; Cardoso, C.; Noel, L.H.; Guerin, A.P.; London, G.M.; et al. Early prediction of IgA nephropathy progression: Proteinuria and AOPP are strong prognostic markers. *Kidney Int.* **2004**, *66*, 1606–1612. [CrossRef] [PubMed]
8. Nakamura, T.; Inoue, T.; Sugaya, T.; Kawagoe, Y.; Suzuki, T.; Ueda, Y.; Koide, H.; Node, K. Beneficial effects of olmesartan and temocapril on urinary liver-type fatty acid-binding protein levels in normotensive patients with immunoglobin A nephropathy. *Am. J. Hypertens.* **2007**, *20*, 1195–1201. [CrossRef] [PubMed]
9. Courtney, A.E.; McNamee, P.T.; Heggarty, S.; Middleton, D.; Maxwell, A.P. Association of functional haem oxygenase-1 gene promoter polymorphism with polycystic kidney disease and IgA nephropathy. *Nephrol. Dial. Transplant. Off. Publ. Eur. Dial. Transpl. Assoc.-Eur. Ren. Assoc.* **2008**, *23*, 608–611. [CrossRef]

10. Chin, H.J.; Cho, H.J.; Lee, T.W.; Na, K.Y.; Yoon, H.J.; Chae, D.W.; Kim, S.; Jeon, U.S.; Do, J.Y.; Park, J.W.; et al. The heme oxygenase-1 genotype is a risk factor to renal impairment of IgA nephropathy at diagnosis, which is a strong predictor of mortality. *J. Korean Med. Sci.* **2009**, *24*, S30–S37. [CrossRef]
11. Merle, N.S.; Grunenwald, A.; Rajaratnam, H.; Gnemmi, V.; Frimat, M.; Figueres, M.L.; Knockaert, S.; Bouzekri, S.; Charue, D.; Noe, R.; et al. Intravascular hemolysis activates complement via cell-free heme and heme-loaded microvesicles. *JCI Insight* **2018**, *3*. [CrossRef]
12. Frimat, M.; Tabarin, F.; Dimitrov, J.D.; Poitou, C.; Halbwachs-Mecarelli, L.; Fremeaux-Bacchi, V.; Roumenina, L.T. Complement activation by heme as a secondary hit for atypical hemolytic uremic syndrome. *Blood* **2013**, *122*, 282–292. [CrossRef]
13. Li, Y.; Ma, K.; Han, Z.; Chi, M.; Sai, X.; Zhu, P.; Ding, Z.; Song, L.; Liu, C. Immunomodulatory Effects of Heme Oxygenase-1 in Kidney Disease. *Front. Med.* **2021**, *8*, 708453. [CrossRef]
14. Gozzelino, R.; Jeney, V.; Soares, M.P. Mechanisms of cell protection by heme oxygenase-1. *Annu. Rev. Pharmacol. Toxicol.* **2010**, *50*, 323–354. [CrossRef]
15. Shepard, M.; Dhulipala, P.; Kabaria, S.; Abraham, N.G.; Lianos, E.A. Heme oxygenase-1 localization in the rat nephron. *Nephron* **2002**, *92*, 660–664. [CrossRef]
16. Detsika, M.G.; Lianos, E.A. Regulation of Complement Activation by Heme Oxygenase-1 (HO-1) in Kidney Injury. *Antioxidants* **2021**, *10*, 60. [CrossRef] [PubMed]
17. Trimarchi, H.; Barratt, J.; Cattran, D.C.; Cook, H.T.; Coppo, R.; Haas, M.; Liu, Z.H.; Roberts, I.S.; Yuzawa, Y.; Zhang, H.; et al. Oxford Classification of IgA nephropathy 2016: An update from the IgA Nephropathy Classification Working Group. *Kidney Int.* **2017**, *91*, 1014–1021. [CrossRef]
18. Rovin, B.H.; Adler, S.G.; Barratt, J.; Bridoux, F.; Burdge, K.A.; Chan, T.M.; Cook, H.T.; Fervenza, F.C.; Gibson, K.L.; Glassock, R.J.; et al. Executive summary of the KDIGO 2021 Guideline for the Management of Glomerular Diseases. *Kidney Int.* **2021**, *100*, 753–779. [CrossRef]
19. Barratt, J.; Floege, J. SGLT-2 inhibition in IgA nephropathy: The new standard of care? *Kidney Int.* **2021**, *100*, 24–26. [CrossRef] [PubMed]
20. Kidney Disease: Improving Global Outcomes Glomerular Diseases Work Group. KDIGO 2021 Clinical Practice Guideline for the Management of Glomerular Diseases. *Kidney Int.* **2021**, *100*, S1–S276. [CrossRef] [PubMed]
21. Keri, K.C.; Blumenthal, S.; Kulkarni, V.; Beck, L.; Chongkrairatanakul, T. Primary membranous nephropathy: Comprehensive review and historical perspective. *Postgrad. Med. J.* **2019**, *95*, 23–31. [CrossRef]
22. Heymann, W.; Hackel, D.B.; Harwood, S.; Wilson, S.G.; Hunter, J.L. Production of nephrotic syndrome in rats by Freund's adjuvants and rat kidney suspensions. *Proc. Soc. Exp. Biol. Medicine. Soc. Exp. Biol. Med.* **1959**, *100*, 660–664. [CrossRef]
23. Farquhar, M.G.; Saito, A.; Kerjaschki, D.; Orlando, R.A. The Heymann nephritis antigenic complex: Megalin (gp330) and RAP. *J. Am. Soc. Nephrol. JASN* **1995**, *6*, 35–47. [CrossRef]
24. Cybulsky, A.V.; Quigg, R.J.; Salant, D.J. Experimental membranous nephropathy redux. *Am. J. Physiol. Ren. Physiol.* **2005**, *289*, F660–F671. [CrossRef]
25. Ma, H.; Sandor, D.G.; Beck, L.H., Jr. The role of complement in membranous nephropathy. *Semin. Nephrol.* **2013**, *33*, 531–542. [CrossRef] [PubMed]
26. Miwa, T.; Song, W.C. Membrane complement regulatory proteins: Insight from animal studies and relevance to human diseases. *Int. Immunopharmacol.* **2001**, *1*, 445–459. [CrossRef] [PubMed]
27. Bao, L.; Spiller, O.B.; St John, P.L.; Haas, M.; Hack, B.K.; Ren, G.; Cunningham, P.N.; Doshi, M.; Abrahamson, D.R.; Morgan, B.P.; et al. Decay-accelerating factor expression in the rat kidney is restricted to the apical surface of podocytes. *Kidney Int.* **2002**, *62*, 2010–2021. [CrossRef]
28. Detsika, M.G.; Goudevenou, K.; Geurts, A.M.; Gakiopoulou, H.; Grapsa, E.; Lianos, E.A. Generation of a novel decay accelerating factor (DAF) knock-out rat model using clustered regularly-interspaced short palindromic repeats, (CRISPR)/associated protein 9 (Cas9), genome editing. *Transgenic Res.* **2021**, *30*, 11–21. [CrossRef] [PubMed]
29. Wu, C.C.; Lu, K.C.; Lin, G.J.; Hsieh, H.Y.; Chu, P.; Lin, S.H.; Sytwu, H.K. Melatonin enhances endogenous heme oxygenase-1 and represses immune responses to ameliorate experimental murine membranous nephropathy. *J. Pineal Res.* **2012**, *52*, 460–469. [CrossRef]
30. Yu, P.; Zhang, X.; Liu, N.; Tang, L.; Peng, C.; Chen, X. Pyroptosis: Mechanisms and diseases. *Signal Transduct. Target. Ther.* **2021**, *6*, 128. [CrossRef]
31. Wang, H.; Lv, D.; Jiang, S.; Hou, Q.; Zhang, L.; Li, S.; Zhu, X.; Xu, X.; Wen, J.; Zeng, C.; et al. Complement induces podocyte pyroptosis in membranous nephropathy by mediating mitochondrial dysfunction. *Cell Death Dis.* **2022**, *13*, 281. [CrossRef]
32. Shi, J.; Zhao, Y.; Wang, K.; Shi, X.; Wang, Y.; Huang, H.; Zhuang, Y.; Cai, T.; Wang, F.; Shao, F. Cleavage of GSDMD by inflammatory caspases determines pyroptotic cell death. *Nature* **2015**, *526*, 660–665. [CrossRef]
33. Fu, L.; Zhang, D.X.; Zhang, L.M.; Song, Y.C.; Liu, F.H.; Li, Y.; Wang, X.P.; Zheng, W.C.; Wang, X.D.; Gui, C.X.; et al. Exogenous carbon monoxide protects against mitochondrial DNAinduced hippocampal pyroptosis in a model of hemorrhagic shock and resuscitation. *Int. J. Mol. Med.* **2020**, *45*, 1176–1186. [CrossRef]
34. Ryter, S.W. Heme Oxgenase-1, a Cardinal Modulator of Regulated Cell Death and Inflammation. *Cells* **2021**, *10*, 515. [CrossRef]
35. Nakatani, Y.; Inagi, R. Epigenetic Regulation Through SIRT1 in Podocytes. *Curr. Hypertens. Rev.* **2016**, *12*, 89–94. [CrossRef] [PubMed]

36. Chen, X.J.; Wu, W.J.; Zhou, Q.; Jie, J.P.; Chen, X.; Wang, F.; Gong, X.H. Advanced glycation end-products induce oxidative stress through the Sirt1/Nrf2 axis by interacting with the receptor of AGEs under diabetic conditions. *J. Cell. Biochem.* **2018**. [CrossRef] [PubMed]
37. Diao, C.; Chen, Z.; Qiu, T.; Liu, H.; Yang, Y.; Liu, X.; Wu, J.; Wang, L. Inhibition of PRMT5 Attenuates Oxidative Stress-Induced Pyroptosis via Activation of the Nrf2/HO-1 Signal Pathway in a Mouse Model of Renal Ischemia-Reperfusion Injury. *Oxidative Med. Cell. Longev.* **2019**, *2019*, 2345658. [CrossRef] [PubMed]
38. Kim, H.; Ronai, Z.A. PRMT5 function and targeting in cancer. *Cell Stress* **2020**, *4*, 199–215. [CrossRef]
39. Fernandez-Juarez, G.; Rojas-Rivera, J.; Logt, A.V.; Justino, J.; Sevillano, A.; Caravaca-Fontan, F.; Avila, A.; Rabasco, C.; Cabello, V.; Varela, A.; et al. The STARMEN trial indicates that alternating treatment with corticosteroids and cyclophosphamide is superior to sequential treatment with tacrolimus and rituximab in primary membranous nephropathy. *Kidney Int.* **2021**, *99*, 986–998. [CrossRef]
40. Kant, S.; Kronbichler, A.; Sharma, P.; Geetha, D. Advances in Understanding of Pathogenesis and Treatment of Immune-Mediated Kidney Disease: A Review. *Am. J. Kidney Dis. Off. J. Natl. Kidney Found.* **2022**, *79*, 582–600. [CrossRef] [PubMed]
41. Segelmark, M.; Hellmark, T. Anti-glomerular basement membrane disease: An update on subgroups, pathogenesis and therapies. *Nephrol. Dial. Transplant. Off. Publ. Eur. Dial. Transpl. Assoc.-Eur. Ren. Assoc.* **2019**, *34*, 1826–1832. [CrossRef] [PubMed]
42. Steblay, R.W. Glomerulonephritis induced in sheep by injections of heterologous glomerular basement membrane and Freund's complete adjuvant. *J. Exp. Med.* **1962**, *116*, 253–272. [CrossRef] [PubMed]
43. Ryan, J.J.; Reynolds, J.; Norgan, V.A.; Pusey, C.D. Expression and characterization of recombinant rat alpha 3(IV)NC1 and its use in induction of experimental autoimmune glomerulonephritis. *Nephrol. Dial. Transplant. Off. Publ. Eur. Dial. Transpl. Assoc.-Eur. Ren. Assoc.* **2001**, *16*, 253–261. [CrossRef]
44. Otten, M.A.; Groeneveld, T.W.; Flierman, R.; Rastaldi, M.P.; Trouw, L.A.; Faber-Krol, M.C.; Visser, A.; Essers, M.C.; Claassens, J.; Verbeek, J.S.; et al. Both complement and IgG fc receptors are required for development of attenuated antiglomerular basement membrane nephritis in mice. *J. Immunol.* **2009**, *183*, 3980–3988. [CrossRef] [PubMed]
45. Sogabe, H.; Nangaku, M.; Ishibashi, Y.; Wada, T.; Fujita, T.; Sun, X.; Miwa, T.; Madaio, M.P.; Song, W.C. Increased susceptibility of decay-accelerating factor deficient mice to anti-glomerular basement membrane glomerulonephritis. *J. Immunol.* **2001**, *167*, 2791–2797. [CrossRef] [PubMed]
46. Cattell, V.; Largen, P.; de Heer, E.; Cook, T. Glomeruli synthesize nitrite in active Heymann nephritis; the source is infiltrating macrophages. *Kidney Int.* **1991**, *40*, 847–851. [CrossRef] [PubMed]
47. Datta, P.K.; Koukouritaki, S.B.; Hopp, K.A.; Lianos, E.A. Heme oxygenase-1 induction attenuates inducible nitric oxide synthase expression and proteinuria in glomerulonephritis. *J. Am. Soc. Nephrol. JASN* **1999**, *10*, 2540–2550. [CrossRef]
48. Datta, P.K.; Lianos, E.A. Nitric oxide induces heme oxygenase-1 gene expression in mesangial cells. *Kidney Int.* **1999**, *55*, 1734–1739. [CrossRef]
49. Liang, M.; Croatt, A.J.; Nath, K.A. Mechanisms underlying induction of heme oxygenase-1 by nitric oxide in renal tubular epithelial cells. *Am. J. Physiol. Ren. Physiol.* **2000**, *279*, F728–F735. [CrossRef]
50. Datta, P.K.; Gross, E.J.; Lianos, E.A. Interactions between inducible nitric oxide synthase and heme oxygenase-1 in glomerulonephritis. *Kidney Int.* **2002**, *61*, 847–850. [CrossRef] [PubMed]
51. Sanchez-Agesta, M.; Rabasco, C.; Soler, M.J.; Shabaka, A.; Canllavi, E.; Fernandez, S.J.; Cazorla, J.M.; Lopez-Rubio, E.; Romera, A.; Barroso, S.; et al. Anti-glomerular Basement Membrane Glomerulonephritis: A Study in Real Life. *Front. Med.* **2022**, *9*, 889185. [CrossRef]
52. Zucchi, D.; Elefante, E.; Schiliro, D.; Signorini, V.; Trentin, F.; Bortoluzzi, A.; Tani, C. One year in review 2022: Systemic lupus erythematosus. *Clin. Exp. Rheumatol.* **2022**, *40*, 4–14. [CrossRef]
53. Ocampo-Piraquive, V.; Nieto-Aristizabal, I.; Canas, C.A.; Tobon, G.J. Mortality in systemic lupus erythematosus: Causes, predictors and interventions. *Expert Rev. Clin. Immunol.* **2018**, *14*, 1043–1053. [CrossRef]
54. Frangou, E.; Georgakis, S.; Bertsias, G. Update on the cellular and molecular aspects of lupus nephritis. *Clin. Immunol.* **2020**, *216*, 108445. [CrossRef]
55. Maria, N.I.; Davidson, A. Renal Macrophages and Dendritic Cells in SLE Nephritis. *Curr. Rheumatol. Rep.* **2017**, *19*, 81. [CrossRef] [PubMed]
56. Kishimoto, D.; Kirino, Y.; Tamura, M.; Takeno, M.; Kunishita, Y.; Takase-Minegishi, K.; Nakano, H.; Kato, I.; Nagahama, K.; Yoshimi, R.; et al. Dysregulated heme oxygenase-1(low) M2-like macrophages augment lupus nephritis via Bach1 induced by type I interferons. *Arthritis Res. Ther.* **2018**, *20*, 64. [CrossRef] [PubMed]
57. Cuitino, L.; Obreque, J.; Gajardo-Meneses, P.; Villarroel, A.; Crisostomo, N.; San Francisco, I.F.; Valenzuela, R.A.; Mendez, G.P.; Llanos, C. Heme-Oxygenase-1 Is Decreased in Circulating Monocytes and Is Associated With Impaired Phagocytosis and ROS Production in Lupus Nephritis. *Front. Immunol.* **2019**, *10*, 2868. [CrossRef] [PubMed]
58. Lech, M.; Anders, H.J. The pathogenesis of lupus nephritis. *J. Am. Soc. Nephrol. JASN* **2013**, *24*, 1357–1366. [CrossRef]
59. Funes, S.C.; Manrique de Lara, A.; Altamirano-Lagos, M.J.; Mackern-Oberti, J.P.; Escobar-Vera, J.; Kalergis, A.M. Immune checkpoints and the regulation of tolerogenicity in dendritic cells: Implications for autoimmunity and immunotherapy. *Autoimmun. Rev.* **2019**, *18*, 359–368. [CrossRef]

60. Funes, S.C.; Rios, M.; Gomez-Santander, F.; Fernandez-Fierro, A.; Altamirano-Lagos, M.J.; Rivera-Perez, D.; Pulgar-Sepulveda, R.; Jara, E.L.; Rebolledo-Zelada, D.; Villarroel, A.; et al. Tolerogenic dendritic cell transfer ameliorates systemic lupus erythematosus in mice. *Immunology* **2019**, *158*, 322–339. [CrossRef]
61. Takeda, Y.; Takeno, M.; Iwasaki, M.; Kobayashi, H.; Kirino, Y.; Ueda, A.; Nagahama, K.; Aoki, I.; Ishigatsubo, Y. Chemical induction of HO-1 suppresses lupus nephritis by reducing local iNOS expression and synthesis of anti-dsDNA antibody. *Clin. Exp. Immunol.* **2004**, *138*, 237–244. [CrossRef]
62. Mackern-Oberti, J.P.; Llanos, C.; Carreno, L.J.; Riquelme, S.A.; Jacobelli, S.H.; Anegon, I.; Kalergis, A.M. Carbon monoxide exposure improves immune function in lupus-prone mice. *Immunology* **2013**, *140*, 123–132. [CrossRef]
63. Mackern-Oberti, J.P.; Obreque, J.; Mendez, G.P.; Llanos, C.; Kalergis, A.M. Carbon monoxide inhibits T cell activation in target organs during systemic lupus erythematosus. *Clin. Exp. Immunol.* **2015**, *182*, 1–13. [CrossRef] [PubMed]
64. Hanly, J.G.; O'Keeffe, A.G.; Su, L.; Urowitz, M.B.; Romero-Diaz, J.; Gordon, C.; Bae, S.C.; Bernatsky, S.; Clarke, A.E.; Wallace, D.J.; et al. The frequency and outcome of lupus nephritis: Results from an international inception cohort study. *Rheumatology* **2016**, *55*, 252–262. [CrossRef]
65. Tektonidou, M.G.; Dasgupta, A.; Ward, M.M. Risk of End-Stage Renal Disease in Patients With Lupus Nephritis, 1971-2015: A Systematic Review and Bayesian Meta-Analysis. *Arthritis Rheumatol.* **2016**, *68*, 1432–1441. [CrossRef] [PubMed]
66. Yap, D.Y.; Tang, C.S.; Ma, M.K.; Lam, M.F.; Chan, T.M. Survival analysis and causes of mortality in patients with lupus nephritis. *Nephrol. Dial. Transplant. Off. Publ. Eur. Dial. Transpl. Assoc.-Eur. Ren. Assoc.* **2012**, *27*, 3248–3254. [CrossRef] [PubMed]
67. Mok, C.C.; Kwok, R.C.; Yip, P.S. Effect of renal disease on the standardized mortality ratio and life expectancy of patients with systemic lupus erythematosus. *Arthritis Rheum.* **2013**, *65*, 2154–2160. [CrossRef]
68. Chen, Y.E.; Korbet, S.M.; Katz, R.S.; Schwartz, M.M.; Lewis, E.J.; Collaborative Study Group. Value of a complete or partial remission in severe lupus nephritis. *Clin. J. Am. Soc. Nephrol. CJASN* **2008**, *3*, 46–53. [CrossRef]
69. Bajema, I.M.; Wilhelmus, S.; Alpers, C.E.; Bruijn, J.A.; Colvin, R.B.; Cook, H.T.; D'Agati, V.D.; Ferrario, F.; Haas, M.; Jennette, J.C.; et al. Revision of the International Society of Nephrology/Renal Pathology Society classification for lupus nephritis: Clarification of definitions, and modified National Institutes of Health activity and chronicity indices. *Kidney Int.* **2018**, *93*, 789–796. [CrossRef]
70. Bolisetty, S.; Zarjou, A.; Agarwal, A. Heme Oxygenase 1 as a Therapeutic Target in Acute Kidney Injury. *Am. J. Kidney Dis. Off. J. Natl. Kidney Found.* **2017**, *69*, 531–545. [CrossRef]
71. Hull, T.D.; Kamal, A.I.; Boddu, R.; Bolisetty, S.; Guo, L.; Tisher, C.C.; Rangarajan, S.; Chen, B.; Curtis, L.M.; George, J.F.; et al. Heme Oxygenase-1 Regulates Myeloid Cell Trafficking in AKI. *J. Am. Soc. Nephrol. JASN* **2015**, *26*, 2139–2151. [CrossRef]
72. Ferenbach, D.A.; Ramdas, V.; Spencer, N.; Marson, L.; Anegon, I.; Hughes, J.; Kluth, D.C. Macrophages expressing heme oxygenase-1 improve renal function in ischemia/reperfusion injury. *Mol. Ther. J. Am. Soc. Gene Ther.* **2010**, *18*, 1706–1713. [CrossRef]
73. Kwong, A.M.; Luke, P.P.W.; Bhattacharjee, R.N. Carbon monoxide mechanism of protection against renal ischemia and reperfusion injury. *Biochem. Pharmacol.* **2022**, *202*, 115156. [CrossRef] [PubMed]
74. Song, R.; Mahidhara, R.S.; Zhou, Z.; Hoffman, R.A.; Seol, D.W.; Flavell, R.A.; Billiar, T.R.; Otterbein, L.E.; Choi, A.M. Carbon monoxide inhibits T lymphocyte proliferation via caspase-dependent pathway. *J. Immunol.* **2004**, *172*, 1220–1226. [CrossRef]
75. Nath, K.A. Heme oxygenase-1 and acute kidney injury. *Curr. Opin. Nephrol. Hypertens.* **2014**, *23*, 17–24. [CrossRef]
76. Eltzschig, H.K.; Eckle, T. Ischemia and reperfusion–from mechanism to translation. *Nat. Med.* **2011**, *17*, 1391–1401. [CrossRef]
77. Salvadori, M.; Rosso, G.; Bertoni, E. Update on ischemia-reperfusion injury in kidney transplantation: Pathogenesis and treatment. *World J. Transplant.* **2015**, *5*, 52–67. [CrossRef]
78. Ferenbach, D.A.; Kluth, D.C.; Hughes, J. Hemeoxygenase-1 and renal ischaemia-reperfusion injury. *Nephron. Exp. Nephrol.* **2010**, *115*, e33–e37. [CrossRef] [PubMed]
79. He, Y.; Li, H.; Yao, J.; Zhong, H.; Kuang, Y.; Li, X.; Bian, W. HO-1 knockdown upregulates the expression of VCAM-1 to induce neutrophil recruitment during renal ischemia-reperfusion injury. *Int. J. Mol. Med.* **2021**, *48*. [CrossRef] [PubMed]
80. Ortega-Trejo, J.A.; Pérez-Villalva, R.; Sánchez-Navarro, A.; Marquina, B.; Rodríguez-Iturbe, B.; Bobadilla, N.A. Repeated Episodes of Ischemia/Reperfusion Induce Heme-Oxygenase-1 (HO-1) and Anti-Inflammatory Responses and Protects against Chronic Kidney Disease. *Int. J. Mol. Sci.* **2022**, *23*, 14573. [CrossRef]
81. Rossi, M.; Thierry, A.; Delbauve, S.; Preyat, N.; Soares, M.P.; Roumeguere, T.; Leo, O.; Flamand, V.; Le Moine, A.; Hougardy, J.M. Specific expression of heme oxygenase-1 by myeloid cells modulates renal ischemia-reperfusion injury. *Sci. Rep.* **2017**, *7*, 197. [CrossRef] [PubMed]
82. Baan, C.; Peeters, A.; Lemos, F.; Uitterlinden, A.; Doxiadis, I.; Claas, F.; Ijzermans, J.; Roodnat, J.; Weimar, W. Fundamental role for HO-1 in the self-protection of renal allografts. *Am. J. Transplant. Off. J. Am. Soc. Transplant. Am. Soc. Transpl. Surg.* **2004**, *4*, 811–818. [CrossRef]
83. Funes, S.C.; Rios, M.; Fernandez-Fierro, A.; Covian, C.; Bueno, S.M.; Riedel, C.A.; Mackern-Oberti, J.P.; Kalergis, A.M. Naturally Derived Heme-Oxygenase 1 Inducers and Their Therapeutic Application to Immune-Mediated Diseases. *Front. Immunol.* **2020**, *11*, 1467. [CrossRef] [PubMed]
84. Detsika, M.G.; Duann, P.; Atsaves, V.; Papalois, A.; Lianos, E.A. Heme Oxygenase 1 Up-Regulates Glomerular Decay Accelerating Factor Expression and Minimizes Complement Deposition and Injury. *Am. J. Pathol.* **2016**, *186*, 2833–2845. [CrossRef]
85. Stec, D.E.; Hinds, T.D., Jr. Natural Product Heme Oxygenase Inducers as Treatment for Nonalcoholic Fatty Liver Disease. *Int. J. Mol. Sci.* **2020**, *21*, 9493. [CrossRef] [PubMed]

86. Nikolic, I.; Saksida, T.; Mangano, K.; Vujicic, M.; Stojanovic, I.; Nicoletti, F.; Stosic-Grujicic, S. Pharmacological application of carbon monoxide ameliorates islet-directed autoimmunity in mice via anti-inflammatory and anti-apoptotic effects. *Diabetologia* **2014**, *57*, 980–990. [CrossRef]
87. Fagone, P.; Mangano, K.; Quattrocchi, C.; Motterlini, R.; Di Marco, R.; Magro, G.; Penacho, N.; Romao, C.C.; Nicoletti, F. Prevention of clinical and histological signs of proteolipid protein (PLP)-induced experimental allergic encephalomyelitis (EAE) in mice by the water-soluble carbon monoxide-releasing molecule (CORM)-A1. *Clin. Exp. Immunol.* **2011**, *163*, 368–374. [CrossRef]
88. Chora, A.A.; Fontoura, P.; Cunha, A.; Pais, T.F.; Cardoso, S.; Ho, P.P.; Lee, L.Y.; Sobel, R.A.; Steinman, L.; Soares, M.P. Heme oxygenase-1 and carbon monoxide suppress autoimmune neuroinflammation. *J. Clin. Investig.* **2007**, *117*, 438–447. [CrossRef]
89. Mangano, K.; Cavalli, E.; Mammana, S.; Basile, M.S.; Caltabiano, R.; Pesce, A.; Puleo, S.; Atanasov, A.G.; Magro, G.; Nicoletti, F.; et al. Involvement of the Nrf2/HO-1/CO axis and therapeutic intervention with the CO-releasing molecule CORM-A1, in a murine model of autoimmune hepatitis. *J. Cell. Physiol.* **2018**, *233*, 4156–4165. [CrossRef]

antioxidants

MDPI

Review

Immune Regulation of Heme Oxygenase-1 in Allergic Airway Inflammation

Zhenwei Xia [1,*] and Wenwei Zhong [2,*]

1 Department of Pediatrics, Ruijin Hospital Affiliated to Shanghai Jiao Tong University School of Medicine, 197 Ruijin 2nd Road, Shanghai 200025, China
2 Department of Pediatrics, Xinhua Hospital Affiliated to Shanghai Jiao Tong University School of Medicine, 1665 Kongjiang Road, Shanghai 200090, China
* Correspondence: xzw10484@rjh.com.cn (Z.X.); zhongwenwei@xinhuamed.com.cn (W.Z.); Tel.: +86-21-64333414 (Z.X.); +86-21-25078384 (W.Z.)

Abstract: Heme oxygenase-1 (HO-1) is not only a rate-limiting enzyme in heme metabolism but is also regarded as a protective protein with an immunoregulation role in asthmatic airway inflammation. HO-1 exerts an anti-inflammation role in different stages of airway inflammation via regulating various immune cells, such as dendritic cells, mast cells, basophils, T cells, and macrophages. In addition, the immunoregulation role of HO-1 may differ according to subcellular locations.

Keywords: heme oxygenase-1; asthma; airway inflammation; immunoregulation

Citation: Xia, Z.; Zhong, W. Immune Regulation of Heme Oxygenase-1 in Allergic Airway Inflammation. *Antioxidants* **2022**, *11*, 465. https://doi.org/10.3390/antiox11030465

Academic Editors: Stanley Omaye, Elias Lianos and Maria G. Detsika

Received: 6 January 2022
Accepted: 23 February 2022
Published: 26 February 2022

1. Introduction

The primary physiological function of heme oxygenase (HO), a rate-limiting enzyme in heme metabolism, is to degrade heme into biliverdin (BV), free ferrous ion (Fe^{2+}), and carbon monoxide (CO). BV is converted to bilirubin (BR), and iron is sequestered into ferritin. HO has two main isoforms, HO-1 and HO-2, of which HO-1 is an inducible isoform with molecular weight 32-kD. HO-1 is considered a protective protein. HO-1 is encoded by HMOX1 and regulated transcriptionally. HO-1 is expressed at low levels or is absent in most tissues except the spleen and liver under homeostatic conditions and highly expressed in response to various stimuli related to cellular stress and pro-oxidant signals, such as reactive oxygen species (ROS), cytokines, inflammatory mediators, and infection. The upregulation of HO-1 or its enzymatic products CO, BV, and BR have been shown to have anti-inflammatory, antioxidant, cell cycle regulation properties in vitro study or animal models of a variety of diseases including asthma [1–3].

Asthma is a chronic inflammatory disease with complex pathogenesis. Different triggers could lead to different types of inflammation. For example, exposure to an allergen in sensitized individuals tends to invoke Th2 high airway inflammation characteristic of increased Th2 cytokine and eosinophil (EOS) infiltration in the airway. On the other hand, respiratory infection or recurrent environmental tobacco smoke exposure tend to induce Th17 or Th1-dominant immune responses and Th2-low airway inflammation characteristic of neutrophil infiltration. A variety of immune cells are involved despite Th2-high or low airway inlfammation. HO-1 expression is significantly upregulated in many immune cells such as dendritic cells (DCs), macrophages [4], mast cells (MCs) [5], basophils (BAs) [6], and T cells [7] in response to cellular stress. Both HO-1 and CO displayed anti-inflammatory, antioxidative stress, and immune regulation properties in an asthma animal model [8,9] and inhibited the proliferation of cultured human airway smooth muscle cells (HASMCs) through downregulation of ERK1/2 activation, which indicated the anti-airway remodeling effects of HO-1. HO-1 has been shown to exert anti-inflammatory effects in both T helper cell type (Th) 2-dominant [6,10,11] and Th17-dominant models of asthmatic airway inflammation [12]. In this article, we will discuss mechanisms by which HO-1 regulates immune responses during allergic airway inflammation with a focus on specific immune cells in different stages of inflammation.

2. Mechanisms of Asthmatic Airway Inflammation and Role of HO-1

Asthma is a heterogeneous disease characterized by chronic airway inflammation and bronchial hyperresponsiveness. As a protective protein, many studies prove that HO-1 expression is upregulated both in asthma patients [13–15] and animal models of asthma [6,16–18]. The pathophysiological significance of this phenomenon is to avoid further deterioration of inflammation since inhibition of endogenous HO-1 further aggravates the inflammation [9,10]. Furthermore, upregulation of HO-1 or exogenous administration of CO and bilirubin which is produced by its degradation of heme has significant protective effects on allergic airway inflammation. These factors can inhibit plasma exudation to the trachea, main bronchi, and segmental bronchi; reduce infiltration of inflammatory cells (such as EOS, neutrophils, lymphocytes, and macrophages) around the airway and in bronchoalveolar lavage fluid (BALF); alleviate airway reactivity and mucus secretion [19–22]; decrease the proportion of antigen-specific Th2 cells [6] and Th17 cells [12,23] in mediastinal lymph nodes and the spleen; and further inhibit allergic airway inflammation. These findings suggest that HO-1 has a protective effect in various types of asthmatic airway inflammation.

The pathogenesis of asthma is complex, and immune imbalance is the most important mechanism of asthma. Th2 immune response-dominant asthma (the most common phenotype) involves EOS inflammation, while Th17 immune response-dominant asthma (the primary refractory phenotype) involves neutrophil inflammation. Regardless of type, this immune disorder proceeds through initiation, Th cell-directed differentiation, amplification, and effective stages. The initiation stage is particularly important because it determines the direction of the immune response. During the initiation of immune responses, antigen-presenting cells (APCs) and T cells interact with each other via the T cell receptor (TCR)–major histocompatibility complex (MHC) peptide complex and costimulatory molecules on the cell surface. In the context of specific cytokine environments, these interactions direct naïve T cell differentiation into antigen-specific Th cells (Th1, Th2, and Th17). Thus, the state and microenvironment of APCs determine the type of T cell immune response. During the effective stage, cytokines secreted by Th2 cells infiltrate around the airway along with EOS, MCs, and BAs, resulting in further release of inflammatory mediators that promote EOS infiltration-domain allergic airway inflammation. Th17 cells, mainly chemotax neutrophils, cause neutrophil-dominant airway inflammation [24–27]. Thus, various immune cells participate in the formation of chronic airway inflammation during asthma onset. HO-1 has important regulatory effects on multiple types of immune cells involved in airway inflammation [6,12,19–23].

3. HO-1 Inhibits Inflammation during the Initial Stage

3.1. *Inhibition of DC Function*

APCs play a key role in the initialization of adaptive immunity via promoting T cell differentiation. APCs capture antigens from the external environment via endocytosis or phagocytosis and degrade them into peptide fragments and binding with MHC II molecules. Antigen peptide–MHCII complex-loaded APCs then contact with naïve T cells via TCRs on the surface and initiate T cell polarization. DCs are the most important professional APCs [28,29]. This function of APCs is regulated by pathogen-associated molecular patterns (PAMPs) via pattern recognition receptors (PRR) [28,30,31]. HO-1 is constitutively expressed in immature DCs (iDCs) such as human monocyte-derived iDCs, freshly isolated rat splenic DC subsets, and rat bone marrow-derived iDCs, and is downregulated during DC maturation [32]. Signaling through PAMPs and its receptor can also regulate HO-1 expression in APCs [28,32,33]. In a mouse model of allergic airway inflammation, reinfusion of DCs highly expressing HO-1 significantly alleviated allergic airway inflammation [34], suggesting a regulatory effect of HO-1 on the antigen-presentation function of DCs. HO-1 can regulate DCs through multiple modes, such as effects on maturation, antigen presentation, and release of cytokines and extracellular vesicles (EVs). EVs are membranous structures loaded with various proteins, lipids, and nucleic acids and play important role in cell–cell communication.

First, HO-1 can inhibit DC maturation. Inhibition of HO-1 in DCs promotes the maturation of DCs [32,35,36], whereas over-expression of HO-1 was shown to inhibit maturation of bone marrow-derived DCs presenting a tolerance phenotype, as well as the presentation of exogenous soluble antigen to naïve T cells [34,37–41], which further affect the polarization of Naïve T cells towards Th1, Th2, and Th17 cells subsets. The function of APCs is regulated by PAMPs, HO-1 and its end-product CO can inhibit DC maturation by interfering with PAMPs and receptor binding. For example, CO can modify the natural conformation of toll-like receptor 4 to reduce DC maturation [42,43]. Thus, similar impairments of key steps required for correct conformational assembly of this complex on the surface of APCs are likely to reduce DC sensitivity to LPS stimulation by interfering with LPS recognition. In addition, upregulation of HO-1 activity renders DCs insensitive to LPS-induced activation of the p38 mitogen-activated protein kinase/cAMP-response element-binding protein/activating transcription factor 1 signaling pathway [38]. Importantly, all of the above factors could influence the LPS-induced maturation of DCs through effects on APCs.

Second, HO-1 and CO can inhibit antigen presentation to regulate DCs. In the process of antigen presentation, APCs first capture antigen components and endocytose them to form early endosomes, late endosomes, and fuse with proteasome/MHC molecules containing endosomes; then, they can fuse with lysosomes to form an MHC peptide complex. HO-1 and CO not only reduce the capability of APCs to identify PAMPs but also impair fusion between late endosomes and lysosomes [40], reduce mitochondrial membrane potential and ATP production in DCs, impairing cargo transport and endosome-to-lysosome fusion [39]. Disrupting fusion between antigen-containing late endosomes and lysosomes further blocks antigen transport by preventing the formation of MHC-II-peptide fragments in lysosomes, thus inhibiting the presentation of soluble antigens by DCs.

Third, HO-1 can regulate patterns of cytokines released by DCs. DCs highly expressing HO-1 secrete high levels of interleukin 10 (IL-10) and TGF-β, and low levels of IL-12 and IL-23, yielding a microenvironment conducive to the differentiation of naïve T cells into regulatory T cells (Tregs) rather than Th2 or Th17 cells [32,34]. In addition, overexpression of HO-1 in DCs can inhibit DCs maturation as we discussed above and direct naïve T cells polarization towards Treg subtypes [34]. The absence of HO-1 in APCs abolished the suppressive activity of Treg cells on effector T cells, indicating that HO-1 activity in APCs is important for the inhibitory function of Tregs [44]. This evidence indicates that the regulatory role of HO-1 on Tregs partly via APCs inhibitory manner.

Finally, HO-1 can regulate immune responses by inhibiting the release of EVs from DCs. DC-derived EVs lead to allergic airway inflammation by presenting allergens and directly contacting $CD4^+$ T cells. Our previous study found that stimulating DCs with dust mite extract expressing MHC II resulted in the concentrated release of EVs, which induced Th2 cell differentiation in vitro. In an animal model of asthma, concentrated EVs were produced following house dust mite stimulation of the airway, indicating typical allergic airway inflammation. In hemin-induced EV-sensitized mice, allergic airway inflammation was significantly alleviated; EOS infiltration and mucus secretion were reduced in the airway; levels of IL-4, IL-5, and IL-13 were decreased in the lung; numbers of Th2 cells in the mediastinum lymph node (MLN) were decreased; numbers of Treg cells in MLN were increased; and numbers of Th17 cells were reduced. These results suggest that the anti-inflammatory effects of EVs are executed through regulation of Th17/Treg balance and inhibition of Th2 and Th17 cell proliferation [45].

3.2. Inhibition of BA Function

In addition to DCs, BAs are an important APC for initiating allergic inflammation. Although DCs have historically been considered an important APC for initiating T cell immune responses and forming memory immune cells, they cannot secrete IL-4 and independently initiate Th2 immune response. Recently, the role of BAs in Th2 immune responses and allergic diseases has attracted increased attention. We and others have confirmed that BAs with antigen-presentation functions express costimulatory molecules and secrete "early IL-4". Moreover, BAs can promote Th2 cell differentiation without exogenous IL-4 in vitro [46–49]. Currently, BAs are considered to both assist APCs (such as DCs) in the initiation of directional differentiation of Th2 cells by secreting Th2 cytokines (such as IL-4) and independently initiate Th2 immune responses as APCs [50]. BAs can also obtain MHC II–peptide complexes from DCs through trogocytosis to exert APC function [47]. Furthermore, our previous study demonstrated HO-1 expression in BAs by immunohistochemistry. Overexpression of HO-1 significantly inhibited the expression of activation marker CD200R and costimulatory factors, inhibited IL-4 release stimulated by DNP-OVA/anti-DNP-IgE, inhibited DQ-OVA up-taken both in the lung-derived BAs from asthma animal models and in cultured bone marrow-derived BAs, and subsequently, inhibited polarization of naïve T cells into Th2 cells in vitro and inhibited OVA-induced allergic airway inflammation and the Th2 immune response.

4. HO-1 Inhibits Inflammation during the Effective Stage

4.1. HO-1 Promotes Treg Cell Function and Inhibits Th2- and Th17 Cell-Mediated Inflammation

The imbalance of the Th cell subgroup plays an important role in the pathogenesis of asthma. HO-1 inhibits Th cell functions via different mechanisms. Firstly, CO, which is one of the end-products of HO-1, can inhibit the proliferation of $CD4^+$ T cells by blocking TCR-dependent IL-2 production [51]. Another end-product, BR, can inhibit $CD4^+$ T cells by inducing apoptosis, suppressing co-stimulatory molecule expression in $CD4^+$ T cells, and inhibiting $CD4^+$ cell proliferation [52].

Secondly, HO-1 can regulate the balance of the Th cell subgroup via Tregs. Tregs are important immune cells to maintain immune homeostasis. Tregs inhibit effector T cells proliferation and function via interactions with negative costimulatory molecules, secrete suppressive cytokines IL-10, and competition for IL-2 [53], and subsequently exert inhibitory effects on Th1, Th2, and Th17 cell-mediated inflammation [54–56]. HO-1 promotes Tregs function, which is regarded as an important mechanism for its immunomodulatory function. HO-1 expression is significantly different between $CD4^+$ $CD25^+$ Treg cells and $CD4^+$ $CD25^+$ T lymphocytes [7], and is consistent with Foxp3 expression in these two cell types. Transfection of Foxp3 into Jurkat T cells significantly upregulated the expression of HO-1 and inhibited their proliferation and cytokine production in a cell contact-dependent manner. Treatment of freshly isolated $CD4^+$ $CD25^{high}$ from the spleen with hemin or transfected with an HO-1 expression vector (pcDNA3HO-1) in vitro not only significantly enhanced Foxp3 expression and IL-10 secretion but also enhanced its ability to inhibit effector T cell proliferation. The regulatory role of HO-1 was significantly inhibited by the addition of an HO-1 activity inhibitor [11,57]. In an animal model of asthmatic allergic airway inflammation, overexpression of HO-1 induced by hemin enhanced proportions and functions of $CD4^+$ $CD25^+$ Treg cells [10,11] and alleviated OVA-induced allergic airway inflammation. On the contrary, inhibition of HO-1 activity with tin-protoporphyrin reversed the above effects of HO-1 [10,11]. These in vivo and in vitro studies show that HO-1 plays an important role in regulating Treg function, However, the direct role of HO-1 in regulating Treg function is challenged since HO-1-deficient mice not only exhibited a significantly higher proportion of Foxp3-expressing cells among total $CD4^+$ and $CD4^+$ $CD25^+$ cells in comparison to wild type mice but also displayed a similar inhibitory role in suppressing the proliferation of effector T cells in vitro. In the same study, HO-1-deficient APCs abolished the suppressive activity of Treg cells [44], indicating that HO-1 may regulate $CD4^+$ $CD25^+$ Treg cells by indirectly promoting Treg differentiation through inhibition of DC maturation.

Considering Tregs have inhibitory effects on T cell subsets of Th1, Th2, and Th17 cells, we speculate that HO-1 enhances Treg function and therefore regulates the balance of Th1, Th2, and Th17 cells.

4.2. HO-1 Inhibits Th17 Cell-Mediated Inflammation

Th17 cells, an important T cell subset in asthma, play key roles in refractory asthma and neutrophil-dominant asthma types by promoting neutrophil growth, development, and chemotactic aggregation in the airway. Th17 cells achieve these effects by secreting cytokines and are essential for inducing neutrophil infiltration-dominant asthma [26,58,59]. Overexpression of HO-1 inhibited the differentiation of naïve T cells into Th17 cells, as well as the secretion of IL-17A in vitro [60]. In an animal model of non-eosinophilic asthma, upregulation of HO-1 expression significantly reduced the proportion of Th17 cells, promoted IL-10 expression, reconstructed the balance of Th17/Treg cells in vivo, and subsequently inhibited Th17 cell-mediated neutrophilic airway inflammation. In contrast, inhibition of HO-1 activity reversed the inhibitory effect of HO-1 on neutrophil airway inflammation and activation of the Th17 cell signaling pathway [12].

4.3. HO-1 Inhibits MC Function

MCs are important effector cells in asthma. Sensitized MCs can be activated and degranulated to release various preformed mediators and pre-synthesized mediators, such as proteases, cytokines, chemokines, and arachidonic acid metabolites. MCs can de novo synthesize lipid mediators by enzymes located in the plasma membrane and synthesize mRNAs encoding cytokines and chemokines. MCs can also regulate inflammation via the secretion of exosomes containing regulatory molecules [61–63]. MCs participate in asthmatic inflammation which is characterized by inflammatory cell infiltration, microvascular leakage, airway hyperresponsiveness, bronchoconstriction by degranulation, and the secretion of various mediators [64–66]. Moreover, MCs participate in allergic inflammation via regulating T cells, DCs, and other inflammatory cells, which results in further chemotaxis, infiltration, and activation of EOS, neutrophils, and other inflammatory cells in the airway [61,67]. Recent studies revealed that MCs can regulate Th17-mediated autoimmune diseases by inducing Tregs [64].

HO-1 has important regulatory effects on MC function. HO-1 is expressed in MC [5] and MC cell lines [68], and can be induced during MCs degranulation [5,68] and upregulation of HO-1 decreased MCs degranulation induced by complex 48/80 and leukocyte adhesion to blood vessels [5]. Upregulation of HO-1 during MCs degranulation was related to cellular oxidative stress since upregulation of HO-1 was inhibited by antioxidant N-acetyl-L-cysteine [68]. HO-1 and its end-products, BR and BV, can inhibit adhesion and degranulation of MCs [68]. HO-1 can also inhibit the production of inflammatory mediators in MCs by selectively inhibiting the DNA-binding activity of the AP-1 transcription factor [69].

In addition, HO-1 regulates MC-mediated immune regulation. Co-culture of MCs and DCs led to a significant release of tumor necrosis factor-α (TNF-α), IL-6, and interferon (IFN), which promoted DC maturation. Upregulation of HO-1 in MCs before co-culture with DCs inhibited expression of costimulatory molecules on DCs and inhibited DCs maturation [70]. In contrast, downregulation of HO-1 expression promoted MC degranulation, DC costimulatory molecule expression, and DC maturation. These findings suggest that upregulation of HO-1 in MCs stabilizes the MC membrane and prevents its degranulation, thereby maintaining the DCs in an immature state to ultimately alleviate the immune response [39].

4.4. HO-1 Regulated Inflammation by Inhibiting NLRP3 Inflammasomes

Factors such as environmental irritants and respiratory infections are common triggers of asthma exacerbation and neutrophilic airway inflammation responds to these situations. Signaling through inflammasome activation plays a key role in neutrophilic airway inflammation. The inflammasome is an intracellular protein complex and activated by ligation of PAMPs and its receptor. Upon ligand sensing, inflammasome components assemble and self-oligomerize, followed by autoactivation of caspase-1 and leading to cleave pro-IL-1β and pro-IL-18 to IL-1β and IL-18, T helper 17 activation, IL-8/IL-6 overproduction, thus initiating or aggravating neutrophilic airway inflammation. Inflammasomes also involved in caspase-1-mediated pyroptosis [71,72]. Among the known inflammasomes, nucleotide-binding domain and leucine-rich repeat protein 3 (NLRP3) are crucially involved in the pathogenesis of asthmatic airway inflammation [73].

HO-1/CO has a potential regulatory role in inflammasome signaling. Li and colleges [74] demonstrate that the induction of HO-1 by hemin inhibited LPS-induced production of IL-1β, inhibited NLRP3 inflammasome activation in human gingival epithelial cells in vivo. Luo and colleges [75] also demonstrated that upregulation of HO-1 by hemin inhibited LPS-induced NLRP3 inflammasome activation, reducing IL-1β and IL-18 production in sepsis-induced acute lung injury. On the contrary, inhibition of HO-1 activity reversed the above results. On the other hand, CO can inhibit LPS and ATP-induced caspase-1 activation and production of IL-1β and IL-18 in bone marrow-derived macrophages. Treatment of CO-releasing molecule-2 (CORM-2) both inhibited NLRP3 inflammasome activation in LPS-induced acute lung injury and ER stress-induced inflammation [76]. CORM2 also can inhibit caspase-1activiation [77–79] and thioredoxin-interacting protein (TXNIP)–NLRP3 complex formation [77,78]. Our previous study showed that HO-1 products, CO and BR, inhibited the NLRP3–RXR axis and NLRP3 inflammasome-mediated apoptosis of AECs, which inhibits subsequent production of IL-25, IL-33, thymic stromal lymphopoietin, and other pro-Th2 epithelial-derived cytokines. In addition, we found that HO-1 bound to the NACHT domain of NLRP3 and RXRα/RXRβ subunits, suggesting that the non-enzymatic action of HO-1 may be involved in the regulation of NLRP3 inflammasomes [80].

4.5. HO-1 Promotes Polarization of Macrophages to M2 Phenotype

Macrophages are one of the important inflammatory cells involved in the pathogenesis of asthma [81]. Macrophages are divided into M1 and M2 subpopulations according to their responses to environmental or inflammatory stimuli. M1 macrophages respond to pro-inflammatory cytokines, such as IFN-γ and TNF-α, and promote a local Th1 environment. M2 macrophages (alternatively referred to as activated macrophages) respond to IL-4 and IL-13, promote the production of the anti-inflammatory cytokine IL-10, and regulate Th2 immune responses [82–84]. HO-1 is considered a regulator of immune responses because it can promote the polarization of M2 macrophages. Indeed, HO-1 is highly expressed in M2 macrophage subsets, and its elevation in response to multiple stimuli can drive phenotypic transfer to M2 macrophages [85–87]. HO-1-knockout bone marrow macrophages (mHO-1-KO) exposed to LPS (M1-inducer) or IL-4 (M2-inducer) exhibited an enhanced M1 phenotype and inhibited M2 phenotype. In contrast, promotion of the M2 phenotype was observed in HO-1-overexpressing (HO-1-Tg) mice [88]. Collectively, these studies support the hypothesis that HO-1 promotes the M2 phenotype. However, there is a lack of studies demonstrating the role of HO-1 on the regulation balance between M1 and M2 subpopulations in clinical studies or animal models of asthma, and further studies are required.

5. Subcellular Localization and Anti-Inflammatory Mechanism of HO-1

It is well accepted for a long time that the biological functions of HO-1 are related to its enzymatic products CO, BR/BV, and ferritin. A recent study revealed that HO-1 exerts its function via interaction between other cellular proteins in an activity-independent manner, which is referred as "non-canonical effects of HO-1" in published studies. Those different fashions may in part relate to specific subcellular locations of HO-1.

HO-1 was initially identified in the endoplasmic reticulum (ER). HO-1 is mainly located in the ER under physiological conditions. Recently, more specific subcellular locations of HO-1 have been revealed. HO-1 was identified in mitochondria, plasma membrane, the caveolae, and the nucleus after various stimuli (e.g., LPS, or hypoxia) [89]. Varying the subcellular localization of HO-1 protein has various effects on its cell-protective functions which may depend on the structural integrity of the HO-1 protein. Except for the nucleus, HO-1 is fixed on the membrane by a transmembrane sequence (TMS) located at the carboxyl terminal of its protein structure, and the rest of the protein structure faces the cytoplasm and colocalizes with cytochrome P450 reductase (CPR) and BVR to facilitate heme degradation [90]. Cleavage of the TMS enables HO-1 relocation. On the other hand, CPR can stabilize the HO-1 protein structure to prevent its relocation and maintain its enzymatic activity by promoting oligomerization [91]. HO-1 locates in the mitochondria, vacuole, and plasma membrane and maintains full protein structure and enzymatic activity and coexists with biliverdin reductase [92,93], indicating that its function is achieved mainly through enzymatic activity. The functional significance of HO-1 in different cellular compartments remains unclear. In caveolae, HO-1 activity is negatively modulated by caveolin-1 (CAV1). It could serve as a brake on HO-1 function and provide a possible approach for the active extracellular transfer of HO-1 [94]. In mitochondria, HO-1 induces increased ROS, which appears important for its regulation of mitochondrial heme content, and plays an important role in apoptosis [89,92].

Unlike HO-1 in the ER, mitochondria, and caveolae, nuclear HO-1(NHO-1) exists in a truncated form (28 kDa) in the COOH terminal with a lack of enzyme activity to degrade heme [89]. Therefore, nuclear HO-1 may exert its roles independent of enzyme activity [89,95]. HO-1 in the ER will be truncated and translocate to the nucleus under pathological conditions or external stimuli, leading to cellular stress [94]. NHO-1 has been reported as a regulator of nuclear transcription factor activities such as NF-κB, AP-1, and Nrf2 [96], the latter of which regulates the antioxidant response. By regulating gene expression levels in the nucleus, HO-1 provides resistance to redox stimulation that can protect cells from dysfunction or death, making it an important part of signaling involved in the cellular response to oxidative stress. As the DNA-binding motif of typical transcription factors is not detected in the HO-1 protein structure and direct recruitment of HO-1 to DNA has not been observed, nuclear HO-1 is unlikely to regulate gene expression by directly binding DNA. It is speculated that HO-1 acts as a transcriptional coregulatory protein that binds transcription factors or complexes to regulate their DNA-binding affinity, thus indirectly regulating transcription of key response genes [89].

Our previous study investigated the non-enzymatic anti-inflammatory effects of HO-1. We found that HO-1 inhibited Th17 cell differentiation mainly by inhibiting IL-6-induced STAT3 phosphorylation and therefore inhibited activation of the STAT3/RORγT signaling pathway. Co-IP results indicated that endogenous HO-1 directly bound STAT3 rather than Jak1, JAK2, or SOCS3, suggesting that HO-1 inhibited STAT3 phosphorylation by interacting with STAT3. This result is consistent with the finding of Elguero and colleagues. In their study, they demonstrated that HO-1 and STAT3 bind to each other and therefore inhibit the phosphorylation of STAT3 and subsequent nuclear translocation of pSTAT3 in prostate cancer cells [97]. Our study demonstrated that co-transfection of 293T cells with plasmids containing HO-1 and STAT3 domains revealed that HO-1 bound all regions of the STAT3 protein except the helical domains, especially the transcriptional activation region (AA 689–770). Further experiments confirmed that HO-1 directly bound STAT3, in particular, the transcriptional activation domain-containing Tyr705 in the STAT3 protein. However, whether HO-1 plays an indirect regulatory role through intermediate proteins (for example, whether HO-1 leads to increased dephosphorylation of STAT3 by promoting SHP-1 activation) or inhibits phosphorylation by directly binding STAT3 requires further study [98].

6. Potential Clinical Application of HO-1

Results from basic research demonstrated that HO-1 alleviated allergic airway inflammation by its anti-inflammatory, antioxidative stress, and immune regulation properties. Moreover, how to apply those results to clinic use has attracted the interest of researchers.

Currently, research focused on the value of HO-1 in both diagnostic and therapeutic applications. In view of diagnosis, clinical studies revealed that exhaled CO elevated in asthma patients and the levels of exhaled CO were associated with the severity of asthma and disease exacerbation [99–101], indicating that exhaled CO has potential value in asthma management as a noninvasive tool. On the other hand, therapeutic applications of drugs targeting HO-1 activity and expression have attracted significant attention, including pharmacologic modulators of HO-1, gene therapy, and enzymatic byproducts of HO-1 such as CO and CO-releasing molecules (CORM) [3,102–104]. Despite the achievement of attempts in animal studies with promising results, pharmacologic use of HO-1 targeted drugs still faces great challenges, especially safety and efficacy.

7. Conclusions

During asthmatic airway inflammation, HO-1 regulates differentiation of Th1/Th2/Th17 cell subsets by inhibiting the functions of APCs (such as DCs and BAs), promotes Treg function, and suppresses allergic airway inflammation characterized by Th2-dominant EOS infiltration and neutrophil infiltration-dominant Th17 immune response. HO-1 directly inhibits airway inflammation by inhibiting BA, MCs, and certain functions during the inflammatory effective stage (summarized in Figure 1). HO-1 not only exerts anti-inflammatory and immunomodulatory effects through its enzymatic products CO and BR/BV but also regulates the transcription of key genes by acting as a transcription coregulatory protein with transcription factors or complexes in the nucleus (summarized in Figure 2). However, studies evaluating the non-enzymatic immunoregulatory mechanism of HO-1 are limited, which represents an important direction for future research.

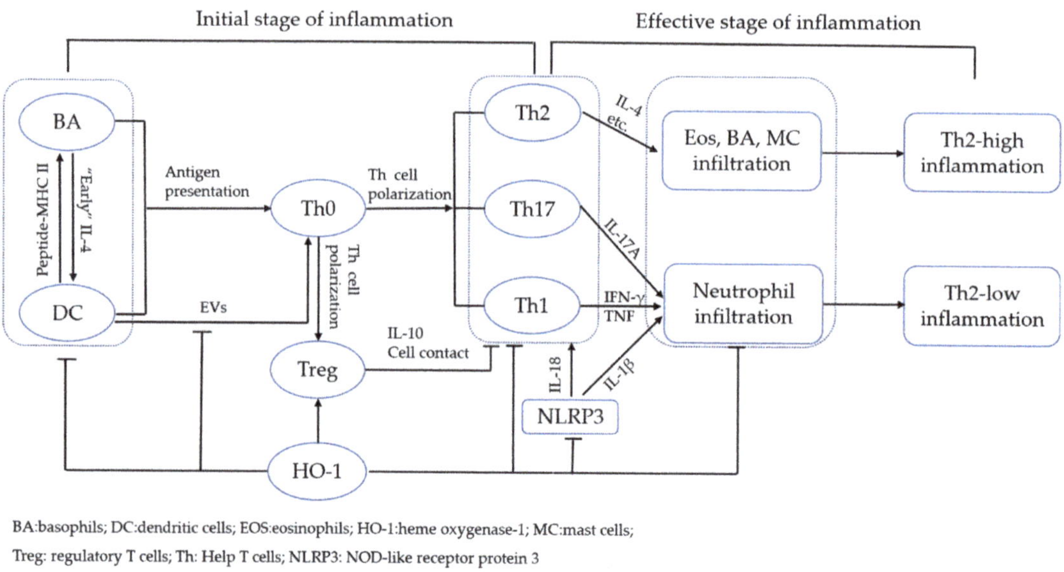

Figure 1. Multi-target effect of HO-1 and its immune regulation role in allergic airway inflammation. 1, HO-1 regulates APCs function and inhibits allergic airway inflammation at the initial stage: a, HO-1 and its end-product CO inhibit DC maturation by interfering with PAMPs and receptor binding, inhibit antigen presentation by impair fusion between late endosomes and lysosome, inhibit release of EVs from DCs and promote Treg polarization; b, BAs assist DCs participate in Th2 cells polarization by secreting "early" IL-4; BAs initiate Th2 polarization independently as APCs or BAs obtain MHC II-peptide complex from DCs through trogocytosis and subsequently initiate Th2 polarization; HO-1

inhibits BAs participate in Th2 cell differentiation by inhibits BAs activation, soluble antigen up-taken, expression of costimulatory molecules and secrete "early IL-4"; 2, HO-1 inhibits allergic airway inflammation at effective stage: a, HO-1 and its end-product inhibit $CD4^{+}$ T cell proliferation and function directly or via promotion of Treg; b, HO-1 inhibits Th17 cell-mediated neutrophilic airway inflammation by inhibiting Th17 cell polarization and IL-17A releasing; c, HO-1 and its end-products suppresses mast cell degranulation and releasing of inflammatory mediators; d, HO-1 and its end-product inhibit NLRP3 inflammasome activation and subsequently inhibit IL-1β and IL-18 mediated airway inflammation.

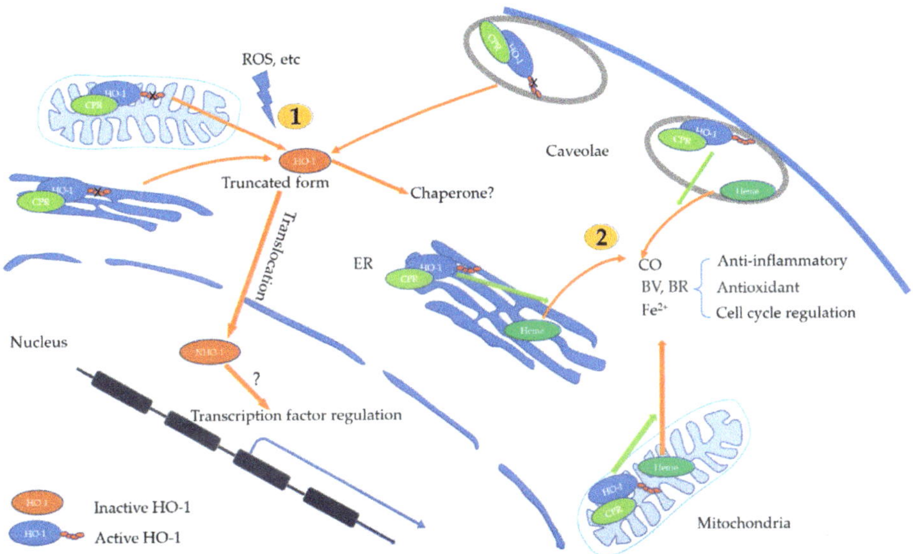

Figure 2. Subcellular localization and roles of HO-1. **1.** HO-1 locates in endoplasmic reticulum (ER), mitochondria, plasma membrane, and the caveolae. HO-1 is fixed on the membrane by a transmembrane sequence (TMS) and stabilized by cytochrome P450 reductase (CPR) to prevent relocation. It remains full protein structure and enzymatic activity and its function is achieved mainly through enzymatic activity. Its enzymatic products CO, BV, and BR have been shown to have anti-inflammatory, antioxidant, cell cycle regulation properties; **2.** HO-1 will be truncated and relocated under pathological conditions or external stimuli which leads to cellular stress. Truncated HO-1 lacks enzyme activity to degrade heme and may act as a chaperone to regulate the activity of signaling pathway proteins or as a transcription factor regulator.

Author Contributions: W.Z. drafted the manuscript; Z.X. edited and revised the manuscript. All authors have read and agreed to the published version of the manuscript.

Funding: This work was supported by grants from the National Natural Science Foundation of China (81970021, 82170035, 82170025), the Shanghai Municipal Science and Technology Commission Foundation (17ZR1418000), the Shanghai Municipal Health Commission (202040164) and Innovative Research Team of High-level Local Universities in Shanghai.

Conflicts of Interest: The authors declare no competing financial interest.

References

1. Consoli, V.; Sorrenti, V.; Grosso, S.; Vanella, L. Heme Oxygenase-1 Signaling and Redox Homeostasis in Physiopathological Conditions. *Biomolecules* **2021**, *11*, 589. [CrossRef]
2. Ryter, S.W. Heme Oxgenase-1, a Cardinal Modulator of Regulated Cell Death and Inflammation. *Cells* **2021**, *10*, 515. [CrossRef]

3. Ryter, S.W. Therapeutic Potential of Heme Oxygenase-1 and Carbon Monoxide in Acute Organ Injury, Critical Illness, and Inflammatory Disorders. *Antioxidants* **2020**, *9*, 1153. [CrossRef] [PubMed]
4. Weis, N.; Weigert, A.; von Knethen, A.; Brune, B. Heme oxygenase-1 contributes to an alternative macrophage activation profile induced by apoptotic cell supernatants. *Mol. Biol. Cell* **2009**, *20*, 1280–1288. [CrossRef] [PubMed]
5. Takamiya, R.; Murakami, M.; Kajimura, M.; Goda, N.; Makino, N.; Takamiya, Y.; Yamaguchi, T.; Ishimura, Y.; Hozumi, N.; Suematsu, M. Stabilization of mast cells by heme oxygenase-1: An anti-inflammatory role. *Am. J. Physiol. Heart Circ. Physiol.* **2002**, *283*, H861–H870. [CrossRef]
6. Zhong, W.; Di, C.; Lv, J.; Zhang, Y.; Lin, X.; Yuan, Y.; Lv, J.; Xia, Z. Heme oxygenase-1 inhibits basophil maturation and activation but promotes its apoptosis in T helper type 2-mediated allergic airway inflammation. *Immunology* **2016**, *147*, 321–337. [CrossRef] [PubMed]
7. Pae, H.O.; Oh, G.S.; Choi, B.M.; Chae, S.C.; Chung, H.T. Differential expressions of heme oxygenase-1 gene in $CD25^-$ and $CD25^+$ subsets of human $CD4^+$ T cells. *Biochem. Biophys. Res. Commun.* **2003**, *306*, 701–705. [CrossRef]
8. Pae, H.O.; Lee, Y.C.; Chung, H.T. Heme oxygenase-1 and carbon monoxide: Emerging therapeutic targets in inflammation and allergy. *Recent Pat. Inflamm. Allergy Drug Discov.* **2008**, *2*, 159–165. [CrossRef]
9. Kuribayashi, K.; Iida, S.; Nakajima, Y.; Funaguchi, N.; Tabata, C.; Fukuoka, K.; Fujimori, Y.; Ihaku, D.; Nakano, T. Suppression of heme oxygenase-1 activity reduces airway hyperresponsiveness and inflammation in a mouse model of asthma. *J. Asthma* **2015**, *52*, 662–668. [CrossRef]
10. Xia, Z.W.; Zhong, W.W.; Xu, L.Q.; Sun, J.L.; Shen, Q.X.; Wang, J.G.; Shao, J.; Li, Y.Z.; Yu, S.C. Heme oxygenase-1-mediated $CD4^+CD25^{high}$ regulatory T cells suppress allergic airway inflammation. *J. Immunol.* **2006**, *177*, 5936–5945. [CrossRef]
11. Xia, Z.W.; Xu, L.Q.; Zhong, W.W.; Wei, J.J.; Li, N.L.; Shao, J.; Li, Y.Z.; Yu, S.C.; Zhang, Z.L. Heme oxygenase-1 attenuates ovalbumin-induced airway inflammation by up-regulation of foxp3 T-regulatory cells, interleukin-10, and membrane-bound transforming growth factor-β1. *Am. J. Pathol.* **2007**, *171*, 1904–1914. [CrossRef] [PubMed]
12. Zhang, Y.; Zhang, L.; Wu, J.; Di, C.; Xia, Z. Heme oxygenase-1 exerts a protective role in ovalbumin-induced neutrophilic airway inflammation by inhibiting Th17 cell-mediated immune response. *J. Biol. Chem.* **2013**, *288*, 34612–34626. [CrossRef] [PubMed]
13. Horvath, I.; Donnelly, L.E.; Kiss, A.; Paredi, P.; Kharitonov, S.A.; Barnes, P.J. Raised levels of exhaled carbon monoxide are associated with an increased expression of heme oxygenase-1 in airway macrophages in asthma: A new marker of oxidative stress. *Thorax* **1998**, *53*, 668–672. [CrossRef] [PubMed]
14. Harju, T.; Soini, Y.; Paakko, R.; Kinnula, V.L. Up-regulation of heme oxygenase-I in alveolar macrophages of newly diagnosed asthmatics. *Respir. Med.* **2002**, *96*, 418–423. [CrossRef]
15. Mo, B.; Zhang, Z.; Xu, Y.; Xiong, W.; Liu, X.A.; Zhen, G. Expression of heme oxygenase-1 in the peripheral blood mononuclear cells from asthmatic patients. *J. Huazhong Univ. Sci. Technol. Med. Sci.* **2005**, *25*, 385–388. [CrossRef]
16. Zhu, Y.; Wang, C.; Luo, J.; Hua, S.; Li, D.; Peng, L.; Liu, H.; Song, L. The protective role of Zingerone in a murine asthma model via activation of the AMPK/Nrf2/HO-1 pathway. *Food Funct.* **2021**, *12*, 3120–3131. [CrossRef] [PubMed]
17. Kim, S.M.; Ryu, H.W.; Kwon, O.K.; Hwang, D.; Kim, M.G.; Min, J.H.; Zhang, Z.; Kim, S.Y.; Paik, J.H.; Oh, S.R.; et al. Callicarpa japonica Thunb. ameliorates allergic airway inflammation by suppressing NF-kappaB activation and upregulating HO-1 expression. *J. Ethnopharmacol.* **2021**, *267*, 113523. [CrossRef]
18. Kitada, O.; Kodama, T.; Kuribayashi, K.; Ihaku, D.; Fujita, M.; Matsuyama, T.; Sugita, M. Heme oxygenase-1 (HO-1) protein induction in a mouse model of asthma. *Clin. Exp. Allergy* **2001**, *31*, 1470–1477. [CrossRef] [PubMed]
19. Jia, Y.X.; Sekizawa, K.; Okinaga, S.; Lie, R.; Sasaki, H. Role of heme oxygenase in pulmonary response to antigen challenge in sensitized rats in vivo. *Int. Arch. Allergy Immunol.* **1999**, *120*, 141–145. [CrossRef] [PubMed]
20. Chapman, J.T.; Otterbein, L.E.; Elias, J.A.; Choi, A.M. Carbon monoxide attenuates aeroallergen-induced inflammation in mice. *Am. J. Physiol. Lung Cell Mol. Physiol.* **2001**, *281*, L209–L216. [CrossRef]
21. Almolki, A.; Taille, C.; Martin, G.F.; Jose, P.J.; Zedda, C.; Conti, M.; Megret, J.; Henin, D.; Aubier, M.; Boczkowski, J. Heme oxygenase attenuates allergen-induced airway inflammation and hyperreactivity in guinea pigs. *Am. J. Physiol. Lung Cell Mol. Physiol.* **2004**, *287*, L26–L34. [CrossRef] [PubMed]
22. He, J.; Jiang, G.; Li, X.; Xiao, Q.; Chen, Y.; Xu, H.; Liu, G.; Lei, A.; Zhou, P.; Shi, K.; et al. Bilirubin represents a negative regulator of ILC2 in allergic airway inflammation. *Mucosal. Immunol.* **2022**, *15*, 314–326. [CrossRef]
23. Lin, X.L.; Lv, J.J.; Lv, J.; Di, C.X.; Zhang, Y.J.; Zhou, T.; Liu, J.L.; Xia, Z.W. Heme oxygenase-1 directly binds STAT3 to control the generation of pathogenic Th17 cells during neutrophilic airway inflammation. *Allergy* **2017**, *72*, 1972–1987. [CrossRef]
24. Grayson, M.H.; Feldman, S.; Prince, B.T.; Patel, P.J.; Matsui, E.C.; Apter, A.J. Advances in asthma in 2017: Mechanisms, biologics, and genetics. *J. Allergy Clin. Immunol.* **2018**, *142*, 1423–1436. [CrossRef]
25. Cevhertas, L.; Ogulur, I.; Maurer, D.J.; Burla, D.; Ding, M.; Jansen, K.; Koch, J.; Liu, C.; Ma, S.; Mitamura, Y.; et al. Advances and recent developments in asthma in 2020. *Allergy* **2020**, *75*, 3124–3146. [CrossRef]
26. Boonpiyathad, T.; Sozener, Z.C.; Satitsuksanoa, P.; Akdis, C.A. Immunologic mechanisms in asthma. *Semin. Immunol.* **2019**, *46*, 101333. [CrossRef] [PubMed]
27. Jeong, J.; Lee, H.K. The Role of $CD4^+$ T Cells and Microbiota in the Pathogenesis of Asthma. *Int. J. Mol. Sci.* **2021**, *22*, 11822. [CrossRef] [PubMed]
28. Holgate, S.T. Innate and adaptive immune responses in asthma. *Nat. Med.* **2012**, *18*, 673–683. [CrossRef]

29. Riquelme, S.A.; Carreno, L.J.; Espinoza, J.A.; Mackern-Oberti, J.P.; Alvarez-Lobos, M.M.; Riedel, C.A.; Bueno, S.M.; Kalergis, A.M. Modulation of antigen processing by haem-oxygenase 1. Implications on inflammation and tolerance. *Immunology* **2016**, *149*, 1–12. [CrossRef]
30. Peron, G.; de Lima Thomaz, L.; da Rosa, L.C.; Thome, R.; Verinaud, L.M.C. Modulation of dendritic cell by pathogen antigens: Where do we stand? *Immunol. Lett.* **2018**, *196*, 91–102. [CrossRef]
31. Ferreira, I.; Liberal, J.; Martins, J.D.; Silva, A.; Neves, B.M.; Cruz, M.T. Inflammasome in Dendritic Cells Immunobiology: Implications to Diseases and Therapeutic Strategies. *Curr. Drug Targets* **2017**, *18*, 1003–1018. [CrossRef]
32. Chauveau, C.; Remy, S.; Royer, P.J.; Hill, M.; Tanguy-Royer, S.; Hubert, F.X.; Tesson, L.; Brion, R.; Beriou, G.; Gregoire, M.; et al. Heme oxygenase-1 expression inhibits dendritic cell maturation and proinflammatory function but conserves IL-10 expression. *Blood* **2005**, *106*, 1694–1702. [CrossRef] [PubMed]
33. Hung, C.C.; Liu, X.; Kwon, M.Y.; Kang, Y.H.; Chung, S.W.; Perrella, M.A. Regulation of heme oxygenase-1 gene by peptidoglycan involves the interaction of Elk-1 and C/EBPalpha to increase expression. *Am. J. Physiol. Lung Cell Mol. Physiol.* **2010**, *298*, L870–L879. [CrossRef] [PubMed]
34. Wong, T.H.; Chen, H.A.; Gau, R.J.; Yen, J.H.; Suen, J.L. Heme Oxygenase-1-Expressing Dendritic Cells Promote Foxp3$^+$ Regulatory T Cell Differentiation and Induce Less Severe Airway Inflammation in Murine Models. *PLoS ONE* **2016**, *11*, e0168919. [CrossRef] [PubMed]
35. Moreau, A.; Hill, M.; Thebault, P.; Deschamps, J.Y.; Chiffoleau, E.; Chauveau, C.; Moullier, P.; Anegon, I.; Alliot-Licht, B.; Cuturi, M.C. Tolerogenic dendritic cells actively inhibit T cells through heme oxygenase-1 in rodents and in nonhuman primates. *FASEB J.* **2009**, *23*, 3070–3077. [CrossRef]
36. Campbell, N.K.; Fitzgerald, H.K.; Malara, A.; Hambly, R.; Sweeney, C.M.; Kirby, B.; Fletcher, J.M.; Dunne, A. Naturally derived Heme-Oxygenase 1 inducers attenuate inflammatory responses in human dendritic cells and T cells: Relevance for psoriasis treatment. *Sci. Rep.* **2018**, *8*, 10287. [CrossRef]
37. Zhao, Y.; Jia, Y.; Wang, L.; Chen, S.; Huang, X.; Xu, B.; Zhao, G.; Xiang, Y.; Yang, J.; Chen, G. Upregulation of Heme Oxygenase-1 Endues Immature Dendritic Cells With More Potent and Durable Immunoregulatory Properties and Promotes Engraftment in a Stringent Mouse Cardiac Allotransplant Model. *Front. Immunol.* **2018**, *9*, 1515. [CrossRef]
38. Al-Huseini, L.M.; Yeang, H.X.A.; Hamdam, J.M.; Sethu, S.; Alhumeed, N.; Wong, W.; Sathish, J.G. Heme oxygenase-1 regulates dendritic cell function through modulation of p38 MAPK-CREB/ATF1 signaling. *J. Biol. Chem.* **2014**, *289*, 16442–16451. [CrossRef]
39. Riquelme, S.A.; Pogu, J.; Anegon, I.; Bueno, S.M.; Kalergis, A.M. Carbon monoxide impairs mitochondria-dependent endosomal maturation and antigen presentation in dendritic cells. *Eur. J. Immunol.* **2015**, *45*, 3269–3288. [CrossRef]
40. Tardif, V.; Riquelme, S.A.; Remy, S.; Carreno, L.J.; Cortes, C.M.; Simon, T.; Hill, M.; Louvet, C.; Riedel, C.A.; Blancou, P.; et al. Carbon monoxide decreases endosome-lysosome fusion and inhibits soluble antigen presentation by dendritic cells to T cells. *Eur. J. Immunol.* **2013**, *43*, 2832–2844. [CrossRef]
41. Jung, I.D.; Lee, J.S.; Lee, C.M.; Noh, K.T.; Jeong, Y.I.; Park, W.S.; Chun, S.H.; Jeong, S.K.; Park, J.W.; Son, K.H.; et al. Induction of indoleamine 2,3-dioxygenase expression via heme oxygenase-1-dependant pathway during murine dendritic cell maturation. *Biochem. Pharmacol.* **2010**, *80*, 491–505. [CrossRef]
42. Riquelme, S.A.; Bueno, S.M.; Kalergis, A.M. Carbon monoxide down-modulates Toll-like receptor 4/MD2 expression on innate immune cells and reduces endotoxic shock susceptibility. *Immunology* **2015**, *144*, 321–332. [CrossRef] [PubMed]
43. Xue, J.; Habtezion, A. Carbon monoxide-based therapy ameliorates acute pancreatitis via TLR4 inhibition. *J. Clin. Investig.* **2014**, *124*, 437–447. [CrossRef] [PubMed]
44. George, J.F.; Braun, A.; Brusko, T.M.; Joseph, R.; Bolisetty, S.; Wasserfall, C.H.; Atkinson, M.A.; Agarwal, A.; Kapturczak, M.H. Suppression by CD4$^+$CD25$^+$ regulatory T cells is dependent on expression of heme oxygenase-1 in antigen-presenting cells. *Am. J. Pathol.* **2008**, *173*, 154–160. [CrossRef]
45. Wu, Y.; Yu, Q.; Zhang, M.; Zhou, Y.; Su, X.; Wu, M.; Lv, J.; Xia, Z. Hemin-primed dendritic cells suppress allergic airway inflammation through releasing extracellular vesicles. *J. Leukoc. Biol.* **2021**, 1–12. [CrossRef] [PubMed]
46. Poddighe, D.; Mathias, C.B.; Freyschmidt, E.J.; Kombe, D.; Caplan, B.; Marseglia, G.L.; Oettgen, H.C. Basophils are rapidly mobilized following initial aeroallergen encounter in naive mice and provide a priming source of IL-4 in adaptive immune responses. *J. Biol. Regul. Homeost. Agents* **2014**, *28*, 91–103.
47. Miyake, K.; Karasuyama, H. Emerging roles of basophils in allergic inflammation. *Allergol. Int.* **2017**, *66*, 382–391. [CrossRef]
48. Sokol, C.L.; Chu, N.Q.; Yu, S.; Nish, S.A.; Laufer, T.M.; Medzhitov, R. Basophils function as antigen-presenting cells for an allergen-induced T helper type 2 response. *Nat. Immunol.* **2009**, *10*, 713–720. [CrossRef]
49. Sarfati, M.; Wakahara, K.; Chapuy, L.; Delespesse, G. Mutual Interaction of Basophils and T Cells in Chronic Inflammatory Diseases. *Front. Immunol.* **2015**, *6*, 399. [CrossRef]
50. Chirumbolo, S.; Bjorklund, G.; Sboarina, A.; Vella, A. The role of basophils as innate immune regulatory cells in allergy and immunotherapy. *Hum. Vaccin. Immunother.* **2018**, *14*, 815–831. [CrossRef]
51. Pae, H.O.; Oh, G.S.; Choi, B.M.; Chae, S.C.; Kim, Y.M.; Chung, K.R.; Chung, H.T. Carbon monoxide produced by heme oxygenase-1 suppresses T cell proliferation via inhibition of IL-2 production. *J. Immunol.* **2004**, *172*, 4744–4751. [CrossRef] [PubMed]
52. Liu, Y.; Li, P.; Lu, J.; Xiong, W.; Oger, J.; Tetzlaff, W.; Cynader, M. Bilirubin possesses powerful immunomodulatory activity and suppresses experimental autoimmune encephalomyelitis. *J. Immunol.* **2008**, *181*, 1887–1897. [CrossRef] [PubMed]

53. Grover, P.; Goel, P.N.; Greene, M.I. Regulatory T Cells: Regulation of Identity and Function. *Front. Immunol.* **2021**, *12*, 750542. [CrossRef] [PubMed]
54. Khan, M.A. Regulatory T cells mediated immunomodulation during asthma: A therapeutic standpoint. *J. Transl. Med.* **2020**, *18*, 456. [CrossRef] [PubMed]
55. Lan, F.; Zhang, N.; Bachert, C.; Zhang, L. Stability of regulatory T cells in T helper 2-biased allergic airway diseases. *Allergy* **2020**, *75*, 1918–1926. [CrossRef] [PubMed]
56. Sakaguchi, S.; Mikami, N.; Wing, J.B.; Tanaka, A.; Ichiyama, K.; Ohkura, N. Regulatory T Cells and Human Disease. *Annu. Rev. Immunol.* **2020**, *38*, 541–566. [CrossRef]
57. Choi, B.M.; Pae, H.O.; Jeong, Y.R.; Kim, Y.M.; Chung, H.T. Critical role of heme oxygenase-1 in Foxp3-mediated immune suppression. *Biochem. Biophys. Res. Commun.* **2005**, *327*, 1066–1071. [CrossRef]
58. Hudey, S.N.; Ledford, D.K.; Cardet, J.C. Mechanisms of non-type 2 asthma. *Curr. Opin. Immunol.* **2020**, *66*, 123–128. [CrossRef]
59. Sze, E.; Bhalla, A.; Nair, P. Mechanisms and therapeutic strategies for non-T2 asthma. *Allergy* **2020**, *75*, 311–325. [CrossRef]
60. Zhang, L.; Zhang, Y.; Zhong, W.; Di, C.; Lin, X.; Xia, Z. Heme oxygenase-1 ameliorates dextran sulfate sodium-induced acute murine colitis by regulating Th17/Treg cell balance. *J. Biol. Chem.* **2014**, *289*, 26847–26858. [CrossRef]
61. Da Silva, E.Z.; Jamur, M.C.; Oliver, C. Mast cell function: A new vision of an old cell. *J. Histochem. Cytochem.* **2014**, *62*, 698–738. [CrossRef] [PubMed]
62. Jimenez, M.; Cervantes-Garcia, D.; Cordova-Davalos, L.E.; Perez-Rodriguez, M.J.; Gonzalez-Espinosa, C.; Salinas, E. Responses of Mast Cells to Pathogens: Beneficial and Detrimental Roles. *Front. Immunol.* **2021**, *12*, 685865. [CrossRef] [PubMed]
63. Skokos, D.; Le Panse, S.; Villa, I.; Rousselle, J.C.; Peronet, R.; David, B.; Namane, A.; Mecheri, S. Mast cell-dependent B and T lymphocyte activation is mediated by the secretion of immunologically active exosomes. *J. Immunol.* **2001**, *166*, 868–876. [CrossRef] [PubMed]
64. Xu, Y.; Chen, G. Mast cell and autoimmune diseases. *Mediat. Inflamm.* **2015**, *2015*, 246126. [CrossRef]
65. Mendez-Enriquez, E.; Hallgren, J. Mast Cells and Their Progenitors in Allergic Asthma. *Front. Immunol.* **2019**, *10*, 821. [CrossRef]
66. Galli, S.J.; Tsai, M. IgE and mast cells in allergic disease. *Nat. Med.* **2012**, *18*, 693–704. [CrossRef]
67. Tsai, M.; Grimbaldeston, M.; Galli, S.J. Mast cells and immunoregulation/immunomodulation. *Adv. Exp. Med. Biol.* **2011**, *716*, 186–211. [CrossRef]
68. Yasui, Y.; Sasao, E.; Sakata, M.; Matsui, N.; Fukuishi, N.; Akagi, R.; Akagi, M. Upregulation of heme oxygenase-1 by degranulation in rat basophilic leukemia cells. *Biol. Pharm. Bull.* **2007**, *30*, 443–446. [CrossRef]
69. Yasui, Y.; Nakamura, M.; Onda, T.; Uehara, T.; Murata, S.; Matsui, N.; Fukuishi, N.; Akagi, R.; Suematsu, M.; Akagi, M. Heme oxygenase-1 inhibits cytokine production by activated mast cells. *Biochem. Biophys. Res. Commun.* **2007**, *354*, 485–490. [CrossRef]
70. Ma, Y.Y.; Yang, M.Q.; Wang, C.F.; Ding, J.; Li, J.Y. Inhibiting mast cell degranulation by HO-1 affects dendritic cell maturation in vitro. *Inflamm. Res.* **2014**, *63*, 527–537. [CrossRef]
71. Lee, T.H.; Song, H.J.; Park, C.S. Role of inflammasome activation in development and exacerbation of asthma. *Asia Pac. Allergy* **2014**, *4*, 187–196. [CrossRef] [PubMed]
72. Pinkerton, J.W.; Kim, R.Y.; Robertson, A.A.B.; Hirota, J.A.; Wood, L.G.; Knight, D.A.; Cooper, M.A.; O'Neill, L.A.J.; Horvat, J.C.; Hansbro, P.M. Inflammasomes in the lung. *Mol. Immunol.* **2017**, *86*, 44–55. [CrossRef] [PubMed]
73. Im, H.; Ammit, A.J. The NLRP3 inflammasome: Role in airway inflammation. *Clin. Exp. Allergy* **2014**, *44*, 160–172. [CrossRef]
74. Li, H.; Zhou, X.; Zhang, J. Induction of heme oxygenase-1 attenuates lipopolysaccharide-induced inflammasome activation in human gingival epithelial cells. *Int. J. Mol. Med.* **2014**, *34*, 1039–1044. [CrossRef] [PubMed]
75. Luo, Y.P.; Jiang, L.; Kang, K.; Fei, D.S.; Meng, X.L.; Nan, C.C.; Pan, S.H.; Zhao, M.R.; Zhao, M.Y. Hemin inhibits NLRP3 inflammasome activation in sepsis-induced acute lung injury, involving heme oxygenase-1. *Int. Immunopharmacol.* **2014**, *20*, 24–32. [CrossRef] [PubMed]
76. Kim, S.; Joe, Y.; Jeong, S.O.; Zheng, M.; Back, S.H.; Park, S.W.; Ryter, S.W.; Chung, H.T. Endoplasmic reticulum stress is sufficient for the induction of IL-1beta production via activation of the NF-kappaB and inflammasome pathways. *Innate Immun.* **2014**, *20*, 799–815. [CrossRef]
77. Jung, S.S.; Moon, J.S.; Xu, J.F.; Ifedigbo, E.; Ryter, S.W.; Choi, A.M.; Nakahira, K. Carbon monoxide negatively regulates NLRP3 inflammasome activation in macrophages. *Am. J. Physiol. Lung Cell Mol. Physiol.* **2015**, *308*, L1058–L1067. [CrossRef]
78. Jiang, L.; Fei, D.; Gong, R.; Yang, W.; Yu, W.; Pan, S.; Zhao, M.; Zhao, M. CORM-2 inhibits TXNIP/$NLRP_3$ inflammasome pathway in LPS-induced acute lung injury. *Inflamm. Res.* **2016**, *65*, 905–915. [CrossRef]
79. Kim, S.K.; Joe, Y.; Chen, Y.; Ryu, J.; Lee, J.H.; Cho, G.J.; Ryter, S.W.; Chung, H.T. Carbon monoxide decreases interleukin-1beta levels in the lung through the induction of pyrin. *Cell Mol. Immunol.* **2017**, *14*, 349–359. [CrossRef]
80. Lv, J.; Su, W.; Yu, Q.; Zhang, M.; Di, C.; Lin, X.; Wu, M.; Xia, Z. Heme oxygenase-1 protects airway epithelium against apoptosis by targeting the proinflammatory NLRP3-RXR axis in asthma. *J. Biol. Chem.* **2018**, *293*, 18454–18465. [CrossRef]
81. Alobaidi, A.H.; Alsamarai, A.M.; Alsamarai, M.A. Inflammation in Asthma Pathogenesis: Role of T Cells, Macrophages, Epithelial Cells and Type 2 Inflammation. *Antiinflamm. Antiallergy Agents Med. Chem.* **2021**, *20*, 317–332. [CrossRef] [PubMed]
82. Saradna, A.; Do, D.C.; Kumar, S.; Fu, Q.L.; Gao, P. Macrophage polarization and allergic asthma. *Transl. Res.* **2018**, *191*, 1–14. [CrossRef] [PubMed]
83. Moreira, A.P.; Hogaboam, C.M. Macrophages in allergic asthma: Fine-tuning their pro- and anti-inflammatory actions for disease resolution. *J. Interferon Cytokine Res.* **2011**, *31*, 485–491. [CrossRef] [PubMed]

84. Sharma, N.; Akkoyunlu, M.; Rabin, R.L. Macrophages-common culprit in obesity and asthma. *Allergy* **2018**, *73*, 1196–1205. [CrossRef]
85. Naito, Y.; Takagi, T.; Higashimura, Y. Heme oxygenase-1 and anti-inflammatory M2 macrophages. *Arch. Biochem. Biophys.* **2014**, *564*, 83–88. [CrossRef]
86. Gwak, S.Y.; Kim, S.J.; Park, J.; Kim, S.H.; Joe, Y.; Lee, H.N.; Kim, W.; Muna, I.A.; Na, H.K.; Chung, H.T.; et al. Potential Role of Heme Oxygenase-1 in the Resolution of Experimentally Induced Colitis through Regulation of Macrophage Polarization. *Gut Liver* **2021**, *16*, 1–13. [CrossRef]
87. Vijayan, V.; Wagener, F.; Immenschuh, S. The macrophage heme-heme oxygenase-1 system and its role in inflammation. *Biochem. Pharmacol.* **2018**, *153*, 159–167. [CrossRef]
88. Zhang, M.; Nakamura, K.; Kageyama, S.; Lawal, A.O.; Gong, K.W.; Bhetraratana, M.; Fujii, T.; Sulaiman, D.; Hirao, H.; Bolisetty, S.; et al. Myeloid HO-1 modulates macrophage polarization and protects against ischemia-reperfusion injury. *JCI Insight* **2018**, *3*, e120596. [CrossRef]
89. Dunn, L.L.; Midwinter, R.G.; Ni, J.; Hamid, H.A.; Parish, C.R.; Stocker, R. New insights into intracellular locations and functions of heme oxygenase-1. *Antioxid. Redox Signal.* **2014**, *20*, 1723–1742. [CrossRef]
90. Tenhunen, R.; Marver, H.S.; Schmid, R. The enzymatic conversion of heme to bilirubin by microsomal heme oxygenase. *Proc. Natl. Acad. Sci. USA* **1968**, *61*, 748–755. [CrossRef]
91. Linnenbaum, M.; Busker, M.; Kraehling, J.R.; Behrends, S. Heme oxygenase isoforms differ in their subcellular trafficking during hypoxia and are differentially modulated by cytochrome P450 reductase. *PLoS ONE* **2012**, *7*, e35483. [CrossRef] [PubMed]
92. Wegiel, B.; Nemeth, Z.; Correa-Costa, M.; Bulmer, A.C.; Otterbein, L.E. Heme oxygenase-1: A metabolic nike. *Antioxid. Redox Signal.* **2014**, *20*, 1709–1722. [CrossRef]
93. Kim, H.P.; Wang, X.; Galbiati, F.; Ryter, S.W.; Choi, A.M. Caveolae compartmentalization of heme oxygenase-1 in endothelial cells. *FASEB J.* **2004**, *18*, 1080–1089. [CrossRef] [PubMed]
94. Wu, J.; Li, S.; Li, C.; Cui, L.; Ma, J.; Hui, Y. The non-canonical effects of heme oxygenase-1, a classical fighter against oxidative stress. *Redox Biol.* **2021**, *47*, 102170. [CrossRef] [PubMed]
95. Mascaro, M.; Alonso, E.N.; Alonso, E.G.; Lacunza, E.; Curino, A.C.; Facchinetti, M.M. Nuclear Localization of Heme Oxygenase-1 in Pathophysiological Conditions: Does It Explain the Dual Role in Cancer? *Antioxidants* **2021**, *10*, 87. [CrossRef] [PubMed]
96. Dennery, P.A. Signaling function of heme oxygenase proteins. *Antioxid. Redox Signal.* **2014**, *20*, 1743–1753. [CrossRef]
97. Elguero, B.; Gueron, G.; Giudice, J.; Toscani, M.A.; De Luca, P.; Zalazar, F.; Coluccio-Leskow, F.; Meiss, R.; Navone, N.; De Siervi, A.; et al. Unveiling the association of STAT3 and HO-1 in prostate cancer: Role beyond heme degradation. *Neoplasia* **2012**, *14*, 1043–1056. [CrossRef] [PubMed]
98. Lin, X.; Lv, J.; Ge, D.; Bai, H.; Yang, Y.; Wu, J. Heme oxygenase-1 alleviates eosinophilic inflammation by inhibiting STAT3-SOCS3 signaling. *Pediatr. Pulmonol.* **2020**, *55*, 1440–1447. [CrossRef] [PubMed]
99. Zhang, J.; Yao, X.; Yu, R.; Bai, J.; Sun, Y.; Huang, M.; Adcock, I.M.; Barnes, P.J. Exhaled carbon monoxide in asthmatics: A meta-analysis. *Respir. Res.* **2010**, *11*, 50. [CrossRef]
100. Ueno, T.; Kataoka, M.; Hirano, A.; Iio, K.; Tanimoto, Y.; Kanehiro, A.; Okada, C.; Soda, R.; Takahashi, K.; Tanimoto, M. Inflammatory markers in exhaled breath condensate from patients with asthma. *Respirology* **2008**, *13*, 654–663. [CrossRef]
101. Jesenak, M.; Banovcin, P.; Havlicekova, Z.; Dobrota, D.; Babusikova, E. Factors influencing the levels of exhaled carbon monoxide in asthmatic children. *J. Asthma* **2014**, *51*, 900–906. [CrossRef] [PubMed]
102. Salerno, L.; Floresta, G.; Ciaffaglione, V.; Gentile, D.; Margani, F.; Turnaturi, R.; Rescifina, A.; Pittala, V. Progress in the development of selective heme oxygenase-1 inhibitors and their potential therapeutic application. *Eur. J. Med. Chem.* **2019**, *167*, 439–453. [CrossRef] [PubMed]
103. Goebel, U.; Wollborn, J. Carbon monoxide in intensive care medicine-time to start the therapeutic application?! *Intensive Care Med. Exp.* **2020**, *8*, 2. [CrossRef] [PubMed]
104. Ling, K.; Men, F.; Wang, W.C.; Zhou, Y.Q.; Zhang, H.W.; Ye, D.W. Carbon Monoxide and Its Controlled Release: Therapeutic Application, Detection, and Development of Carbon Monoxide Releasing Molecules (CORMs). *J. Med. Chem.* **2018**, *61*, 2611–2635. [CrossRef] [PubMed]

antioxidants

MDPI

Article

Exploiting Interdata Relationships in Prostate Cancer Proteomes: Clinical Significance of HO-1 Interactors

Sofia Lage-Vickers [1,2], Pablo Sanchis [1,2], Juan Bizzotto [1,2], Ayelen Toro [1,2], Agustina Sabater [1,2], Rosario Lavignolle [1,2], Nicolas Anselmino [3], Estefania Labanca [3], Alejandra Paez [1,2], Nora Navone [3], Maria P. Valacco [1,2], Javier Cotignola [1,2], Elba Vazquez [1,2] and Geraldine Gueron [1,2,*]

1 Laboratorio de Inflamación y Cáncer, Departamento de Química Biológica, Facultad de Ciencias Exactas y Naturales, Universidad de Buenos Aires, Buenos Aires C1428EGA, Argentina; sofilage@qb.fcen.uba.ar (S.L.-V.); pablosanchis@qb.fcen.uba.ar (P.S.); jbizzotto@qb.fcen.uba.ar (J.B.); ayelentoro@qb.fcen.uba.ar (A.T.); asabater@qb.fcen.uba.ar (A.S.); rlavignolle@qb.fcen.uba.ar (R.L.); alejandravpaez@qb.fcen.uba.ar (A.P.); pvalacco@qb.fcen.uba.ar (M.P.V.); jcotignola@qb.fcen.uba.ar (J.C.); elba@qb.fcen.uba.ar (E.V.)

2 CONICET—Universidad de Buenos Aires, Instituto de Química Biológica de la Facultad de Ciencias Exactas y Naturales (IQUIBICEN), Buenos Aires C1428EGA, Argentina

3 Department of Genitourinary Medical Oncology and the David H. Koch Center for Applied Research of Genitourinary Cancers, The University of Texas MD Anderson Cancer Center, Houston, TX 77030, USA; nanselmino@mdanderson.org (N.A.); elabanca@mdanderson.org (E.L.); nnavone@mdanderson.org (N.N.)

* Correspondence: ggueron@iquibicen.fcen.uba.ar; Tel.: +54-9114-408-7796; Fax: +54-9114-788-5755

Citation: Lage-Vickers, S.; Sanchis, P.; Bizzotto, J.; Toro, A.; Sabater, A.; Lavignolle, R.; Anselmino, N.; Labanca, E.; Paez, A.; Navone, N.; et al. Exploiting Interdata Relationships in Prostate Cancer Proteomes: Clinical Significance of HO-1 Interactors. *Antioxidants* **2022**, *11*, 290. https://doi.org/10.3390/antiox11020290

Academic Editors: Elias Lianos, Maria G. Detsika and Alessandra Napolitano

Received: 6 December 2021
Accepted: 27 January 2022
Published: 31 January 2022

Abstract: Prostate cancer (PCa) cells display abnormal expression of proteins resulting in an augmented capacity to resist chemotherapy and colonize distant organs. We have previously shown the anti-tumoral role of heme oxygenase 1 (HO-1) in this disease. In this work, we undertook a mass spectrometry-based proteomics study to identify HO-1 molecular interactors that might collaborate with its modulatory function in PCa. Among the HO-1 interactors, we identified proteins with nuclear localization. Correlation analyses, using the PCa GSE70770 dataset, showed a significant and positive correlation between *HMOX1* and 6 of those genes. Alternatively, *HMOX1* and *YWHAZ* showed a negative correlation. Univariable analyses evidenced that high expression of *HNRNPA2B1, HSPB1, NPM1, DDB1, HMGA1, ZC3HAV1,* and *HMOX1* was associated with increased relapse-free survival (RFS) in PCa patients. Further, PCa patients with high *HSPB1/HMOX1, DDB1/HMOX1,* and *YWHAZ/HMOX1* showed a worse RFS compared with patients with lower ratios. Moreover, a decrease in RFS for patients with higher scores of this signature was observed using a prognostic risk score model. However, the only factor significantly associated with a higher risk of relapse was high *YWHAZ*. Multivariable analyses confirmed *HSPB1, DDB1,* and *YWHAZ* independence from PCa clinic-pathological parameters. In parallel, co-immunoprecipitation analysis in PCa cells ascertained HO-1/14-3-3ζ/δ (protein encoded by *YWHAZ*) interaction. Herein, we describe a novel protein interaction between HO-1 and 14-3-3ζ/δ in PCa and highlight these factors as potential therapeutic targets.

Keywords: *YWHAZ; HMOX1*; prostate cancer; proteomics; transcriptomics; protein interactions

1. Introduction

Prostate cancer (PCa) is the second most common cancer type in men and the sixth leading cause of cancer-related death in men worldwide [1]. The discovery of new therapeutic avenues in PCa, and the development of effective drugs in the era of personalized medicine, would greatly benefit from the field of proteomics. The proteomics approach for a high-throughput study of biological samples by mass spectrometry has emerged as one of the main analytical strategies from the last years, and proteomic-based studies have greatly improved cancer research. Thus, proteomics represents an important tool for the identification of new molecular targets for PCa's tailored therapy.

Inflammation is widely recognized as a hallmark of cancer [2]. Cell proliferation is enhanced in an inflammatory microenvironment rich in cytokines, growth factors, and agents that cause DNA damage [3]. This combination of factors makes the risk of developing a tumor much higher [3]. Further, inflammatory cells release reactive oxygen species (ROS), which generate oxidative stress and damage the DNA of neighboring epithelial cells, thus accelerating the evolution towards a more malignant phenotype [4]. Elevated intracellular ROS levels might affect several signaling pathways, resulting in the activation or repression of processes related to cell proliferation, motility, and survival [5,6]. Although the cause of prostatic inflammation is uncertain in most cases, it is believed that viral or bacterial infections, physical trauma from urinary reflux, dietary factors, estrogens, or a combination of these factors, could contribute to the establishment of an inflammatory microenvironment [7].

Heme oxygenase 1 (HO-1), the rate-limiting enzyme that catalyzes heme degradation, is a key player in cellular responses to pro-oxidative and pro-inflammatory insults [8,9]. HO-1 participates in cell homeostasis by attenuating inflammation, reducing oxidative injury, and regulating cell proliferation [10]. Reports from our laboratory documented that HO-1 has a strong anti-tumoral effect in vivo and in vitro in PCa [8,11–16]. Moreover, HO-1 induction in PCa cells impairs cell proliferation, invasion, and migration in vitro, and angiogenesis and tumor growth in vivo [11,12]. HO-1 is recognized as an integral smooth endoplasmic reticulum membrane protein; however, it has been detected in other subcellular compartments, including the nucleus [13,17,18]. It has been suggested that HO-1 undergoes proteolytic degradation at its carboxy-terminal hydrophobic end, which would facilitate its entry into the nucleus [13]. Interestingly, this truncated form of HO-1 does not possess catalytic activity [17]. It has been proposed that HO-1 has a non-canonical function in the nucleus, participating in the regulation of the activity of nuclear transcription factors and even regulating its own expression [19]. In line with this, we have previously documented that HO-1 binds to the proximal promoter of genes involved in PCa, such as the prostate-specific antigen (PSA), and represses androgen receptor (AR) activation revealing an undescribed function for HO-1 in the nucleus [12].

Due to the pleiotropic actions of HO-1, we hypothesized that its multiple functions could be mediated by interactions with several other relevant proteins associated with the carcinogenic process. Through co-immunoprecipitation assays, we previously verified that HO-1 interacts with STAT3, producing its retention in the cytoplasm of PCa cells [13]. Another HO-1 interactor protein in PCa cells identified by our group was Annexin A2 (ANXA2) [20], a key molecule in the adhesion process of PCa cells to the bone microenvironment. We found that HO-1 modulation in tumor cells interferes with ANXA2-mediated signaling [20]. These results clearly suggest that HO-1 is involved in cellular processes beyond the degradation of the heme group. However, further research into the mechanisms associated with HO-1 non-canonical functions is needed. Given that HO-1 does not show DNA binding motifs, it is possible that HO-1 needs to interact with transcription factors to fulfill its regulatory function in the cell nuclei.

To further our analysis, in this work, we undertook a proteomics approach to assess whether in PCa cells and under oxidative stress conditions, HO-1 could interact with proteins previously documented to have nuclear localization. Further, we evaluated the clinical relevance of such a network in PCa patients and performed correlation analyses among HO-1 and its partners, selecting those with higher correlation and building a risk score model. Taking into account all of our results, we report novel interactions between HO-1 and HSPB1, DDB1, and 14-3-3ζ/δ, highlighting their clinical relevance in PCa.

2. Materials and Methods

2.1. Cell Culture, Treatments, Reagents, and Antibodies

PC3 cells were obtained from the American Type Culture Collection (Manassas, VA, USA) and were routinely cultured in RPMI 1640 (Invitrogen, Grand Island, NY, USA) supplemented with 10% fetal bovine serum (FBS, Internegocios, Mercedes, Buenos Aires,

Argentina), penicillin 100 U/mL, streptomycin 100 μg/mL, and amphotericin 0.5 μg/mL. Cells were cultured at 37 °C and 5% CO_2.

For PC3 H_2O_2 treatment, cells were treated with 200 μM for 30 min, prepared in sterile PBS 1X. After treatment, cells were incubated in complete medium for 24 h and were harvested for the different experiments performed.

2.2. *Antibodies*

Monoclonal rabbit anti-human HO-1, anti-human 14-3-3ζ/δ and anti-human IgG antibodies were obtained from Cell Signaling (Danvers, MA, USA). Monoclonal mouse anti-human HO-1 antibody was obtained from Abcam (Cambridge, UK). Horseradish peroxidase (HRP) conjugated secondary anti-mouse antibody was obtained from Cell Signaling (Danvers, MA, USA). Secondary antibodies associated with the Alexa 555 and Alexa 647 fluorophores were obtained from Molecular Probes, Invitrogen (Carlsbad, CA, USA).

2.3. *PEBG-GST-HO-1 Cloning*

The vector pEBG-GST-HO-1 was generated by cloning the copy DNA sequence (cDNA) encoding the human HO-1 gene (*HMOX1*) into the restriction sites *BamHI* and *NotI* of the mammalian expression vector pEBG-GST (Addgene, Watertown, MA, USA). This strategy results in the fusion of the GST peptide at the N-terminus of HO-1. The sequences of primers used were: forward—5′-GCCGGATCCATGGAGCGTCCGCAAC-3′; reverse—5′-GCCGCGGCCGCCATTCACATGGCATAAAGC-3′.

2.4. *Transfection with PEBG-GST and PEBG-GST-HO1*

PC3 cells were transiently transfected with the HO-1 expression plasmid (pEBG-GST-HO-1) or the empty vector as control (pEBG-GST). Each 10 cm diameter plate was transfected using 10 μg of plasmid and 20 μL of polyethylene glycol (PEI) (Sigma-Aldrich, Gillingham, UK) in a final volume of 200 μL of RPMI 1640 culture medium. Transfections were performed on plates with 3 mL of RPMI without FBS or antibiotics. After 5 h of transfection, the culture medium was replaced by complete RPMI with 10% v/v FBS and antibiotics at the previously mentioned concentrations. For the proteomics tests, 40 plates of 10 cm in diameter of each experimental condition (GST-HO-1 vs. GST) were transfected and incubated for 48 h to carry out the different experiments.

2.5. *GST Immunoprecipitation Strategy*

48 h after transfection with pEBG-GST-HO-1 or the empty vector pEBG-GST and subsequent treatment with H_2O_2, proteins were extracted using a low concentration NaCl buffer (20 mM Tris, 150 mM NaCl, 5 mM MgCl2, 0.5% NP40, pH 7.5) to avoid disruption of protein-protein interactions. Protein extracts were incubated for 2 h at 4 °C with beads coated with glutathione-S-agarose.

2.6. *Separation of Peptides and Mass Spectrometry Analysis*

Recombinant GST-HO-1 protein complexes were reduced (200 mM DTT), alkylated (200 mM iodoacetamide), and digested with trypsin in-solution overnight, using an estimated 1:30 enzyme to substrate ratio. The peptides were desalted and concentrated in a C18 resin (Zip-Tips, Waters Technologies Corporation, Milford, MA, USA) before analysis by LC ESI-MS/MS at the Center for Metabolomics and Mass Spectrometry (The Scripps Research Institute, La Jolla, CA, USA). Peptides were separated by reverse-phase chromatography before mass spectrometry analysis using the following method: nanoelectrospray capillary column tips were made in-house by using a P-100 laser puller (Sutter Instruments, Novato, CA, USA). The columns were packed with Zorbax SB-C18 stationary phase (Agilent Technologies, Santa Clara, CA, USA) purchased in bulk (5 mm particles, with a 15 cm length and a 75 mm inner diameter). The reverse-phase gradient separation was performed by using water and acetonitrile (0.1% formic acid) as the mobile phases. The gradient consisted of 5% acetonitrile for 10 min followed by a gradient to 8% acetonitrile for 5 min,

35% acetonitrile for 113 min, 55% acetonitrile for 12 min, 95% acetonitrile for 15 min, and re-equilibrated with 5% acetonitrile for 15 min.

Data-dependent MS/MS data were obtained with an LTQ linear ion trap mass spectrometer using a home-built nanoelectrospray source at 2 kV at the tip. One MS spectrum was followed by 4 MS/MS scans on the most abundant ions after the application of a dynamic exclusion list. Tandem mass spectra were extracted using the Xcalibur software (Thermo Scientific, Waltham, MA, USA). All MS/MS samples were analyzed by using Mascot (version 2.1.04; Matrix Science, London, UK) with *H. Sapiens* proteins contained in the NCBI protein database (NCBInr2 20080628 (6655203 sequences; 2281585098 residues). Taxonomy: *Homo sapiens* (human) (202147 sequences), assuming the digestion enzyme trypsin with a maximum of 1 miscleavage. Mascot was searched with a fragment ion mass tolerance of 0.80 Da and a parent ion tolerance of 2.0 Da, and fixed modifications: carbamidomethyl (C) and variable modifications: oxidation (M). Identification was carried out at the 95% confidence level with a calculated false-positive rate of <1% as determined by using a reversed concatenated protein database. The minimum score for a nonrandom identification with more than 95% confidence was 45 for the GST control search and 44 for the GST-HO-1 search. At least one unique peptide with MS/MS data was identified for each protein hit. To control for nonspecific binding, we compared GST-HO-1 co-purifying proteins with those immunoprecipitated in cells transfected with a GST empty vector. Only differential GST-HO-1 binding proteins compared with GST-binding proteins were considered further.

2.7. Co-Immunoprecipitation (Co-IP)

After treatment with H_2O_2, total proteins were extracted from PC3 cells and quantified using the BCA method (Sigma-Aldrich, Gillingham, UK) (98% BCA + 2% $CuSO_4$). 500 μg of proteins from the cell lysates and 10 μg of the specific rabbit anti-human HO-1 antibody were diluted in a final volume of 500 μL in RIPA buffer (150 μM NaCl, 20 μM EDTA 1% *v*/*v* sodium deoxycholate, 0.1% *v*/*v* sodium dodecyl sulfate (SDS), 1% Triton x-100Tris, pH 7.4). As a specificity control, samples were incubated with 10 μg of nonspecific rabbit anti-human IgG antibody. All samples were supplemented with MPI protease inhibitors (Sigma-Aldrich, Gillingham, UK), and were incubated overnight at 4 °C with orbital shaking. Protein A/G PLUS-Agarose beads (Santa Cruz Biotechnology Inc., Santa Cruz, CA, USA) were washed; 20 μL were added to each sample and incubated overnight at 4 °C with orbital shaking. Antibody-protein-beads complexes were washed with 500 μL of RIPA buffer and re-suspended in 40 μL of RIPA 20% loading buffer (5% B-Mercaptoethanol). Samples were heated for 5 min at 95 °C. 60 μg of the total protein lysate was used as input. The immune complexes were analyzed by Western blot as previously described [21], using mouse anti-human HO-1 or rabbit anti-human 14-3-3ζ/δ antibodies. Three independent experiments were performed.

2.8. Immunofluorescence (IF) Experiment

60,000 PC3 cells were cultured on coverslips placed in 12-well plates and the respective treatments were carried out. Afterwards, cells were washed three times with PBS and then fixed with 1 mL of 100% methanol for 20 min at −20 °C. After removing the methanol, cells were incubated at room temperature in a permeabilizing solution of 1 mL PBS-triton 0.1% *v*/*v* for 10 min. Subsequently, coverslips were placed in a humid and dark chamber and were incubated with a blocking solution (40 μL of PBS-BSA 3% m/v for 1 h). Coverslips were then incubated overnight at 4 °C with 40 μL of the corresponding primary antibodies (mouse anti-human HO-1 and/or rabbit anti-human 14-3-3ζ/δ antibodies), prepared in PBS-BSA 3% m/v. After performing three washes of 10 min with PBS, the coverslips were incubated for 1 h at room temperature with the Alexa Fluor 555 and/or Alexa Fluor 647 secondary antibodies in a 1:5000 dilution prepared in PBS-BSA 3% m/v.

After three washes of 10 min with PBS, the coverslips were mounted on a slide with 2 μL of Mowiol (Sigma-Aldrich, Gillingham, UK) and stored at 4 °C in the dark until use.

Confocal images were acquired by confocal microscopy (FV1000, Olympus, Tokyo, Japan), using a UPlanSApo 100X oil immersion objective (NA 1/41.35; Olympus), a diode laser of 543 and 635 nm as the excitation sources. Images were obtained with a Qimaging EXI Aqua camera; >20 cells were analyzed for the IF assay for each condition.

2.9. Image Processing for Presentation

Confocal images were processed for presentation using ImageJ software (NIH, Bethesda, MD, USA). The background of each channel was subtracted. The JACoP plugin was used in order to evaluate co-localization, and Manders and Pearson's coefficients were estimated. GraphPad Prism software (La Jolla, CA, USA) was used to assess statistical significance, which was set at $p < 0.05$.

2.10. Bioinformatics Analysis

2.10.1. Identification of HO-1 Interactor Proteins with Nuclear Localization and GO Enrichment Analysis

Protein–protein interactions and gene ontology (GO) analyses were performed using the STRING webtool [22], using a minimum combined interaction score of 0.400. Plots were generated using Cytoscape [23] and enrichplot package [24] in R. Statistical significance was set at false discovery rate (FDR) $p < 0.05$.

2.10.2. Information Source and Eligibility Criteria (GEO: Gene Expression Omnibus)

To study the impact of the expression of the selected genes on the survival of patients, we selected the following dataset:

Ross-Adams 2015 (GSE70770) GPL10558 series [25]: a PCa patient's cohort with 206 tumor tissue samples from men with PCa who had undergone radical prostatectomy and a clinical follow-up of 9 years, including biochemical relapse (BCR) information, defined according to European Guidelines as a persistent rise of PSA above 0.2 ng/mL. Tumor sample expression of 31,000 transcripts was measured by 47,000 probes using the Illumina Human HT-12 V4.0 platform. A descriptive table regarding patient characteristics at baseline (start of the follow-up for survival analyses) is depicted in Supplementary Table S1.

2.10.3. Gene Correlation

Pairwise gene correlation between *HMOX1* and all genes encoding for the HO-1 interactor proteins of interest was analyzed with the ggcorrplot package in R. Correlation coefficients were classified as weak ($|r| \leq 0.30$), intermediate ($0.30 < |r| < 0.66$), and strong ($|r| \geq 0.66$). Statistical significance was set at $p < 0.05$.

2.10.4. Risk Scoring System Analysis

Based on the expression of genes with a strong/intermediate gene correlation with *HMOX1*, a risk score model was created based on the coefficients of a Cox logistic regression analysis. Then, the sum of the product of the coefficient (Coef) values of all genes and their dichotomic expression (Expr) values was calculated as the patient risk score (risk score = $\sum_{i=1}^{n}(Coef_i \times Expr_i)$). Using Cutoff Finder software [26], patients were divided into high-risk and low-risk groups. Kaplan–Meier survival analysis was used to determine whether relapse-free survival (RFS) was significantly different between high-risk and low-risk patients.

2.10.5. Survival Analyses

Kaplan–Meier curves showing the RFS of patients with PCa were plotted using the survminer package [27] in R. To find the optimal cutoff value to stratify patients into two groups based on the expression levels of each gene, we used the Cutoff Finder tool. Multivariable analysis was performed in Stata (StataCorp LLC, College Station, TX, USA) and plotted in GraphPad Prism software (La Jolla, CA, USA). For univariable and multivariable analyses of prognostic factors, the Log-rank test and Cox proportional hazard model regression were employed. Statistical significance was set at $p < 0.05$.

3. Results

3.1. Proteomics Profile of HO-1 Interactors in PCa Cells

We have previously identified HO-1 molecular partners involved in cell–cell communication and cell adhesion through an integrative "omics" approach [28], establishing four molecular pathways (ANXA2/HMGA1/POU3F1; PLAT/PLAU; TMOD3/RAI14/VWF; NFRSF13/GSN). Further, we also described ANXA2, a protein associated with both bone physiology and in PCa bone progression [29,30], as an HO-1 interactor [20].

In light of the strong anti-tumoral role of HO-1 in PCa, we deepened our analyses in the search for other HO-1 interactors with clinical relevance for the disease. We carried out a proteomics analysis, in which PC3 cells, derived from a bone metastasis of PCa [31], were transfected with the HO-1 expression plasmid (pEBG-GST-HO-1) or with the empty vector as control (pEBG-GST) (Figure 1A). After 48 h, cells were treated with 200 μM H_2O_2 for 30 min, with the aim of generating an inflammatory and oxidative microenvironment similar to the one which characterizes PCa [7]. Subsequently, GST-HO-1 was immunoprecipitated together with its interactors. The eluates were digested for analysis by LC ESI-MS/MS (Figure 1A). We identified 41 HO-1 interactor proteins, including GSN (Gelsolin), FLNB (filamin B), 14-3-3 family proteins, TES (testin), TRIM28 (tripartite motif-containing 28), and SRSF3 (splicing factor rich in serine and arginine 3), with clinical relevance in PCa [32–37]. Supplementary Table S2 shows the differential proteins identified (GST-HO-1 vs. GST) together with the number of unique peptides, the name of the encoding gene, the score, and the percentage of coverage with respect to the complete sequence. Next, an interaction network was built using STRING [22] and Cytoscape [23], which is shown in Figure 1B. Significant protein hits (Supplementary Table S2) identified with more than 2 PSMs were selected for further analysis (Figure 1B, larger spheres). This filter resulted in the selection of 14 proteins. Among those, 11 proteins had been previously reported with nuclear localization, including tripartite motif-containing 28 (TRIM28), heterogeneous nuclear ribonucleoprotein A2/B1 (HNRNPA2B1), heat shock protein family B (small) member 1 (HSPB1), chromobox 1 (CBX1), chromobox 3 (CBX3), matrin 3 (MATR3), nucleophosmin 1 (NPM1), damage specific DNA binding protein 1 (DDB1), high mobility group AT-hook 1 (HMGA1), 14-3-3 zeta delta-like protein (14-3-3ζ/δ) and zinc finger CCCH-type containing, antiviral 1 (ZC3HAV1) [22] (Figure 1B, pink spheres). The gene ontology (GO) for the top biological processes (BP) categories of the genes encoding for the HO-1 interactor proteins with nuclear localization included response to stress and cellular response to DNA damage stimulus (Figure 1C).

3.2. Gene Correlation between HMOX1 and the Genes Encoding for HO-1 Interactors with Nuclear Localization in PCa Cells

To study the relevance of the HO-1 interactors in PCa, we assessed the clinical significance of these nuclear factors by using the GSE70770 dataset [25], which gathers transcriptomic and clinical data from PCa patients who had undergone radical prostatectomy and clinical follow-up of 9 years, including biochemical relapse. Next, we performed pairwise correlation analyses between *HMOX1* and the genes encoding for the 11 selected nuclear localized HO-1 interactors (Figure 2B). Results show a significant and positive intermediate/strong Spearman correlation between *HMOX1* and 6 of those genes ($r = 0.4$, $p < 0.0001$ for *HNRNPA2B1*; $r = 0.3$, $p < 0.0001$ for *HSPB1*; $r = 0.4$, $p < 0.0001$ for *NPM1*; $r = 0.5$, $p < 0.0001$ for *DDB1*; $r = 0.6$, $p < 0.0001$ for *HMGA1*; and $r = 0.4$, $p < 0.0001$ for *ZC3HAV1*). Interestingly, *HMOX1* only correlated negatively with *YWHAZ* ($r = -0.4$, $p < 0.0001$) (Figure 2A,B).

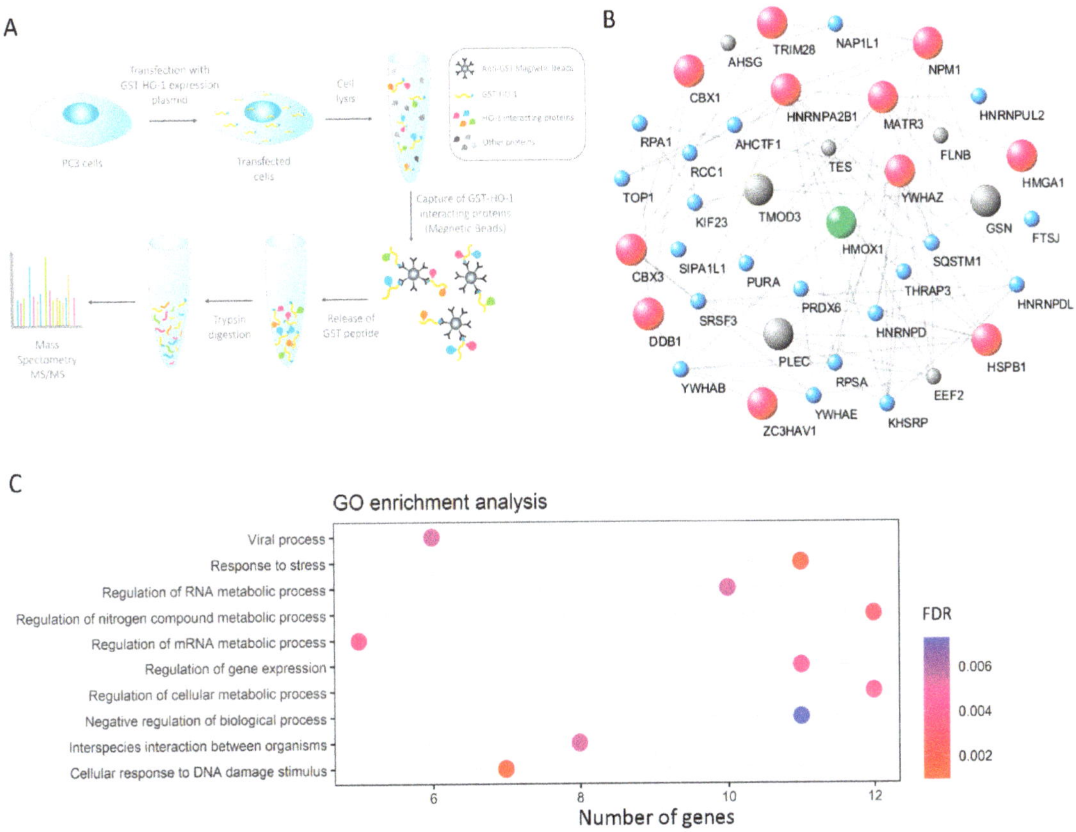

Figure 1. Construction of the HO-1 interactome in PCa cells. (**A**) Simplified schematic workflow of the construction of the HO-1 interactome in PC3 cells. GST-HO-1 immunoprecipitation assays were performed from PC3 cell extracts that had been previously treated with H_2O_2 (200 μM, 1 h). For the LC ESI-MS/MS analysis, peptides were desalted and concentrated on a C18 resin. Tandem mass spectra were extracted using the Xcalibur software. All MS/MS samples were analyzed by using Mascot. (**B**) Interactions network for the GST-HO-1 binding proteins. The network was built with Cytoscape 3.7.0. using the interactions data obtained with STRING, using a minimum combined interaction score of 0.400. The protein size increases proportionally to the PSMs obtained in the LC ESI-MS/MS analysis. Proteins in pink are proteins identified with PSM > 2 and previously reported in the cell nucleus. (**C**) Gene ontology (GO) (biological processes) categories significantly dysregulated for genes encoding for HO-1 interactor proteins with nuclear localization. The color gradient shows the adjusted *p*-value. FDR = false discovery rate.

3.3. Clinical Relevance of HO-1 Interactors with Nuclear Localization in PCa

In order to evaluate the clinical relevance of the HO-1 interactors that have been previously reported in the nucleus, we analyzed the biochemical relapse-free survival (RFS) of PCa patients associated with the expression of the genes encoding for those proteins. The analyzed genes were the ones which showed a significant ($p < 0.05$) and intermediate ($0.30 < |r| < 0.66$) or strong ($|r| \geq 0.66$) mRNA correlation with *HMOX1* (*HNRNPA2B1*, *HSPB1*, *NPM1*, *DDB1*, *HMGA1*, *ZC3HAV1*, and *YWHAZ*). We also included *HMOX1* expression in this analysis. We plotted KM curves for each individual gene assessed (Figure 3). Patients were stratified into two groups based on the expression levels for each gene: high and low expression. The analysis showed that high expression of

HNRNPA2B1, *HSPB1*, *NPM1*, *DDB1*, *HMGA1*, *ZC3HAV1*, and *HMOX1* was associated with an increased RFS in PCa patients (HR = 0.467, Cox $p = 0.003$ for *HNRNPA2B1* (Figure 3A); HR = 0.364, Cox $p = 0.001$ for *HSPB1* (Figure 3B); HR = 0.377, Cox $p < 0.0001$ for *NPM1* (Figure 3C); HR = 0.52, Cox $p = 0.01$ for *DDB1* (Figure 3D); HR = 0.361, Cox $p < 0.0001$ for *HMGA1* (Figure 3E); HR = 0.266, Cox $p < 0.0001$ for *ZC3HAV1* (Figure 3F); and HR = 0.505, Cox $p = 0.018$ for *HMOX1* (Figure 3H)). However, high expression of *YWHAZ* was associated with a lower RFS (HR = 3.942, Cox $p < 0.0001$) (Figure 3G).

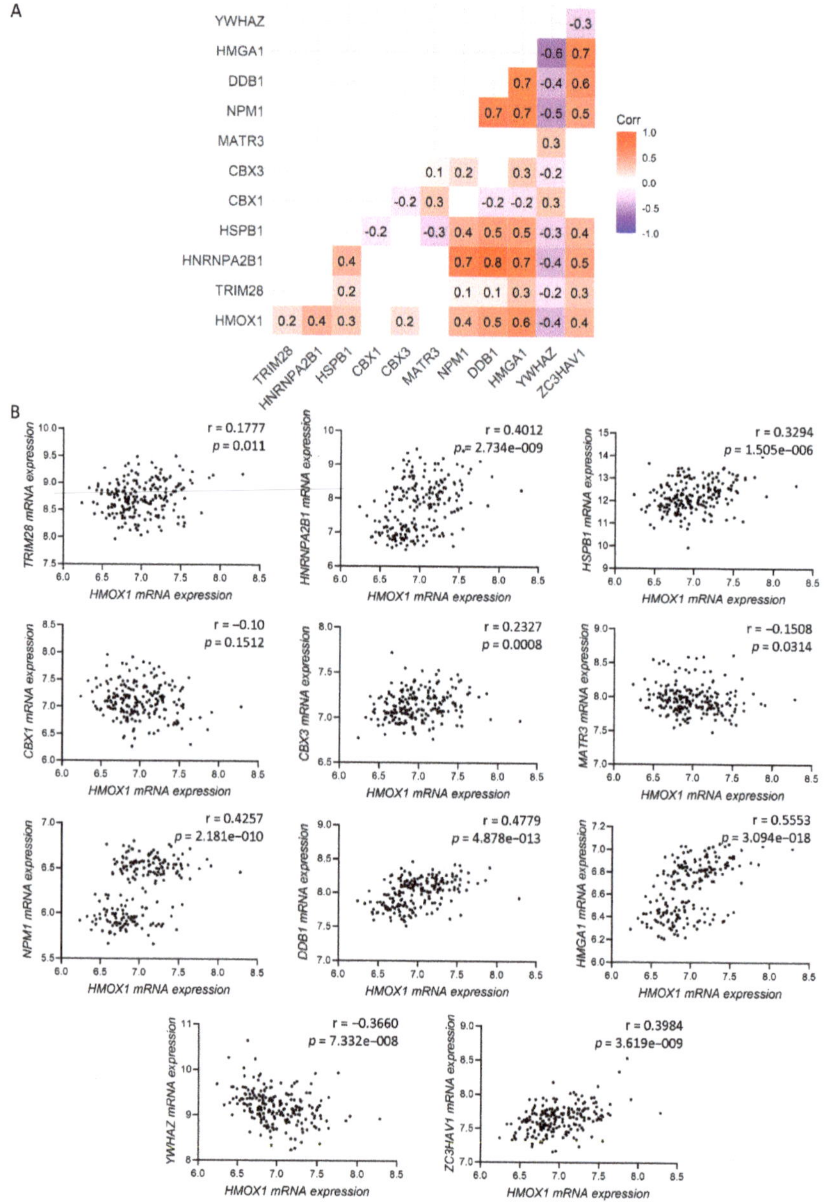

Figure 2. Gene correlation. (**A**) Pairwise Spearman correlation matrix analysis between *HMOX1* and all of the genes encoding for the HO-1 interactor proteins which were previously reported in the cell

nucleus (*TRIM28, HNRNPA2B1, HSPB1, CBX1, CBX3, MATR3, NPM1, DDB1, HMGA1, YWHAZ*, and *ZC3HAV1*). Rounded Spearman correlation values are included inside each color box. Color scale ranges from blue (r = −1) to white (r = 0) to red (r = 1). (**B**) Spearman correlation analysis for *HMOX1* and the genes encoding for the HO-1 interactor proteins, which were previously reported in the cell nucleus (*TRIM28, HNRNPA2B1, HSPB1, CBX1, CBX3, MATR3, NPM1, DDB1, HMGA1, YWHAZ*, and *ZC3HAV1*). Statistical significance was set at $p < 0.05$.

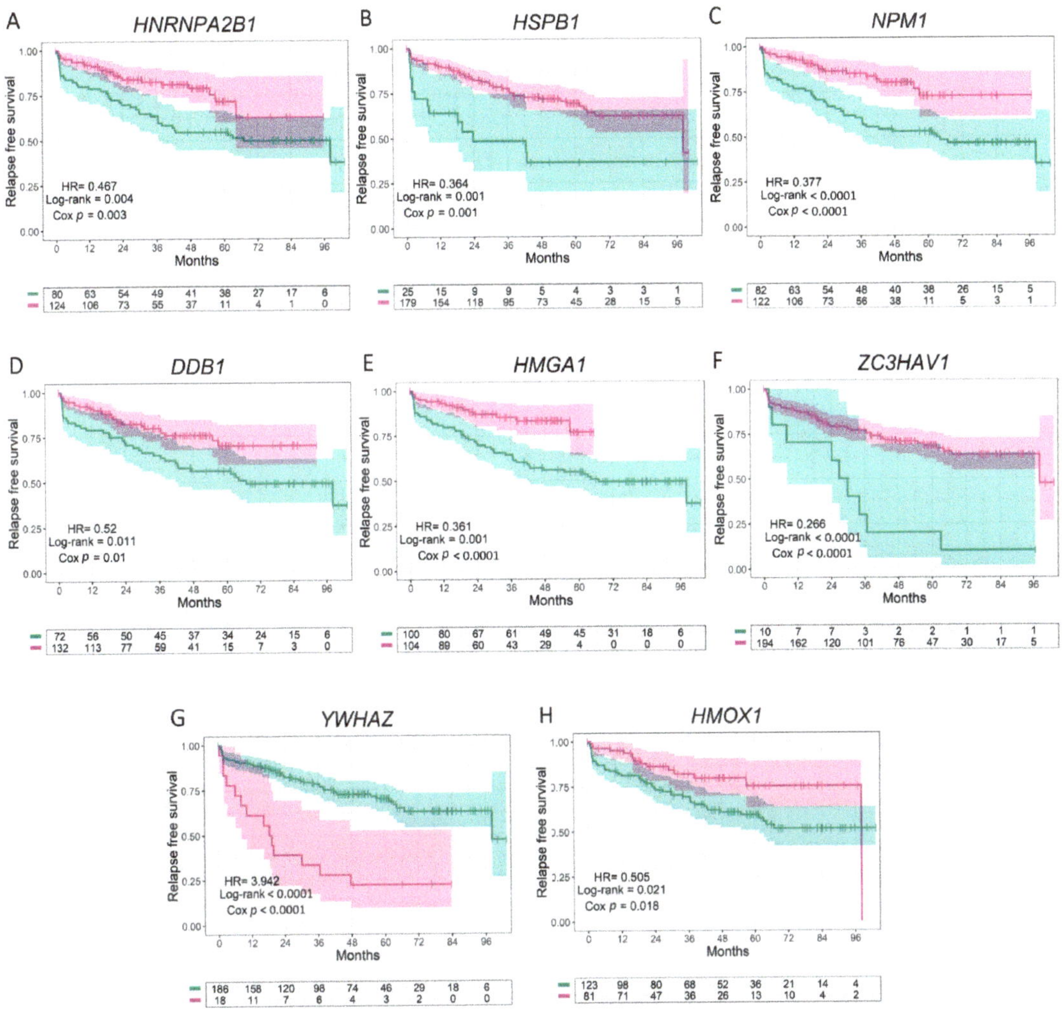

Figure 3. Relapse-free survival (RFS) of PCa patients using the Ross–Adams dataset, GSE70770, n = 206. (**A–H**) Kaplan–Meier curves for RFS of PCa patients segregated based on the gene expression levels for *HNRNPA2B1* (**A**), *HSPB1* (**B**), *NPM1* (**C**), *DDB1* (**D**), *HMGA1* (**E**), *ZC3HAV1* (**F**), *YWHAZ* (**G**), and *HMOX1* (**H**). RFS of patients with high (purple lines) vs. low (green lines) expression for each gene. HR = hazard ratios (95% confidence interval). All comparisons consider low expression patients as the reference group. Cox p = Cox proportional hazard model p-value. Statistical significance was set at Cox $p < 0.05$.

Next, we furthered our analysis by plotting the gene expression ratio between each of these seven genes and *HMOX1* in biochemical relapse (BCR) patients compared with non-BCR. Results show that *HNRNPA2B1/HMOX1*, *NPM1/HMOX1*, and *YWHAZ/HMOX1* were significantly higher in BCR compared with non-BCR patients (p = 0.028 for *HNRNPA2B1/HMOX1* (Figure 4(Ai)); p = 0.018 for *NPM1/HMOX1* (Figure 4(Ci)); p < 0.0001 for *YWHAZ/HMOX1* (Figure 4(Gi))). Next, we plotted the RFS stratifying patients according to their gene expression ratio. PCa patients with higher *HSPB1/HMOX1*, *DDB1/HMOX1*, and *YWHAZ/HMOX1* showed a worse RFS compared with patients with lower ratios (HR = 2.291, Cox p = 0.006 for *HSPB1/HMOX1* (Figure 4(Bii)), HR = 1.888, Cox p = 0.014 for *DDB1/HMOX1* (Figure 4(Dii)), and HR = 3.764, Cox p < 0.0001 for *YWHAZ/HMOX1* (Figure 4(Gii))). Results evidenced that increased *HMOX1* expression in combination with high expressions of *HSPB1*, *DDB1*, and *YWHAZ*, improves RFS for PCa patients.

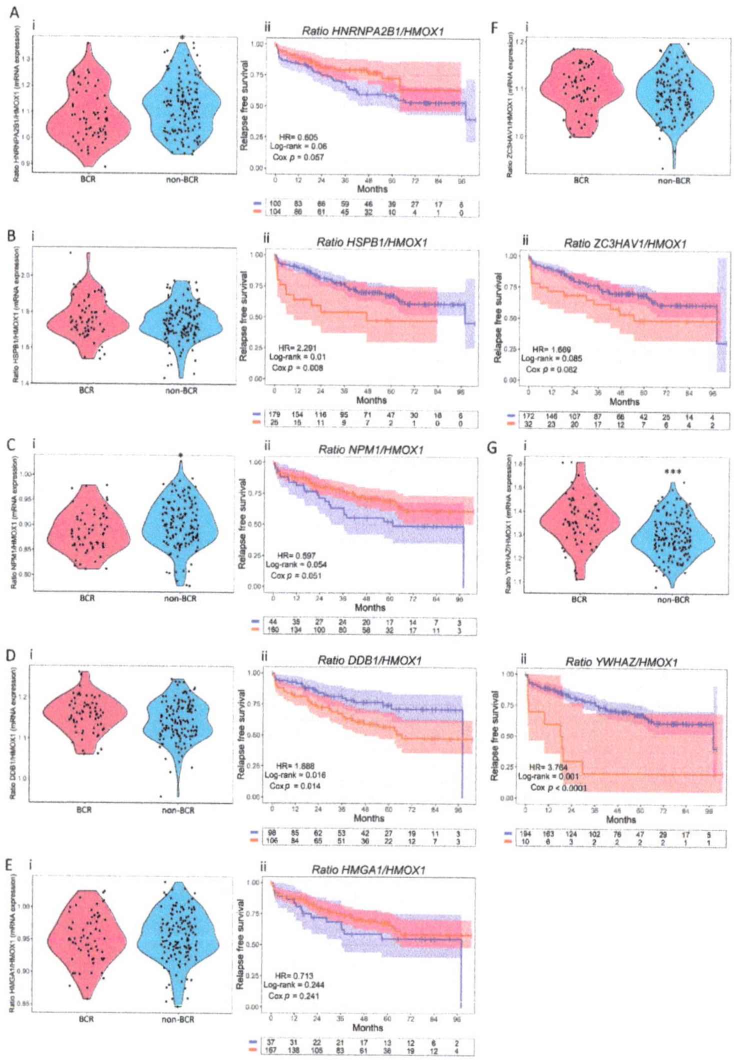

Figure 4. Expression ratios between *HNRNPA2B1*, *HSPB1*, *NPM1*, *DDB1*, *HMGA1*, *ZC3HAV1*, *YWHAZ*, and *HMOX1* and their association with biochemical relapse (BCR) and relapse-free survival

(RFS) using the Ross-Adams dataset, GSE70770, n = 206. **(i)** Violin plots depicting *HNRNPA2B1/HMOX1* (**A**), *HSPB1/HMOX1* (**B**), *NPM1/HMOX1* (**C**), *DDB1/HMOX1* (**D**), *HMGA1/HMOX1* (**E**), *ZC3HAV1/HMOX1* (**F**), and *YWHAZ/HMOX1* (**G**) expressions in BCR vs. non-BCR patients. **(ii)** Kaplan–Meier curves for RFS of PCa patients segregated based on the gene expression *HNRNPA2B1/HMOX1* (**A**), *HSPB1/HMOX1* (**B**), *NPM1/HMOX1* (**C**), *DDB1/HMOX1* (**D**), *HMGA1/HMOX1* (**E**), *ZC3HAV1/HMOX1* (**F**), and *YWHAZ/HMOX1* (**G**) ratios. RFS of patients with high (red lines) vs. low (blue lines) expression for each ratio. HR = hazard ratios [95% confidence interval]. All comparisons consider low expression patients as the reference group. Cox p = Cox proportional hazard model p-value. Statistical significance was set at Cox $p < 0.05$. * $p < 0.05$; *** $p < 0.001$.

We then performed a Cox regression analysis and built a prognostic model for predicting the RFS, based on the expression (Expr) and coefficient (Coef) values of *HMOX1* and each of the three genes whose higher ratios with *HMOX1* showed a significant decrease in RFS compared with patients with lower ratios (*YWHAZ*, *DDB1*, and *HSPB1*) (Table 1). The risk score was calculated as follows: $= \sum_{i=1}^{n} (Coef_i \times Expr_i)$. On the basis of the result for each PCa patient, the GSE70770 dataset was divided into two groups (high-risk group and low-risk group) according to the optimal cutoff value. Interestingly, we observed that patients with higher risk scores had a worse clinical outcome than patients with lower risk scores (HR = 4.807, Cox $p < 0.0001$) (Figure 5A).

Table 1. Detailed information of HO-1 interactors for the risk score model.

Gene	Coeff.	[95% CI]	p-Value
HMOX1	−0.44	−1.04–0.16	0.147
YWHAZ	1.43	0.82–2.04	<0.001
DDB1	−0.48	−1.01–0.05	0.08
HSPB1	−0.86	−1.47––0.25	0.006

To validate *HSPB1*, *DDB1*, *YWHAZ*, and *HMOX1*'s clinical relevance in PCa, multivariable analyses were performed in the presence of clinic-pathological parameters previously associated with increased PCa relapse risk. These parameters included Gleason score (GS), PSA levels, patients' clinical and pathological stages, and *HMOX1* expression. *HSPB1* behaved independently from the patients' GS, PSA levels, clinical and pathological stages, and *HMOX1* expression ($p = 0.03$ for GS; $p = 0.001$ for PSA; $p = 0.002$ for the clinical stage; $p = 0.005$ for the pathological stage; $p = 0.002$ for *HMOX1* (Figure 5B)). *DDB1* behaved independently from the patients' GS, PSA levels, and clinical and pathological stages ($p = 0.035$ for GS; $p = 0.018$ for PSA; $p = 0.014$ for the clinical stage; and $p = 0.001$ for the pathological stage (Figure 5C)). When we further adjusted the model to include all variables simultaneously, the associations remained significant ($p = 0.022$) (Figure 5C). Further, *YWHAZ* behaved independently from the patients' GS, PSA levels, clinical and pathological stages, and *HMOX1* expression (Figure 5D). When analyzing all variables simultaneously, the associations remained significant ($p = 0.01$) (Figure 5D). Figure 5E depicts *HMOX1* independence from all of the clinic-pathological parameters previously analyzed ($p = 0.027$ for GS; $p = 0.019$ for PSA; $p = 0.019$ for the clinical stage; $p = 0.004$ for the pathological stage; $p = 0.023$ for all the variables simultaneously (Figure 5E)). Altogether, the multivariable analyzes add support to the independence of variables to predict the patient outcome.

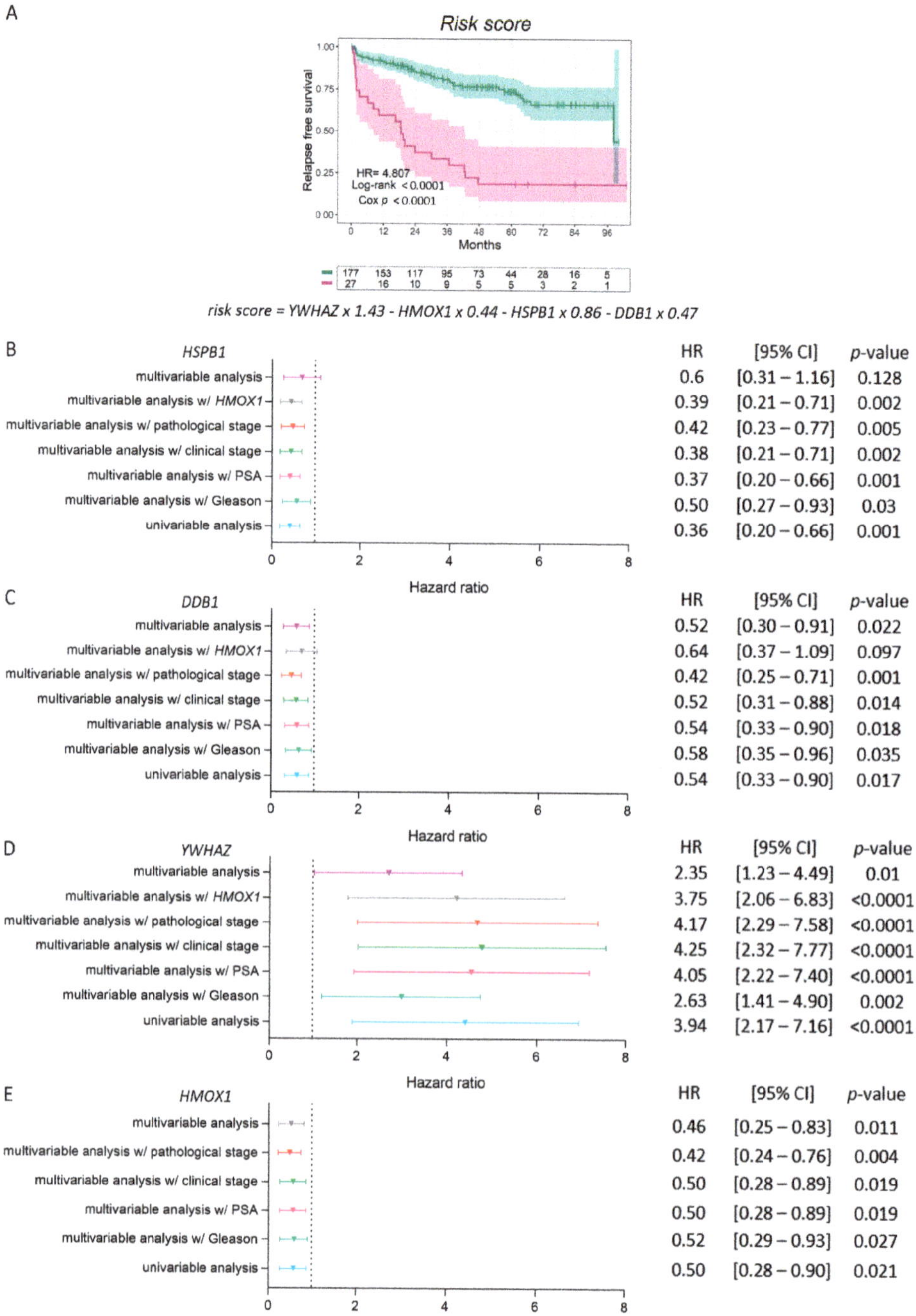

Figure 5. Risk score regression model and multivariable analyses for *HSPB1*, *DDB1*, *YWHAZ*, and *HMOX1* in PCa patients. (**A**) Kaplan–Meier curve for RFS in high-risk (purple lines) and low-risk

(green lines) groups, according to a risk score model based on the expression of *HSPB1*, *DDB1*, *YWHAZ*, and *HMOX1* in PCa patients. (**B–D**) Multivariable analyses presented by forest plots between each gene (*HSPB1* (**B**), *DDB1* (**C**), *YWHAZ* (**D**), and *HMOX1* (**E**)) and GS, PSA, clinical and pathological stage, *HMOX1*'s expression, or all the variables together. Univariable analysis (light blue); multivariable analysis with GS (light green) = adjusted for the GS (6; 7 (3 + 4); 7 (4 + 3); 8-10); multivariable analysis with PSA (pink) = adjusted for the PSA serum levels at diagnosis (PSA (ng/mL): <4; 4-10; > 10); multivariable analysis with the clinical stage (dark green) = adjusted for the clinical stage; multivariable analysis with the pathological stage (red) = adjusted for the pathological stage; multivariable analysis with *HMOX1* (grey) = adjusted for the expression of *HMOX1*; multivariable analysis (purple) = adjusted for all the variables simultaneously. HR = hazard ratios (95% confidence interval). All comparisons consider low expression patients as the reference group. Cox p = Cox proportional hazard model p-value. Statistical significance was set at Cox $p < 0.05$.

3.4. Validation of the Interaction between HO-1 and 14-3-3ζ/δ in PCa Cells

Considering that: (1) *YWHAZ* has been reported to be an independent and strong predictor of aggressiveness in PCa [38]; (2) its expression showed a significant negative correlation with *HMOX1* expression; (3) PCa patients with high *YWHAZ/HMOX1* showed the highest HR; and (4) high *YWHAZ* was the only factor significantly associated with a higher risk of relapse; we validated the interaction between HO-1 and 14-3-3ζ/δ. PC3 cells were treated with H_2O_2 or vehicle, and protein eluates were subjected to co-immunoprecipitation. As seen in Supplementary Figure S1A, HO-1 and 14-3-3ζ/δ interact in PC3 cells.

In addition, by confocal microscopy, HO-1 and 14-3-3ζ/δ co-localization was evaluated in PC3 cells treated with H_2O_2 or vehicle. The images displayed in Supplementary Figure S1B showed that HO-1 and 14-3-3ζ/δ co-localize in the cell nucleus under the induction of HO-1 with H_2O_2 (Supplementary Figure S1B). When analyzing the Manders and Pearson co-localization coefficients, a significant increase is observed in PC3 cells treated with H_2O_2 compared with controls (Supplementary Figure S1C).

Altogether, we confirmed for the first time the interaction between HO-1 and 14-3-3ζ/δ, highlighting them as critical players in PCa, and potential targets for clinical intervention.

4. Discussion

HO-1 is a key player in the cellular defense system against pro-oxidative and pro-inflammatory insults [10]. Regarding its role in pathological conditions, this protein is commonly considered as a survival molecule that plays an important role in cancer [10]. However, there is controversy about its role in tumor development and progression, possibly because its expression profile is associated with the type of tissue in question and depends, in turn, on the context or the tumor microenvironment.

Previous reports from our laboratory documented for the first time that HO-1 is expressed in human primary prostate carcinomas and is localized in the cell nucleus [18]. In PCa cell lines, we found that the pharmacological and genetic induction of HO-1 inhibits proliferation, migration, and invasion in vitro; further, it slows down tumor growth, limits tumor-associated angiogenesis [12] and neovascularization [15], and boosts the antitumor response in vivo [15].

Given the pleiotropic anti-tumoral role of HO-1 in PCa, in this work, we set out to evaluate whether HO-1 interacted with proteins previously described with nuclear localization, enabling it to reprogram prostate tumor cells fate, favoring the acquisition of a less aggressive phenotype. After generating oxidative stress conditions, HO-1 co-immunoprecipitates were subjected to LC ESI-MS/MS, identifying 11 proteins reported with nuclear localization. Interestingly, GO analyses showcased response to stress and cellular response to DNA damage stimulus as the top significant biological processes categories, highlighting the non-canonical HO-1 potential nuclear function in PCa.

Our next aim was to evaluate the clinical relevance of such interactors in association with HO-1 in PCa patients that had undergone radical prostatectomy (GSE70770) [25]. Gene expression correlation analyses showed a significant and positive Spearman correlation between *HMOX1* and *HNRNPA2B1*, *HSPB1*, *NPM1*, *DDB1*, *HMGA1*, and *ZC3HAV1*. Of note, *HMOX1* and *YWHAZ* showed a significant negative correlation.

We next set out to study whether the ratios of *HNRNPA2B1/HMOX1*, *HSPB1/HMOX1*, *NPM1/HMOX1*, *DDB1/HMOX1*, *HMGA1/HMOX1*, *ZC3HAV1/HMOX1*, and *YWHAZ/HMOX1* might affect patients RFS. Results confirmed that PCa patients with higher *HSPB1/HMOX1*, *DDB1/HMOX1*, and *YWHAZ/HMOX1* showed a worse RFS, highlighting the protective role of HO-1.

Further, we computed a risk score constituted by *HSPB1*, *DDB1*, *YWHAZ*, and *HMOX1*, evidencing a decrease in the RFS of patients with higher risk scores. Multivariable analyses supported the independence of variables to predict the patient outcome.

Interestingly, it has been reported that HSPB1 correlates with the overall survival of patients with several types of cancer. Particularly, HSPB1 induction is associated with highly aggressive disease and poor clinical outcomes in PCa. At early tumor stages, HSPB1 expression is inhibited, but it is re-expressed during PCa progression, leading to a more aggressive phenotype [39,40]. On the other hand, Zoubeidi et al. described a novel cooperative interaction between AR and HSPB1 that enhances AR stability and transcriptional activity, thereby increasing prostate cancer cell survival [41].

DDB1 is involved in DNA repairing and has been related to tumor suppression [42,43]. Regarding its role as a member of the CUL4A-DDB1 E3 ligase complex, it promotes ubiquitination-dependent AR degradation. Accordingly, DDB1 and AR protein levels negatively correlate in PCa cells [44]. 14-3-3ζ/δ is an adapter protein encoded by *YWHAZ*. Members of the 14-3-3 protein family are involved in the regulation of a wide spectrum of signaling pathways by binding to various proteins [45] and contributing to the regulation of crucial cellular processes such as protein trafficking, malignant transformation, and differentiation [46,47]. Particularly, the ζ/δ isoform constitutes a potential prognostic and therapeutic target since its high expression correlates with the progression of different cancers [38,48,49]. Previous studies from our group demonstrated *YWHAZ* relevance as a prognostic factor independent from the clinic-pathological parameters associated with the disease, such as age, GS, and TMPRSS2-ERG fusion [38]. Moreover, we observed that PCa patients with amplification, or increased mRNA or protein levels for *YWHAZ*, have significant alterations in key DNA repair genes [38].

In terms of PCa immunotherapy, different therapeutic mechanisms have been described for these 3 HO-1 interactors: HSPB1, DDB1, and 14-3-3ζ/δ. It has been reported a novel immune escape mechanism mediated by HSPB1, in which this protein expressed in the breast tumor microenvironment, promotes the differentiation of monocytes to macrophages with immune-tolerogenic phenotypes, which, in turn, trigger severe anergy in T-cells [50]. DDB1 has been proposed as a key factor for immunomodulatory drug sensitivity [51]. Moreover, DDB1 has been identified in high-throughput analyses as a potential target, showing sensitivity to Poly(ADP-ribose) polymerase (PARP) inhibition [52]. This therapeutic avenue might be combined together with different immunomodulatory drugs, enhancing their effect [53]. In the case of 14-3-3ζ/δ, Yu et al. [54] reported that immune-associated genes involved in interferon signaling, TLR-4 signaling, inflammasome network, antigen presentation/TCR recognition, and CD28 co-stimulation were found significantly downregulated in patients with urothelial carcinomas of the urinary bladder (UCUBs) presenting *YWHAZ* amplification/overexpression. However, there is no evidence of *YWHAZ* immunomodulation in PCa. Further studies are required in order to determine whether an anti-*YWHAZ* approach might be a useful strategy for improving the therapeutic efficacy of immunotherapy in PCa.

Remarkably, 14-3-3 proteins can affect various physical and functional aspects of their targets, such as: blocking nuclear localization or export signals affecting their subcellular localization, blocking their binding to other proteins, affecting their stability, and modulating their catalytic activity by modifying their conformation [47,55–58]. In most cases, the binding of 14-3-3 sequesters the target protein in a particular subcellular compartment, and the release of 14-3-3 allows the protein to relocate. Likewise, recently Chen et al. [58] revealed through in vitro and in vivo studies that 14-3-3ζ/δ promotes invasion and metastasis of non-small-cell lung cancer by binding to the soluble form of β-catenin phosphorylated at Ser552, protecting it from ubiquitin-mediated degradation, suggesting a novel mechanism by which β-catenin accumulates in the cytoplasm and remains protected from degradation in tumoral cells. Hence, 14-3-3ζ/δ promotes the activation of the Wnt pathway and the transcriptional activity of β-catenin, which enters the nucleus and interacts with the TCF/LEF complexes, inducing epithelial-mesenchymal transition, proliferation, and cell migration. Furthermore, it is widely accepted that the canonical Wnt pathway modulates osteoblast function and participates in the induction of the osteoblastic phenotype in PCa bone metastasis [59]. These results show the importance of the 14-3-3ζ/δ/β-catenin axis in PCa progression.

In this work, we confirmed the HO-1/14-3-3ζ/δ interaction, performing a co-immunop recipitation assay. Further, through an immunofluorescence assay, we determined that HO-1 and 14-3-3ζ/δ co-localize in the nucleus of PCa cells under oxidative stress conditions. Remarkably, this is the first time that this interaction has been reported in this type of cell. Although Song et al. reported this interaction in hepatocellular carcinoma [55], they only observed co-localization of both proteins in the cell cytoplasm, demonstrating that the interaction inhibited the ubiquitination and subsequent degradation of HO-1, facilitating its stability.

5. Conclusions

In summary, the results obtained in this study describe for the first time the interaction between HO-1 and HSPB1, DDB1, and 14-3-3ζ/δ in PCa cells. Further work will be necessary to identify the fine molecular mechanisms tuning HO-1 towards the acquisition of a less aggressive phenotype and to delineate the role of its interactions in PCa.

Supplementary Materials: The following supporting information can be downloaded at: https://www.mdpi.com/article/10.3390/antiox11020290/s1, Figure S1: Validation of the interaction between 14-3-3ζ/δ and HO-1 by co-immunoprecipitation and fluorescence microscopy; Table S1: Ross-Adams patients' characteristics at baseline (start of the follow-up survival analyses); Table S2: Differential proteins identified by mass spectrometry in GST-HO-1 vs. GST.

Author Contributions: Conceptualization, S.L.-V., P.S., J.C., E.V. and G.G.; methodology, S.L.-V., P.S., J.B., A.S., R.L., N.A., E.L., A.P., M.P.V., J.C., E.V. and G.G.; software, P.S., J.B. and M.P.V.; validation, S.L.-V., P.S., J.B., M.P.V., J.C., E.V. and G.G.; formal analysis, S.L.-V., P.S., J.B., M.P.V., J.C., E.V. and G.G.; investigation, S.L.-V., P.S., J.B., M.P.V., J.C., E.V. and G.G.; resources, N.N., M.P.V., J.C., E.V. and G.G.; data curation, S.L.-V., P.S., J.B., M.P.V., J.C., E.V. and G.G.; writing—original draft preparation, S.L.-V., P.S., J.C., E.V. and G.G.; writing—review and editing, S.L.-V., P.S., J.B., A.T., A.S., R.L., N.A., E.L., A.P., N.N., M.P.V., J.C., E.V. and G.G.; visualization, S.L.-V., P.S., J.C., E.V. and G.G.; supervision, J.C., E.V. and G.G.; project administration, J.C., E.V. and G.G.; funding acquisition, N.N., J.C., E.V. and G.G. All authors have read and agreed to the published version of the manuscript.

Funding: This research was funded by the Agencia Nacional de Promocion de la Investigacion, el Desarrollo Tecnologico y la Innovación (ANPCyT), Argentina: PICT-2016-0056, PICT-RAICES-2018-02639; PICT-2019-2019-03215; and Universidad de Buenos Aires, Argentina: 20020170100585BA.

Institutional Review Board Statement: Not applicable.

Informed Consent Statement: Not applicable.

Data Availability Statement: The data is contained in the manuscript and supplementary file.

Conflicts of Interest: The authors declare no conflict of interest.

References

1. Sung, H.; Ferlay, J.; Siegel, R.L.; Laversanne, M.; Soerjomataram, I.; Jemal, A.; Bray, F. Global Cancer Statistics 2020: GLOBOCAN Estimates of Incidence and Mortality Worldwide for 36 Cancers in 185 Countries. *CA Cancer J. Clin.* **2021**, *71*, 209–249. [CrossRef] [PubMed]
2. Hanahan, D.; Weinberg, R.A. Hallmarks of cancer: The next generation. *Cell* **2011**, *144*, 646–674. [CrossRef] [PubMed]
3. Coussens, L.M.; Werb, Z. Inflammation and cancer. *Nature* **2002**, *420*, 860–867. [CrossRef] [PubMed]
4. Grivennikov, S.I.; Greten, F.R.; Karin, M. Immunity, inflammation, and cancer. *Cell* **2010**, *140*, 883–899. [CrossRef] [PubMed]
5. Sauer, H.; Wartenberg, M.; Hescheler, J. Reactive oxygen species as intracellular messengers during cell growth and differentiation. *Cell. Physiol. Biochem.* **2001**, *11*, 173–186. [CrossRef] [PubMed]
6. Bolton, D.M.; Papa, N.; Ta, A.D.; Millar, J.; Davidson, A.-J.; Pedersen, J.; Syme, R.; Patel, M.I.; Giles, G.G. Predictors of prostate cancer specific mortality after radical prostatectomy: 10 year oncologic outcomes from the Victorian Radical Prostatectomy Registry. *BJU Int.* **2015**, *116*, 66–72. [CrossRef] [PubMed]
7. De Marzo, A.M.; Platz, E.A.; Sutcliffe, S.; Xu, J.; Gronberg, H.; Drake, C.G.; Nakai, Y.; Isaacs, W.B.; Nelson, W.G. Inflammation in prostate carcinogenesis. *Nat. Rev. Cancer* **2007**, *7*, 256–269. [CrossRef] [PubMed]
8. Labanca, E.; De Luca, P.; Gueron, G.; Paez, A.; Moiola, C.P.; Massillo, C.; Porretti, J.; Giudice, J.; Zalazar, F.; Navone, N.; et al. Association of HO-1 and BRCA1 Is Critical for the Maintenance of Cellular Homeostasis in Prostate Cancer. *Mol. Cancer Res.* **2015**, *13*, 1455–1464. [CrossRef]
9. Gueron, G.; De Siervi, A.; Vazquez, E. Advanced prostate cancer: Reinforcing the strings between inflammation and the metastatic behavior. *Prostate Cancer Prostatic Dis.* **2012**, *15*, 213–221. [CrossRef]
10. Jozkowicz, A.; Was, H.; Dulak, J. Heme oxygenase-1 in tumors: Is it a false friend? *Antioxid. Redox Signal.* **2007**, *9*, 2099–2117. [CrossRef]
11. Gueron, G.; De Siervi, A.; Ferrando, M.; Salierno, M.; De Luca, P.; Elguero, B.; Meiss, R.; Navone, N.; Vazquez, E.S. Critical role of endogenous heme oxygenase 1 as a tuner of the invasive potential of prostate cancer cells. *Mol. Cancer Res.* **2009**, *7*, 1745–1755. [CrossRef] [PubMed]
12. Ferrando, M.; Gueron, G.; Elguero, B.; Giudice, J.; Salles, A.; Leskow, F.C.; Jares-Erijman, E.A.; Colombo, L.; Meiss, R.; Navone, N.; et al. Heme oxygenase 1 (HO-1) challenges the angiogenic switch in prostate cancer. *Angiogenesis* **2011**, *14*, 467–479. [CrossRef] [PubMed]
13. Elguero, B.; Gueron, G.; Giudice, J.; Toscani, M.A.; De Luca, P.; Zalazar, F.; Coluccio-Leskow, F.; Meiss, R.; Navone, N.; De Siervi, A.; et al. Unveiling the association of STAT3 and HO-1 in prostate cancer: Role beyond heme degradation. *Neoplasia* **2012**, *14*, 1043–1056. [CrossRef] [PubMed]
14. Gueron, G.; Giudice, J.; Valacco, P.; Paez, A.; Elguero, B.; Toscani, M.; Jaworski, F.; Leskow, F.C.; Cotignola, J.; Marti, M.; et al. Heme-oxygenase-1 implications in cell morphology and the adhesive behavior of prostate cancer cells. *Oncotarget* **2014**, *5*, 4087–4102. [CrossRef] [PubMed]
15. Jaworski, F.M.; Gentilini, L.D.; Gueron, G.; Meiss, R.P.; Ortiz, E.G.; Berguer, P.M.; Ahmed, A.; Navone, N.; Rabinovich, G.A.; Compagno, D.; et al. In Vivo Hemin Conditioning Targets the Vascular and Immunologic Compartments and Restrains Prostate Tumor Development. *Clin. Cancer Res.* **2017**, *23*, 5135–5148. [CrossRef]
16. Ferrando, M.; Wan, X.; Meiss, R.; Yang, J.; De Siervi, A.; Navone, N.; Vazquez, E. Heme oxygenase-1 (HO-1) expression in prostate cancer cells modulates the oxidative response in bone cells. *PLoS ONE* **2013**, *8*, e80315. [CrossRef] [PubMed]
17. Lin, Q.; Weis, S.; Yang, G.; Weng, Y.-H.; Helston, R.; Rish, K.; Smith, A.; Bordner, J.; Polte, T.; Gaunitz, F.; et al. Heme oxygenase-1 protein localizes to the nucleus and activates transcription factors important in oxidative stress. *J. Biol. Chem.* **2007**, *282*, 20621–20633. [CrossRef]
18. Sacca, P.; Meiss, R.; Casas, G.; Mazza, O.; Calvo, J.C.; Navone, N.; Vazquez, E. Nuclear translocation of haeme oxygenase-1 is associated to prostate cancer. *Br. J. Cancer* **2007**, *97*, 1683–1689. [CrossRef]
19. Lin, Q.S.; Weis, S.; Yang, G.; Zhuang, T.; Abate, A.; Dennery, P.A. Catalytic inactive heme oxygenase-1 protein regulates its own expression in oxidative stress. *Free Radic. Biol. Med.* **2008**, *44*, 847–855. [CrossRef]
20. Anselmino, N.; Bizzotto, J.; Sanchis, P.; Lage-Vickers, S.; Ortiz, E.; Valacco, P.; Paez, A.; Labanca, E.; Meiss, R.; Navone, N.; et al. HO-1 Interactors Involved in the Colonization of the Bone Niche: Role of ANXA2 in Prostate Cancer Progression. *Biomolecules* **2020**, *10*, 467. [CrossRef]
21. Sacca, P.; Caballero, F.; Batlle, A.; Vazquez, E. Cell cycle arrest and modulation of HO-1 expression induced by acetyl salicylic acid in hepatocarcinogenesis. *Int. J. Biochem. Cell Biol.* **2004**, *36*, 1945–1953. [CrossRef] [PubMed]
22. Szklarczyk, D.; Franceschini, A.; Wyder, S.; Forslund, K.; Heller, D.; Huerta-Cepas, J.; Simonovic, M.; Roth, A.; Santos, A.; Tsafou, K.P.; et al. STRING v10: Protein-protein interaction networks, integrated over the tree of life. *Nucleic Acids Res.* **2015**, *43*, D447–D452. [CrossRef] [PubMed]
23. Shannon, P.; Markiel, A.; Ozier, O.; Baliga, N.S.; Wang, J.T.; Ramage, D.; Amin, N.; Schwikowski, B.; Ideker, T. Cytoscape: A software environment for integrated models of biomolecular interaction networks. *Genome Res.* **2003**, *13*, 2498–2504. [CrossRef]
24. Yu, G.; Hu, E. Package 'Enrichplot': Visualization of Functional Enrichment Result. 2021. Available online: https://www.bioconductor.org/packages/devel/bioc/manuals/enrichplot/man/enrichplot.pdf (accessed on 10 October 2021).

25. Ross-Adams, H.; Lamb, A.; Dunning, M.; Halim, S.; Lindberg, J.; Massie, C.; Egevad, L.; Russell, R.; Ramos-Montoya, A.; Vowler, S.; et al. Integration of copy number and transcriptomics provides risk stratification in prostate cancer: A discovery and validation cohort study. *EBioMedicine* **2015**, *2*, 1133–1144. [CrossRef] [PubMed]
26. Budczies, J.; Klauschen, F.; Sinn, B.V.; Gyorffy, B.; Schmitt, W.D.; Darb-Esfahani, S.; Denkert, C. Cutoff Finder: A comprehensive and straightforward Web application enabling rapid biomarker cutoff optimization. *PLoS ONE* **2012**, *7*, e51862. [CrossRef] [PubMed]
27. Kassambara, A.; Kosinski, M.; Biecek, P. Survminer: Drawing Survival Curves Using "ggplot2". 2019. Available online: https://cran.r-project.org/web/packages/survminer/survminer.pdf (accessed on 10 October 2021).
28. Paez, A.V.; Pallavicini, C.; Schuster, F.; Valacco, M.P.; Giudice, J.; Ortiz, E.G.; Anselmino, N.; Labanca, E.; Binaghi, M.; Salierno, M.; et al. Heme oxygenase-1 in the forefront of a multi-molecular network that governs cell-cell contacts and filopodia-induced zippering in prostate cancer. *Cell Death Dis.* **2016**, *7*, e2570. [CrossRef]
29. Kan, C.; Vargas, G.; Pape, F.L.; Clezardin, P. Cancer Cell Colonisation in the Bone Microenvironment. *Int. J. Mol. Sci.* **2016**, *17*, 1674. [CrossRef]
30. Christensen, M.V.; Høgdall, C.K.; Jochumsen, K.M.; Høgdall, E.V.S. Annexin A2 and cancer: A systematic review. *Int. J. Oncol.* **2018**, *52*, 5–18. [CrossRef]
31. Kaighn, M.E.; Narayan, K.S.; Ohnuki, Y.; Lechner, J.F.; Jones, L.W. Establishment and characterization of a human prostatic carcinoma cell line (PC-3). *Investig. Urol.* **1979**, *17*, 16–23.
32. Nishimura, K.; Ting, H.-J.; Harada, Y.; Tokizane, T.; Nonomura, N.; Kang, H.-Y.; Chang, H.-C.; Yeh, S.; Miyamoto, H.; Shin, M.; et al. Modulation of androgen receptor transactivation by gelsolin: A newly identified androgen receptor coregulator. *Cancer Res.* **2003**, *63*, 4888–4894.
33. Narain, N.R.; Diers, A.R.; Lee, A.; Lao, S.; Chan, J.Y.; Schofield, S.; Andreazi, J.; Ouro-Djobo, R.; Jimenez, J.J.; Friss, T.; et al. Identification of Filamin-A and -B as potential biomarkers for prostate cancer. *Future Sci. OA* **2016**, *3*, FSO161. [CrossRef] [PubMed]
34. Fan, X.; Cui, L.; Zeng, Y.; Song, W.; Gaur, U.; Yang, M. 14-3-3 Proteins Are on the Crossroads of Cancer, Aging, and Age-Related Neurodegenerative Disease. *Int. J. Mol. Sci.* **2019**, *20*, 3518. [CrossRef] [PubMed]
35. Popiel, A.; Kobierzycki, C.; Dzięgiel, P. The Role of Testin in Human Cancers. *Pathol. Oncol. Res.* **2019**, *25*, 1279–1284. [CrossRef] [PubMed]
36. Fong, K.; Zhao, J.C.; Song, B.; Zheng, B.; Yu, J. TRIM28 protects TRIM24 from SPOP-mediated degradation and promotes prostate cancer progression. *Nat. Commun.* **2018**, *9*, 5007. [CrossRef] [PubMed]
37. Jiménez-Vacas, J.M.; Herrero-Aguayo, V.; Montero-Hidalgo, A.J.; Gómez-Gómez, E.; Fuentes-Fayos, A.C.; León-González, A.J.; Sáez-Martínez, P.; Alors-Pérez, E.; Pedraza-Arévalo, S.; González-Serrano, T.; et al. Dysregulation of the splicing machinery is directly associated to aggressiveness of prostate cancer. *EBioMedicine* **2020**, *51*, 102547. [CrossRef] [PubMed]
38. Lage-Vickers, S.; Bizzotto, J.; Valacco, M.P.; Sanchis, P.; Nemirovsky, S.; Labanca, E.; Scorticati, C.; Mazza, O.; Mitrofanova, A.; Navone, N.; et al. The expression of YWHAZ and NDRG1 predicts aggressive outcome in human prostate cancer. *Commun. Biol.* **2021**, *4*, 103. [CrossRef] [PubMed]
39. Cornford, P.A.; Dodson, A.R.; Parsons, K.F.; Desmond, A.D.; Woolfenden, A.; Fordham, M.; Neoptolemos, J.P.; Ke, Y.; Foster, C.S. Heat shock protein expression independently predicts clinical outcome in prostate cancer. *Cancer Res.* **2000**, *60*, 7099–7105.
40. Foster, C.S.; Dodson, A.R.; Ambroisine, L.; Fisher, G.; Møller, H.; Clark, J.; Attard, G.; De-Bono, J.; Scardino, P.; Reuter, V.E.; et al. Hsp-27 expression at diagnosis predicts poor clinical outcome in prostate cancer independent of ETS-gene rearrangement. *Br. J. Cancer* **2009**, *101*, 1137–1144. [CrossRef]
41. Zoubeidi, A.; Zardan, A.; Beraldi, E.; Fazli, L.; Sowery, R.; Rennie, P.; Nelson, C.; Gleave, M. Cooperative interactions between androgen receptor (AR) and heat-shock protein 27 facilitate AR transcriptional activity. *Cancer Res.* **2007**, *67*, 10455–10465. [CrossRef]
42. Yoon, T.; Chakrabortty, A.; Franks, R.; Valli, T.; Kiyokawa, H.; Raychaudhuri, P. Tumor-prone phenotype of the DDB2-deficient mice. *Oncogene* **2005**, *24*, 469–478. [CrossRef]
43. Alekseev, S.; Kool, H.; Rebel, H.; Fousteri, M.; Moser, J.; Backendorf, C.; de Gruijl, F.R.; Vrieling, H.; Mullenders, L.H.F. Enhanced DDB2 expression protects mice from carcinogenic effects of chronic UV-B irradiation. *Cancer Res.* **2005**, *65*, 10298–10306. [CrossRef] [PubMed]
44. Chang, S.-W.; Su, C.-H.; Chen, H.-H.; Huang, C.-W.; Tsao, L.-P.; Tsao, Y.-P.; Chen, S.-L. DDB2 is a novel AR interacting protein and mediates AR ubiquitination/degradation. *Int. J. Biochem. Cell Biol.* **2012**, *44*, 1952–1961. [CrossRef] [PubMed]
45. Morrison, D.K. The 14-3-3 proteins: Integrators of diverse signaling cues that impact cell fate and cancer development. *Trends Cell Biol.* **2009**, *19*, 16–23. [CrossRef] [PubMed]
46. Dougherty, M.K.; Morrison, D.K. Unlocking the code of 14-3-3. *J. Cell Sci.* **2004**, *117*, 1875–1884. [CrossRef] [PubMed]
47. Tzivion, G.; Gupta, V.S.; Kaplun, L.; Balan, V. 14-3-3 proteins as potential oncogenes. *Semin. Cancer Biol.* **2006**, *16*, 203–213. [CrossRef] [PubMed]
48. Neal, C.L.; Yu, D. 14-3-3ζ as a prognostic marker and therapeutic target for cancer. *Expert Opin. Ther. Targets* **2010**, *14*, 1343–1354. [CrossRef]
49. Matta, A.; Siu, K.W.M.; Ralhan, R. 14-3-3 zeta as novel molecular target for cancer therapy. *Expert Opin. Ther. Targets* **2012**, *16*, 515–523. [CrossRef]

50. Banerjee, S.; Lin, C.-F.L.; Skinner, K.A.; Schiffhauer, L.M.; Peacock, J.; Hicks, D.G.; Redmond, E.M.; Morrow, D.; Huston, A.; Shayne, M.; et al. Heat shock protein 27 differentiates tolerogenic macrophages that may support human breast cancer progression. *Cancer Res.* **2011**, *71*, 318–327. [CrossRef]
51. Barrio, S.; Munawar, U.; Zhu, Y.X.; Giesen, N.; Shi, C.-X.; Da Viá, M.; Sanchez, R.; Bruins, L.; Demler, T.; Müller, N.; et al. IKZF1/3 and CRL4(CRBN) E3 ubiquitin ligase mutations and resistance to immunomodulatory drugs in multiple myeloma. *Haematologica* **2020**, *105*, e237–e241. [CrossRef]
52. Chan, N.; Pires, I.M.; Bencokova, Z.; Coackley, C.; Luoto, K.R.; Bhogal, N.; Lakshman, M.; Gottipati, P.; Oliver, F.J.; Helleday, T.; et al. Contextual Synthetic Lethality of Cancer Cell Kill Based on the Tumor Microenvironment. *Cancer Res.* **2010**, *70*, 8045–8054. [CrossRef]
53. Teyssonneau, D.; Margot, H.; Cabart, M.; Anonnay, M.; Sargos, P.; Vuong, N.-S.; Soubeyran, I.; Sevenet, N.; Roubaud, G. Prostate cancer and PARP inhibitors: Progress and challenges. *J. Hematol. Oncol.* **2021**, *14*, 51. [CrossRef] [PubMed]
54. Yu, C.-C.; Li, C.-F.; Chen, I.-H.; Lai, M.-T.; Lin, Z.-J.; Korla, P.K.; Chai, C.-Y.; Ko, G.; Chen, C.-M.; Hwang, T.; et al. YWHAZ amplification/overexpression defines aggressive bladder cancer and contributes to chemo-/radio-resistance by suppressing caspase-mediated apoptosis. *J. Pathol.* **2019**, *248*, 476–487. [CrossRef] [PubMed]
55. Song, J.; Zhang, X.; Liao, Z.; Liang, H.; Chu, L.; Dong, W.; Zhang, X.; Ge, Q.; Liu, Q.; Fan, P.; et al. 14-3-3zeta inhibits heme oxygenase-1 (HO-1) degradation and promotes hepatocellular carcinoma proliferation: Involvement of STAT3 signaling. *J. Exp. Clin. Cancer Res.* **2019**, *38*, 3. [CrossRef] [PubMed]
56. Aitken, A. Post-translational modification of 14-3-3 isoforms and regulation of cellular function. *Semin. Cell Dev. Biol.* **2011**, *22*, 673–680. [CrossRef] [PubMed]
57. Dar, A.; Wu, D.; Lee, N.; Shibata, E.; Dutta, A. 14-3-3 proteins play a role in the cell cycle by shielding cdt2 from ubiquitin-mediated degradation. *Mol. Cell. Biol.* **2014**, *34*, 4049–4061. [CrossRef] [PubMed]
58. Chen, C.-H.; Chuang, S.-M.; Yang, M.-F.; Liao, J.-W.; Yu, S.-L.; Chen, J.J.W. A novel function of YWHAZ/beta-catenin axis in promoting epithelial-mesenchymal transition and lung cancer metastasis. *Mol. Cancer Res.* **2012**, *10*, 1319–1331. [CrossRef] [PubMed]
59. Li, Z.G.; Yang, J.; Vazquez, E.S.; Rose, D.; Vakar-Lopez, F.; Mathew, P.; Lopez, A.; Logothetis, C.J.; Lin, S.-H.; Navone, N.M. Low-density lipoprotein receptor-related protein 5 (LRP5) mediates the prostate cancer-induced formation of new bone. *Oncogene* **2008**, *27*, 596–603. [CrossRef]

MDPI
St. Alban-Anlage 66
4052 Basel
Switzerland
Tel. +41 61 683 77 34
Fax +41 61 302 89 18
www.mdpi.com

Antioxidants Editorial Office
E-mail: antioxidants@mdpi.com
www.mdpi.com/journal/antioxidants

www.ingramcontent.com/pod-product-compliance
Lightning Source LLC
Chambersburg PA
CBHW061441210926
51850CB00008B/1571